AF567020

Nanomedicine – Basic and Clinical Applications in Diagnostics and Therapy

Else Kröner-Fresenius Symposia

Vol. 2

Series Editor

S. Pahernik Heidelberg

Nanomedicine – Basic and Clinical Applications in Diagnostics and Therapy

Volume Editor

Christoph Alexiou Erlangen

80 figures, 56 in color, and 11 tables, 2011

Basel · Freiburg · Paris · London · New York · New Delhi ·
Bangkok · Beijing · Tokyo · Kuala Lumpur · Singapore · Sydney

Univ.-Prof. Dr. med. Christoph Alexiou

Department of Otorhinolaryngology, Head and Neck Surgery
Else Kröner-Fresenius-Stiftung Professorship for Nanomedicine
Section for Experimental Oncology and Nanomedicine (SEON)
University Hospital Erlangen
Glückstrasse 10, DE–91054 Erlangen (Germany)

This book is sponsored by the Else Kröner-Fresenius-Stiftung.

Library of Congress Cataloging-in-Publication Data

Nanomedicine : basic and clinical applications in diagnostics and therapy / volume editor, Christoph Alexiou.
p. ; cm. -- (Else Kröner-Fresenius symposia, ISSN 1663-0114 ; v. 2)
Includes bibliographical references and indexes.
ISBN 978-3-8055-9818-7 (hard cover : alk. paper) -- ISBN 978-3-8055-9819-4 (e-ISBN)
I. Alexiou, Christoph. II. Series: Else Kröner-Fresenius symposia ; v. 2. 1663-0114
[DNLM: 1. Nanomedicine--Congresses. 2. Nanostructures--diagnostic use--Congresses. 3. Nanostructures--therapeutic use--Congresses. QT 36.5]
LC classification not assigned
616.07--dc23

2011029822

Cover art by C. Alexiou and the Else Kröner-Fresenius-Stiftung Professorship for Nanomedicine

www.karger.com
Printed in Germany on acid-free and non-aging paper (ISO 9706) by Kraft Druck, Ettlingen
ISSN 1663–0114
ISBN 978–3–8055–9818–7
e-ISBN 978–3–8055–9819–4

Contents

Preface

This book series features the proceedings of the Else Kröner-Fresenius Symposia, which is intended to cover clinically relevant topics at the forefront of biomedical research. The meetings should give experts the opportunity to discuss the most recent findings of evolving fields of biomedicine and outline future research strategies.

Today's research is characterized by accelerated generation of biomedical data, an increasingly interdisciplinary and translational nature of biomedical science, as well as efforts to integrate the new results into health care.

In this context nanomedicine, i.e. applying nanotechnology to health, is one of the promising fields of biomedical research. The technology aims, at the level of molecules and atoms, to discover and develop new material for use in medicine. Nanotechnology is regarded as one of the key technologies of the future.

Nanomedicine has the potential to yield intelligent solutions to many of the current medical problems. Research in nanomedicine will provide a better insight into pathophysiology at the molecular and nanometric level. Novel promising solutions for intervention in chronic or acute pathologies are to be expected. Nanotechnology already impacts biomedical research. Thus, the first nanotechnology-based targeted drug delivery systems are already being used in clinical settings, and some are currently being investigated in clinical trials or are under development. Another highly important field of nanotechnology in medicine is diagnostics with the aim of yielding precise information at a nanoscale at a very early stage of the disease. Furthermore, new concepts using nanotechnology are being investigated to selectively target tumor diseases. Another promising related area of biomedical research is regenerative medicine leading to the development of artificial skin, bone, cartilage, and other tissue to restore organ function.

A wide variety of potential diagnostic and therapeutic applications of nanotechnology, as well as their basic effects on pathophysiology and toxicity, was discussed at the 2nd Else Kröner-Fresenius Symposia. The Else Kröner-Fresenius-Stiftung together with the Medical Faculty of Erlangen established a Professorship for Nanomedicine in 2009 headed by Dr. Christoph Alexiou. The foundation thanks him for organizing this excellent meeting.

The Else Kröner-Fresenius-Stiftung

In 1983, Else Kröner (1925–1988) founded the Else Kröner-Fresenius-Stiftung, a nonprofit foundation dedicated to promoting medical science, supporting medical education, and providing humanitarian aid.

Else Kröner, born Fernau, was born on May 15, 1925, in Frankfurt am Main, Germany. At the age of 3 years, her father died. After his death, she lived with her mother in the home of Dr. Eduard Fresenius, a pharmacist and owner of the Hirsch Pharmacy in Frankfurt, who founded the Fresenius pharmaceutical company in 1912. Dr. Eduard Fresenius, whose marriage was childless, took care of Else Fernau. In 1944, she started an internship at the Hirsch Pharmacy and decided to study pharmacy, which was supported by Dr. Eduard Fresenius. In 1946, Dr. Eduard Fresenius unexpectedly died. At this time, Else Fernau had not completed her pharmaceutical education. However, Dr. Eduard Fresenius bequeathed the Hirsch Pharmacy and the Fresenius company to her.

At the age of 21 years, Else Fernau decided, against the council of many, to take responsibility for the Hirsch Pharmacy and the Fresenius company. At that time, the Fresenius company was completely indebted. Of the original 400 employees, all but 30 had to be laid off. She joined efforts in the early 1950s with her later husband, Hans Kröner, a lawyer and economist, to ensure the survival and force the expansion of the company. The company was progressively rebuilt. At this time, the need to maintain the company determined all activities.

Entrepreneurial important and far-sighted business decisions in the 1950s and 1960s ensured the successful future development of the company. Thereafter, decades of growth followed, in particular within the fields of dialysis, nutrition, and intensive care, leading to an internationally competitive enterprise and market leader in special areas of health care.

Until 1981, Else Kröner led the company. After the transformation of Fresenius into a stock company, Else Kröner remained chairman of the Supervisory Board until her death on June 5, 1988. From 1981 to 1992, her husband Hans Kröner led the company as CEO. Thereafter, he significantly influenced the policy of the Else Kröner-Fresenius-Stiftung, of which he was chairman of the board from 1995 to 2005.

Today, the Fresenius group, of which the Else Kröner-Fresenius-Stiftung is the leading share holder, is an international health care group with products and services for dialysis, hospital, and medical care of patients. The Fresenius group consists of 5 business areas that operate globally: Fresenius Medical Care, Fresenius Kabi, Fresenius Helios, Fresenius Vamed, and Fresenius Biotech. Fresenius Medical Care is the world's first provider of products and services for patients with chronic renal failure. Fresenius Kabi is the leader in infusion therapy and clinical nutrition in Europe. Fresenius Helios is one of the largest and leading medical clinic operators in Germany. Fresenius Vamed is a health care project and financial management operator. Fresenius Biotech is a biotechnology company focusing on the development and commercialization of biopharmaceutical therapies. The Fresenius group currently employs more than 140,000 people in over 100 countries and generates annual sales of over EUR 15 billion.

Else Kröner transferred nearly her entire fortune to the foundation for the pursuit of promoting medical science and providing humanitarian aid. It corresponds to her requests and represents the continuation of her concept that her fortune be used for nonprofit purposes. The symposia are published as part of the foundation's commitment to the advancement of medical research and treatment.

Sascha Pahernik, Heidelberg
Member of the Scientific Commitee of the Else Kröner-Fresenius-Stiftung
Series Editor

Alexiou C (ed): Nanomedicine – Basic and Clinical Applications in Diagnostics and Therapy.
Else Kröner-Fresenius Symp. Basel, Karger, 2011, vol 2, pp 1–2

Introduction

Christoph Alexiou

Else Kröner-Fresenius-Stiftung Professorship for Nanomedicine, Section for Experimental Oncology and Nanomedicine, Department of Otorhinolaryngology, Head and Neck Surgery, University Hospital Erlangen, Erlangen, Germany

ELSE KRÖNER-FRESENIUS-STIFTUNG

Forschung fördern. Menschen helfen.

Second Else Kröner-Fresenius-Symposium September, 3rd-5th 2010
Nanomedicine - Basic and Clinical Application in Diagnostics and Therapy
Organized by Univ.- Prof. Dr. med. Christoph Alexiou

University Hospital Erlangen

Nanomedicine stands at the boundary between physical, chemical, biological, and medical sciences. Recent advances have made it possible to analyze biological systems at cellular and subcellular levels, offering numerous promising possibilities to significantly improve medical diagnosis and therapy. It is expected that nanomedicine will have a great impact, especially on drug delivery and imaging. In this context, the development of targeted, highly specific nanoparticles is of pivotal importance. The results of these and other advances will offer personalized diagnostic tools and treatment in the future. Acknowledging the scientific and clinical potential of this field, the Else Kröner-Fresenius-Stiftung initiated a competitive call for the establishment of a professorship for nanomedicine in 2008. The concept submitted by the University of Erlangen-Nuremberg was selected as the most promising, and a professorship for nanomedicine was established at the Department of Otorhinolaryngology, Head and Neck Surgery in 2009. The Else Kröner-Fresenius Symposium, in September 2010, focused on recent advances in nanomedicine and offered an excellent platform to discuss this exciting field from different perspectives and to promote an open scientific exchange with leading experts. Under the topic 'Nanomedicine – Basic and Clinical Applications in Diagnostics and Therapy', a broad spectrum ranging from nanoscale drug delivery/drug design, nanotoxicity, diagnostics and imaging, therapeutic applications, and antibody therapies was intensely discussed and contributed to a scientific amendment in this emerging field. This will lead to new developments to translate research results from the laboratory into clinical trials and to achieve in the end the acceptance of industrial organizations to place it on the market.

This book is a meeting report and summarizes the current knowledge of nanomedicine and reflects the major topics discussed at the symposium.

Disclosure Statement

The author has nothing to disclose.

Univ. Prof. Dr. med. Christoph Alexiou
Department of Otorhinolaryngology, Head and Neck Surgery
Else Kröner-Fresenius-Stiftung Professorship for Nanomedicine
Section for Experimental Oncology and Nanomedicine (SEON)
University Hospital Erlangen
Glückstrasse 10, DE–91054 Erlangen (Germany)
Tel. +49 9131 853 4769, E-Mail c.alexiou@web.de

Alexiou C (ed): Nanomedicine – Basic and Clinical Applications in Diagnostics and Therapy.
Else Kröner-Fresenius Symp. Basel, Karger, 2011, vol 2, pp 3–14

Speakers at the Symposium

Christoph Alexiou, MD

C.A., born in 1967, received his PhD in 1995 from Munich Technical University Medical School. After finishing his internship in the Department of Gastroenterology at the University Hospital Munich he started as a physician and researcher at the Department of Otorhinolaryngology, Head and Neck Surgery. There he founded a research group working in the field of local cancer therapy with magnetic nanoparticles (magnetic drug targeting).

In 2000, he received his degree as an ENT specialist and changed to the Department of Otorhinolaryngology, Head and Neck Surgery in Erlangen (Germany) in 2002. There he performed his postdoctoral lecture qualification ('habilitation') and now works as an assistant medical director in the hospital as well as leading the Section for Experimental Oncology and Nanomedicine (SEON). Since 2009, he has held the Else Kröner-Fresenius-Stiftung Professorship for Nanomedicine at the University Hospital Erlangen. His research focuses on the application of functionalized magnetic nanoparticles in diagnostics and therapy of cancer and the translation into human trials. He is a member of the editorial board of the journals *Nanomedicine: Nanotechnology, Biology and Medicine* (Elsevier), and *OncoTargets and Therapy* and has received several national and international awards for his research.

Hanene Ali-Boucetta, PhD

H.A.-B. graduated with a first-class (honors) MPharm degree from the School of Pharmacy at the University of London. In October 2006, she joined the Nanomedicine Lab for a PhD under the supervision of Prof. Kostas Kostarelos.

Her research focused on studying and elucidating the important and critical parameters in carbon nanotube pharmacokinetics and toxicology and establishing effective CNT nanovectors for cancer therapeutics. Upon completing her PhD, she was awarded a fellowship by the University of London and rejoined the Nanomedicine Lab as a C.W. Maplethorpe Research and Teaching Fellow working on the engineering of carbon nanotubes as multimodal modalities for cancer therapy.

Lajos P. Balogh, PhD

L.P.B. received his PhD in chemical technology from Kossuth L. University (KLTE Debrecen) in Hungary and was invited to the University of Massachusetts Lowell in 1991. He later worked at the Michigan Molecular Institute as a senior scientist and had various faculty appointments at the University of Michigan Ann Arbor and at the University at Buffalo, SUNY. He is the former co-director of the NanoBiotechnology Center and director of Nanotechnology Research in the Department of Radiation Medicine at the Roswell Park Cancer Institute, Buffalo, N.Y. (USA). He is an internationally renown scientist with an expertise in dendrimers and dendrimer nanocomposites who has authored or coauthored over 150 scientific publications and has 12 patents in various disciplines.

He is the editor in chief of the journal *Nanomedicine: Nanotechnology, Biology and Medicine* (Elsevier) and the principal of AA Nanotechnology Consulting (AANTC). AANTC provides expert advice to private enterprises and government agencies in nanotechnology and nanomedicine.

L.P.B. is a member of numerous US government, NGO, and international expert committees, including the NIH NANO and EPA Nanotechnology study sections. He is also a voting member of the Steering Committee of the American National Standards Institute Nanotechnology Panel and the ASTM E56 Committee on Nanotechnology and serves on the US Technical Advisory Committee to the International Standard Organization on Nanotechnology (TC-229). He is one of the 5 founders of the American Society for Nanomedicine (www.amsocnanomed.org).

His research interests involve the design, synthesis, and characterization of multifunctional hybrid nanodevices for the targeted delivery of smart contrast agents and therapeutic materials, as well as the fabrication of nanosized tools to better understand fundamental biologic processes and interactions of nanodevices with cells and tissues including toxicity, biodistribution, and pharmacokinetics.

Thorsten Buzug, PhD

T.B., born in 1963, received his PhD in Applied Physics from the University of Kiel in 1993. After a postdoctoral position at the German Federal Armed Forces Underwater Acoustics and Marine Geophysics Research Institute where he worked on signal processing for SONAR systems, he joined Philips Research Laboratories Hamburg at the end of 1994. As leader of the Philips research cluster Medical Image Processing he was responsible for several projects in that field. In 1998, T.B. was appointed professor of physics and medical engineering at RheinAhrCampus Remagen. In 2006, he was appointed director of the Institute of Medical Engineering at the University of Lübeck.

Esther H. Chang, PhD

E.H.C. is a member of the Departments of Oncology and Otolaryngology at the Lombardi Comprehensive Cancer Center of Georgetown University Medical Center. Before joining Georgetown University, she held positions in the National Cancer Institute, as a professor in the Department of Surgery at Stanford University, and as a professor in the Departments of Pathology and Surgery at the Uniformed Services University of the Health Sciences. E.H.C. has been a pioneer and major contributor to understanding genetic influences on both the development and the suppression of cancerous tumors, as well as to understanding tumors' resistance to radio/chemotherapy. She was the first to identify human versions of a mouse viral oncogene and to demonstrate the role of these '*ras* oncogenes' in causing cancer. Her research group was also the first to demonstrate, simultaneously with another group, that a mutated p53 gene inherited in a cancer-prone family plays a role in the early stages of cancer development. As determined by *Science Watch,* E.H.C.'s seminal papers in both of these fields were among the top 100 and top 10 most cited publications in *Life Sciences and Medicine* in their respective years of publication.

Recently, her research on the application of nanotechnology to cancer therapy, diagnosis, and prevention was recognized in *Science News* magazine as being among the most promising efforts in the field. This tumor-targeted nanodelivery platform technology is in clinical trials. Her group has also been evaluating the combination of this tumor-targeted gene therapy and more conventional radiotherapy or chemotherapy for the treatment of cancers.

E.H.C. has over 130 publications and has served as a member of a number of scientific advisory boards for the National Cancer Institute, NASA, the US Military Cancer Institute, and the Department of Energy. Her scientific findings have been published in prominent journals including *Nature, Science, Cancer Research, Human Gene Therapy* and *Cell,* as well as in many other leading journals. She has also dedicated much time to the education of graduate and medical students and has held peer-reviewed grants from the NIH for more than 25 years without interruption.

Mike Eaton, PhD

M.E. has worked in research in the pharmaceutical industry for more than 35 years. He was a founding member of Celltech (later acquired by UCB) in 1980. He has worked on a number of marketed drugs – Mylotarg in 2000, the first antibody drug conjugate, and certolizumab pegol in 2009, the first PEGylated antibody. Unusually, he has worked with both small molecules and large molecules, including DNA. He built the first automated DNA synthesizer in Europe, which is now owned by the Science Museum in London. M.E. is a special professor at Nottingham and has been an executive board member of the European Technology Platform for Nanomedicine since its inception in 2005. He left UCB in February 2010 and is now a strategic and technical adviser to a number of companies.

Rutledge G. Ellis-Behnke, PhD

R.G.E.-B. received his PhD in Neuroscience from MIT and his BS from Rutgers University and graduated from Harvard Business School's International Senior Management Program (AMP/ISMP). He is associate director of the Technology Transfer Office at the University of Hong Kong, as well as professor at the university's Li Ka Shing Faculty of Medicine (Department of Anatomy, State Key Lab for Brain and Cognitive Sciences and Research Centre of Heart, Brain, Hormone and Healthy Aging). He is also a research affiliate in the Brain and Cognitive Sciences Department at MIT. His primary interest is using nanobiotechnology to reconnect disconnected parts of the central nervous system.

Prior to returning to school to pursue his PhD, R.G.E.-B. held various management positions including senior vice-president of a public company for testing and consulting services and co-founder/CEO in 1995 of one of the first internet companies to do online commerce in computer memory. He is associate editor (neurology) for the journal *Nanomedicine: Nanotechnology, Biology and Medicine,* member of the board of directors and the scientific advisory board of the Glaucoma Foundation, and member of the China Spinal Cord Clinical Trial Network, the Society for Neuroscience, the American Chemical Society, the Association for Research in Vision and Ophthalmology, and Sigma Xi. *Technology Review* named his nanotechnology discoveries one of the 'Top 10 Emerging Technologies of 2007'.

Bengt Fadeel, MD, PhD

B.F. holds MD and PhD degrees from Karolinska Institutet in Stockholm (Sweden), and he is professor of toxicology and Head of the Division of Molecular Toxicology. He is also Vice-Chairman of the Institute of Environmental Medicine at Karolinska Institutet. He acts as project coordinator of NANOMMUNE, a research consortium funded by the Seventh Framework Programme of the European Commission, including 13 research teams at 10 different universities/agencies in Europe and the USA. The project focuses on studies of the hazardous effects of engineered nanomaterials on the immune system. B.F. was the main organizer of the 1st Nobel Forum Mini-Symposium on Nanotoxicology, Karolinska Institutet, Stockholm (2006), and Chairman of the organizing committee of the 6th Key Symposium on Nanomedicine, Grand Hophone Saltsjöbaden, Stockholm (2009), organized in collaboration with The Royal Swedish Academy of Sciences. He is also organizer of the 2nd Nobel Forum Mini-Symposium on Nanotoxicology, Karolinska Institutet, Stockholm (2010), which is an official part of the Karolinska Institutet Bicentennial Celebration.

The focus of his research has been on programmed cell death/apoptosis and in particular on cell death in the immune system, with an emphasis in more recent years on interactions of engineered nanomaterials with immunocompetent cells. He has published more than 110 peer-reviewed articles. He is a member of the editorial board of the journal *Apoptosis*, and served as guest editor of a special issue on nanomedicine in the *Journal of Internal Medicine* (2010). He was awarded the Alvarenga Prize by the Swedish Medical Society in 2003 and ranked as one of the top 5 most cited young cancer investigators in the country by the Swedish Cancer Foundation (2007). He is the former director (2002–2006) of the Summer Research School in Biomedicine at Karolinska Institutet. He has also served as the project coordinator (2005–2008) of an interdisciplinary research project sponsored by the Human Frontier Science Program (HFSP). B.F. was awarded a national Senior Investigator Award (6-year position) by the Swedish Cancer Foundation (2009) and a national Senior Investigator Award in Inflammation Research (6-year position) by the Swedish Research Council (2010).

Urs Häfeli, PhD

U.H. received a pharmacy degree from the Federal Institute of Technology in Zurich (Switzerland) and his PhD from the Paul Scherrer Institute in Villigen (Switzerland). He spent 1.5 years as a postdoctoral fellow at the Joint Center of Radiation Therapy at Harvard University, followed by 11 years as a research scientist in the Radiation Oncology Department of the Cleveland Clinic Foundation.

Urs Häfeli is an assistant professor at the Faculty of Pharmaceutical Sciences of the University of British Columbia in Vancouver (Canada). Every 2 years since 1996 he has organized and cochaired the International Conference on the Scientific and Clinical Applications of Magnetic Carriers. At its last meeting in May 2010, the conference attracted more than 370 participants from 43 countries to Rostock (for more information see www.magneticmicrosphere.com).

U.H.'s research has concentrated on magnetic drug targeting with microspheres and nanospheres and on the preparation of biocompatible and biodegradable monosized particles. In order to make uniform particles, he is investigating flow focusing based on silica and PDMS nanotechnologies. The overall aim is to use drug-releasing targeted particles for radiopharmaceutical cancer therapy.

Mukesh G. Harisinghani, MD

M.G.H. has been an associate professor of radiology at Harvard Medical School, Boston, Mass. (USA) since 2006. Additionally he has been director of the Abdominal MRI unit at Massachusetts General Hospital (MGH), Boston, Mass. (USA) since 2002 and director of the Clinical Discovery Program at the Center for Molecular Imaging and Research (CMIR), Charlestown, Mass. (USA) since 2004, as well as section editor of GU Radiology, *American Journal of Roentgenology,* since 2008.

He obtained his diploma in medical radiodiagnosis in 1991 and the diplomate from the National Board of Radio Diagnosis in 1992 at Bombay University (India). From 1992 to 1993, he was a clinical assistant at the Department of Radiology and Imaging, Hinduja National Hospital and Research Center, Bombay (India). Additionally, from 1992 to 1994 he was a lecturer in the Department of Radiology and Imaging, Lokmanya Tilak Municipal Medical College and Lokmanya Tilak Municipal General Hospital, Bombay (India). From 1993 to 1994, he worked as a specialist in the Department of Radiology and Imaging, Dr. Khalid Idriss Hospital, Jeddah (Kingdom of Saudi Arabia), and from 1994 to 1995 he worked as a medical officer in the Department of Radiology and Imaging, First Referral Unit Hospital, New Bombay Municipal Corporation, New Bombay (India). In 1997, he obtained the Certificate of Merit at the RSNA Annual Meeting, Chicago, Ill. (USA), and in 1998 he received a Roentgen Resident Research Award.

From 1995 to 1996, he was a research fellow in the Department of Abdominal Imaging and Intervention, from 1996 to 2000 he was a clinical fellow in radiology, from 1996 to 2000 he was a resident, and from 1998 to 1999 he was chief resident. In 2000, he was board certified by the American Board of Radiology. From 2000 to 2001, he passed fellowship training in the Division of Abdominal Imaging and Intervention, from 2000 to 2001 he worked as a clinical assistant in Radiology, from 2001 to 2003 he was an instructor, and from 2003 to 2006 he was assistant professor at MGH/Harvard Medical School.

Furthermore, he received several of awards, such as the 2003 Best Translational Research Award, MGH; the 2003 Magna Cum Laude, RSNA Annual Meeting, Chicago, Ill. (USA), 2004 Melvin M. Figley Fellowships in Radiology Journalism; the 2007 Society of Uroradiology Research Award, and the 2008 Best Departmental Research Award, MGH.

Christopher Heeschen, MD, PhD

Following his university studies in Germany (Munich and Berlin), C.H. obtained his MD in 1997 at the University of Hamburg (Germany). After 3 years of clinical training in internal medicine at the same university, he joined the Falk Cardiovascular Research Center at Stanford University (USA) in 1999 where he worked on basic mechanisms of angiogenesis and vasculogenesis. He obtained his PhD in 2001 and then joined Johann Wolfgang Goethe University in Frankfurt/Main as a staff scientist (he was promoted to junior group leader in 2003). In 2004, he became a professor of experimental oncology and transplantation at Ludwig Maximilian University in Munich (Germany).

He has published more than 80 articles in leading scientific journals such as the *New England Journal of Medicine, Nature Medicine, The Lancet, Cancer Cell, Cell Stem Cell,* the *Journal of Experimental Medicine,* and *PNAS,* among others. He has made major contributions to the field of stem and vascular cell biology that have advanced our understanding of the basic processes of stem cell function and trafficking. His research has been recognized through numerous international awards (most recently the Richtzenhain Award from the German Cancer Research Centre) and has merited him an ERC Advanced Investigator Grant for pancreatic cancer research. He moved to the CNIO with his lab in January 2009 as Head of the Stem Cells and Cancer Clinical Research Group in the newly founded Clinical Research Programme.

Ingrid Hilger, PhD

I.H. is professor and leader of the Department of Experimental Radiology at Jena University Hospital. She is one of the main researchers worldwide bringing magnetic heating closer to clinical practice. She brings extensive experience in the field, particularly in the analysis of NMP for magnetic heating purposes, MNP functionalization, the interaction of MNP with biological systems, MNP pharmacokinetics, the design of magnetic field application systems for clinical practice, and MRI/molecular imaging approaches.

Patrick Hunziker, MD

P.H. studied medicine at the University of Zurich (Switzerland). He received a doctoral degree based on his thesis work in experimental immunology from the University of Zurich and did further research in experimental hematology at the University Hospital in Zurich (Switzerland). He earned specialist degrees in internal medicine, cardiology, and intensive care medicine. As a fellow at Massachusetts General Hospital, Harvard Medical School, he worked on cardiac imaging in a joint project with the Massachusetts Institute of Technology, Cambridge. His professional activities in Europe, the USA, Africa, and China gave him a broad insight into the needs for the medicine of the future in a variety of settings. P.H. became involved in medical applications of nanoscience in the late 90s and has been the pioneer physician in nanomedicine in Switzerland since then. With improved prevention, diagnosis, and cure of cardiovascular disease as his main research topic, he worked in the nanoscience fields of atomic force microscopy, nanoptics, micro/nanofluidics, nanomechanical sensors, and polymer nanocarriers for targeting. He is the founding president of the European Society of Nanomedicine, cofounder of the European Foundation for Clinical Nanomedicine, and coinitiator of the European Conference for Clinical Nanomedicine and is clinically active as Deputy Head of the Clinic for Intensive Care Medicine at the University Hospital Basel (Switzerland). In November 2008, P.H. became professor of cardiology and intensive care medicine at the University of Basel.

Udo Jeschke, PhD

U.J., born in 1963, started his chemistry studies at the University of Rostock in 1984. In 1988, he gained work experience in the synthesis of X-ray contrast media at the pharmaceutical company Fahlberg-List in Magdeburg (Germany). In the year 1989, he finished his chemistry studies with the diploma and diploma thesis entitled: 'Maurer oxidation on unprotected carbohydrates'. From 1989 to 1992 he worked as a scientific assistant in the research group of Prof. Dr. C. Vogel at the University of Rostock. He received his PhD (thesis: 'Syntheses of D-galakturonic-acid-ester, -thioester, -amides and -mercaptales and amides of the 6-amino-6-deoxy-D-galac-tose') in 1992. After that, he worked in a postdoctoral position (DFG grant) in the research group of Prof. Dr. Hindsgaul at the University of Alberta (Canada). The title of this work was 'Enzyme synthesis of sugar nucleotide analogs'. From December 1993 to December 1994, he worked as a scientific assistant in the Chemistry and Biochemistry Department at the University of Rostock. In 1995, he applied for a postdoctoral position (DFG grant) in the research group of Prof. Dr. Vliegenthart at the University of Utrecht; the study was entitled: 'Isolation and analysis of human glycoproteins and lipopolysaccharides'.

He became a research group leader at the Department of Obstetrics and Gynecology of the University of Rostock in 1996, and from November 2002 to the present he has held the position of Head of the research laboratories at the Department of Obstetrics and Gynecology of Ludwig Maximilians University in Munich. He finished his 'habilitation' in 2003 and received the 'venia legendi' in biochemistry. In February 2010, he obtained the title of 'Extraordinary Professor' at the faculty for Medicine of Ludwig Maximilians University of Munich.

Challa S.S.R. Kumar, PhD

C.S.S.R.K. is currently the director of Nanofabrication and Nanomaterials at the Center for Advanced Microstructures and Devices (CAMD), Baton Rouge, La. (USA). He is also the president and CEO of Magnano Technologies, a company established to commercialize nanomaterials for applications in life sciences. His research interests are in developing novel synthetic methods, including those based on micro fluidic reactors, for multifunctional nanomaterials. He has also been involved in the development of innovative therapeutic and diagnostic tools based on nanotechnology. He has 8 years of industrial research and development experience working for Imperial Chemical Industries and United Breweries. He is the founding editor in chief of the *Journal of Biomedical Nanotechnology,* published by American Scientific Publishers, and series editor for the 20-volume book series *Nanotechnologies for the Life Sciences* (NtLS) and *Nanomaterials for the Life Sciences* (NmLS), published by Wiley-VCH. He has coauthored more than 45 scientific publications and is coinventor of over 10 patentable technologies.

Luis M. Liz-Marzán, PhD

L.M.L.-M. was born in Lugo (Spain) and graduated with a BS in chemistry from the University of Santiago de Compostela in 1988 where he also completed a PhD in the field of nanoparticle synthesis using microemulsions. He opened a new research line on metal nanoparticles and silica coating at Van't Hoff Laboratory, Utrecht (The Netherlands) with Prof. Albert Philipse as a postdoctoral research associate. After a short appointment at the University of Santiago de Compostela, L.M.L.-M. joined the University of Vigo in 1995 and was promoted to associate professor in 1999 and to professor (in the Department of Physical Chemistry) in 2006. He has also been a visiting professor at Tohoku University and The University of Michigan. He is currently research leader of the Colloid Chemistry Group, which has grown to have more than 30 research staff and students. L.M.L.-M.'s research has always been related to the study of nanomaterials on the basis of colloid chemistry. The main topics underlying his work include the control of metal nanoparticle morphology and tuning of their optical response, the ability to tailor the nanoparticles' surface chemistry through capping ligand exchange and inorganic coatings, the ability to organize colloidal nanoparticles into well-defined nanoparticle assemblies in 2 and 3 dimensions, and the exploitation of the optical (plasmonic) response of the nanoparticles and their assemblies for useful applications, such as diagnosis and detection. L.M.L.-M. has authored over 200 articles and presented over 75 invited lectures on this subject at international conferences. He is a fellow of the Royal Society of Chemistry and has been recognized with several research awards, including the Physical Chemistry Award of the Spanish Royal Society of Chemistry, the Alexander von Humboldt Research Award, and the Wilsmore Fellowship from the University of Melbourne. Additionally, he was recently elected President of the Colloid and Interface Science Group of the Spanish

Royal Societies of Chemistry and Physics. L.M.L.-M. is senior editor of the American Chemical Society Journal for Colloids and Surfaces *Langmuir.* He serves on the editorial board of *Nano Today,* the *Journal of Materials Chemistry,* the *Journal of Physical Chemistry,* and *Science of Advanced Materials.* He was guest editor for *MRS Bulletin* and the *Journal of Materials Chemistry.* He served on the editorial board of the *Journal of Colloid and Interface Science.* He was a member of the Particulate Fluids Processing Centre of the University of Melbourne (Australia).

Beat Löffler

B.L. studied communications science, philosophy, and political science. He received his MA from Freie Universität Berlin. In 1983, he started his first company for concepts and new media. Six years later he became director of the International High-Tech Forum of Messe Basel. After working a further 6 years in the new technology sector as developer and conference organizer and creating concepts for emerging technology events, he created in 1994 his present company, L&A Concept Engineering, and specialized in the fields of the development of innovation concepts and the development of science and knowledge promotion initiatives as well as leadership training and interdisciplinary bridging events. His fields of work are computational fluid dynamic materials science, energy technology, and life sciences. B.L. had numerous mandates for projects developed by his company and was mandated for 4 years as leader of the life sciences business development EMEA for the Japanese company NEC High Performance Computing. In 2005, he conceived and realized the European Summit for New Materials in Energy and Mobility in Essen (Germany) for 'Initiativkreis Ruhrgebiet'. He coconceived and realized the first World Summit for New Materials in Energy Technology in Lisbon (Portugal). In October 2006, he started the development of a concept for a conference on applied nanomedicine. In 2007, together with Patrick Hunziker, MD, he established the European Foundation for Clinical Nanomedicine and started up the European Society for Nanomedicine. The company has focused since then on novel materials with mandates of the European Materials Forum and the European Materials Research Society predominantly in nanotechnology. Since 2007, he has been CEO of the CLINAM Foundation and Secretary General of the European Society for Nanomedicine. He is responsible for the European Conference for Clinical Nanomedicine.

Scott McNeil, PhD

S.M. serves as director of the Nanotechnology Characterization Laboratory of the National Cancer Institute (NCI) at Frederick where he conducts characterization of nanomaterials and facilitates their translation to clinical applications. He advises industries and state and US governments on the development of nanotechnology and is a member of several governmental working groups related to nanotechnology policy and commercialization. Prior to joining NCI-Frederick (i.e. SAIC-Frederick), he served as senior scientist at Science Applications International Corp. (SAIC) where he transitioned basic research into government and commercial markets. The professional career of S.M. includes tenure as a combat arms officer and as chief of biochemistry at Tripler Army Medical Center. He is an invited speaker to numerous nanotechnology-related conferences and has several patents pending related to nanotechnology and biotechnology.

Jan Mollenhauer, PhD

J.M., born in 1968, received his PhD in 1998 from the University of Heidelberg (Germany). In 2003, he received his 'habilitation' in molecular medicine from the University Heidelberg, which was mentored by the Nobel laureate in Medicine, Prof. Harald zur Hausen. Since 2008, he has worked as a professor of molecular oncology at the University of Southern Denmark, Odense (Denmark). J.M. received the 2005 Future Award in Health Sciences and was listed in the 2007 edition of the Who Is Who of Emerging Leaders. Since 2010, he has been the director of the Lundbeckfonden Center of Excellence in Nanomedicine NanoCAN. His research focuses on pattern recognition molecules, synthetic biology, drug target discovery, and – since 2010 – on nanomedicine.

Stefan Odenbach, PhD

S.O. is a professor of magnetofluiddynamics at the Faculty of Mechanical Engineering of Technische Universität Dresden. His PhD in physics is from Ludwig-Maximilians-Universität Munich (1993); after a postdoc at the University of Wuppertal, he spent 9 years at the Center of Applied Space Technology and Microgravity of the University of Bremen where he was appointed with a professorship for complex fluids in 2004. He moved to Dresden in November 2005 where he is responsible for magnetofluiddynamics as well as for measuring technology.

Most of his work has been on the fluid mechanics of complex fluids, especially suspensions of magnetic nanoparticles but also liquid metals and metallic foams. In the past years, his group has started additionally extensive experimental work on X-ray microtomography and on biomedical applications of magnetic nanoparticles.

S.O. was the coordinator of the DFG Priority Program on Magnetic Fluids (2000–2006) and became the coordinator of DFG's Collaborative Research Centre on Electromagnetic Flow Control in Dresden in 2007. He is the German representative in the International Steering Committee on Magnetic Fluids, member of the editorial board of various journals, and one of the editors in chief of the *Journal of Applied Mathematics and Mechanics* (ZAMM).

He has written and edited several books and has roughly about 90 refereed publications.

Quentin Pankhurst, PhD

Q.P. is director of the Davy Faraday Research Laboratory and Wolfson Professor of Natural Philosophy – a post once held by Lord Ernest Rutherford – at the Royal Institution of Great Britain in London.

He runs research programs in bio- and nanomagnetism aimed at making practical advances in the use of magnetic nanoparticles in healthcare. These include medical imaging devices, targeted regenerative medicine, molecular imaging microscopy for living cells, and the development of multifunctional nanoparticles for therapy and diagnostics. He is a founder and CTO of Endomagnetics Ltd., a spin-out company which is currently running clinical trials on the SentiMag™, an intraoperative device for breast cancer surgery.

Alexander Pfeifer, PhD

A.P. works as a professor and research group leader at the Institute of Pharmacology and Toxicology, University Hospital Bonn. He has a long-standing interest in viral vectors, especially lentiviral vectors. An important aspect of his work is the targeting of viral vectors. Within the German Research Council (DFG) research unit Nanoparticle-Based Targeting of Gene- and Cell-Based Therapies (Nanoguide/FOR 917), he is developing approaches to target viral vectors as well as transduced cells with magnetic nanoparticles.

Annika Scheynius, PhD

A.S. was born in 1948 in Sweden and is the Head of the Clinical Allergy Research Unit at the Department of Medicine of Karolinska Institutet. After studying at Umeå University she completed her specialization in clinical bacteriology and 1983 her PhD at the University Hospital, Uppsala (Sweden). In 1986 she became associate professor ('Docent'). She has worked since 1990 at Karolinska Hospital (specializing in clinical immunology) and since 1995 as a professor of clinical allergy research. She obtained her professional experience from the Institute for Anatomy and Histology, Umeå University, as assistant instructor (1970–1972), from MD appointments at Umeå University Hospital (1972–1976), and from MD appointments at the Department of Clinical Microbiology, University Hospital Uppsala (1976–1988). She was named assistant professor by the Swedish Medical Research Council (1985–1989), assistant professor (1990–1992) and associated head of the Department of Clinical Immunology at Karolinska Institutet and Hospital (1990–1994), and assistant professor by the Swedish Medical Research Council in immunological disease mechanisms (1995). She won the Nicholson Research Fellowship for senior investigators (1991–1992) and the 2004 Swedish Research Council Excellent Investigator Award. She is a member of numerous very important committees, societies, and advisory boards, e.g. the Nobel Committee for Physiology or Medicine, the Nobel Assembly at Karolinska Institutet, the board of medicine and health within the Swedish Research Council, and the advisory board of *The Open Allergy Journal,* and she is a reviewer of manuscripts for scientific journals in the area of immunology, allergy, mycology, and dermatology.

Benjamin Shapiro, PhD

B.S., born in Jerusalem in 1973, received his bachelor's degree from the Aerospace Engineering Department at Georgia Tech and his PhD from the Control and Dynamical Systems option at Caltech. He has been at the University of Maryland for 9 years. His research is focused on the modeling, design, and control of micro-scale systems for chemical, biological, and now clinical applications. His primary appointment (as of January 2010) is with the Fischell Department of Bioengineering, he has a joint appointment with the Institute for Systems Research at the Nanocenter, and he is affiliated with the Applied Math and Scientific Computation Program. He is the recipient of a 2003 NSF CAREER award, has filed 16 patents (2 of which were awarded 1st and 3rd places as inventions of the year at Maryland), and is a Fulbright scholar (to Germany).

Matthias Taupitz, MD

M.T. was born in 1959. He studied medicine at Freie Universität Berlin from 1981 to 1986 and physics from 1982 to 1989. In 1986, he got his license to practice as a physician and in 1999 he received the qualification for diagnostic radiology from the Berlin Medical Council.

His scientific activities including clinical research comprise more than 20 clinical trials as coinvestigator, e.g. AMI-227 for lymph node assessment (altogether 3 phase III trials), gadobenate dimeglumine in MR angiography of the neck, abdominal and pelvic vessels (3 clinical phase II trials), gadobenate dimeglumine in breast MRI (phase II trials), Gd-EOB-DTPA in liver MRI (1 phase I trial, 1 phase II trail, and 1 phase III trial), and VSOP-C184 phase I, phase Ib, and phase II clinical trial. As principal investigator he supervises projects like the whole-heart contrast enhanced coronary MR angiography after injection of Vasovist® in patients with coronary artery disease (phase IIIb, duration <2 years) and noninvasive detection of prostate cancer using functional MRI in patients with histopathologically confirmed prostate cancer prostatectomy (phase II).

Since 2006, he has supported the Bundesamt für Arzneimittel und Medizinprodukte (BfArM; Federal Institute for Drugs and Medical Products) as a consultant. M.T. is a professor of radiology at the Department of Radiology, Charité Centrum 6, and Head of the Magnetic Resonance Imaging section and of the Experimental Radiology section at the Charité – Universitätsmedizin Berlin.

Vladimir P. Torchilin, PhD

V.P.T., PhD, DSc., is a distinguished professor of pharmaceutical sciences and director of the Center for Pharmaceutical Biotechnology and Nanomedicine, Northeastern University, Boston, Mass. (USA). He graduated from Moscow University with an MS in chemistry and also obtained there his PhD and DSc in polymer chemistry, chemical kinetics and catalysis, and the chemistry of physiologically active compounds in 1971 and 1980, respectively. In 1991, V.P.T. joined Massachusetts General Hospital and Harvard Medical School as Head of the Chemistry Program of the Center for Imaging and Pharmaceutical Research and as associate professor of radiology. Since 1998, he has been with Northeastern University, where he is Chairman of the Department of Pharmaceutical Sciences (1998–2008). His research interests have focused on biomedical polymers, polymeric drugs, immobilized medicinal enzymes, drug delivery and targeting, pharmaceutical nanocarriers for diagnostic and therapeutic agents, and experimental cancer immunology. He has published more than 300 original papers and more than 100 reviews and book chapters, has written and edited 10 books, including *Immobilized Enzymes in Medicine, The Handbook on Targeted Delivery of Imaging Agents, Liposomes: A Practical Approach, Nanoparticulates as Pharmaceutical Carriers, Multifunctional Pharmaceutical Nanocarriers, Biomedical Aspects of Drug Targeting,* and *Delivery of Protein and Peptide Drugs in Cancer,* and holds more than 40 patents. He is editor in chief of *Current Drug Discovery Technologies* and coeditor in chief of *Drug Delivery,* and he is on the editorial boards of many leading journals in the field including the *Journal of Controlled Release* (review editor), *Bioconjugate Chemistry, Advanced Drug Delivery Reviews,* the *European Journal of Pharmaceutics and Biopharmaceutics,* the *Journal of Drug Targeting, Molecular Pharmaceutics,* the *Journal of Biomedical Nanotechnology,* and a few others. Among his many awards, V.P.T. was the recipient of the 1982 Lenin Prize in Science and Technology (the highest scientific award in the former USSR). He was elected as a member of the European Academy of Sciences. He is also a fellow of the American Institute of Medical and Biological Engineering and of the American Association of Pharmaceutical Scientists (AAPS) and received the 2005 Research Achievements in Pharmaceutics and Drug Delivery Award from the AAPS, the 2007 Research Achievements Award from the Pharmaceutical Sciences World Congress, the 2009 AAPS Journal Award, and the 2009 International Journal of Nanomedicine Distinguished Scientist Award. From 2005 to 2006, he served as a president of the Controlled Release Society.

Lutz Trahms, PhD

L.T. was educated in physics at the Technical University of Aachen and at the Free University of Berlin, where he received his PhD in 1982. After a few years of postdoctoral activity he joined the Physikalisch-Technische Bundesanstalt Berlin in 1986 where he now is Head of the biosignal department. In his lab, 3 high-end multichannel SQUID devices are operated in magnetically shielded rooms, one of them being a unique low-noise home-made 304 channel vector magnetometer installed in the world's best magnetically shielded room, BMSR-2.

Throughout his career, his research fields have been nuclear magnetic resonance, biosignal processing, the SQUID-based biomagnetic measurement technique, and biomedical applications of magnetic particles. Most of his research and development activity is part of interdisciplinary research projects funded by national or European research programs, which include both academic and industrial partners. L.T. is coauthor or author of about 150 peer-reviewed research and review articles. Since 2000, he has been Chairman of the section Magnetic Methods in Medicine of the German Society of Biomedical Engineering.

Ladislaus Vekas, PhD

L.V. was born in 1945 in Arad (Romania). He was educated in physics at the University of Timisoara (graduated in 1968) and obtained his doctorate in physics (theoretical physics) in 1983 at A.I. Cuza Iasi University (Romania). From 1968 to 1970, he worked as a scientific researcher in the Center for Technical Research of the Romanian Academy-Timisoara Branch. He then worked for 4 years in the Research Center for Hydraulic Machinery at Politehnica University in Timisoara. He became senior researcher (3rd degree) and associate professor in the Laboratory of Magnetic Fluids at the same university (course on magnetohydrodynamics of magnetic fluids and applications; 1974–1991). From 1991 to 1997, L.V. was senior researcher (1st degree) and Head of the Laboratory of Magnetic Fluids, Research Center of Hydrodynamics, Cavitation and Magnetic Fluids, University Timisoara, and from 1994 to 1998 he was associate professor at the Faculty of Physics, West University, Timisoara (course on ferrofluids and applications). In 1996, he became director of the Institute of Condensed Matter Research, Timisoara. From 1997 to the present, L.V. conducted his research at the Laboratory of Magnetic Fluids, Center of Fundamental and Advanced Technical Research, Romanian Academy-Timisoara Branch, as a senior researcher (1st degree) and Head of the laboratory until 1997 and as the director from 2009. During the course of the DAAD and EU grants, he spent 3 months at the Physics section of Ludwig Maximilians University, Munich (1993–1994) and at the ZARM, University of Bremen (1997–1998).

His scientific interests and main research topics are: magnetic fluids: synthesis, structure, and properties; engineering and biomedical applications and magnetorheological fluids, and liquid-vapor phase transition: cavitation and boiling phenomena. He is responsible for 51 multiannual national research projects and is the Romanian coordinator of international projects [3 bilateral (2 in Hungary and 1 in Cyprus), 1 ERA-NET, and 1 FP7-NMP-2008-Large (MagPro^{2}Life)]. As an example of his cooperation with the industry one could emphasize the transfer of magnetic fluid sealing technology to ROSEAL Co. (Odorheiu Secuies, Romania; http://roseal.topnet.ro/ang/index1.html). Ninety-seven papers can be found in international refereed journals, as well as 6 book chapters, 1 book, and 27 papers in national journals, and 12 patents (RO). In 1984, he won the Dragomir Hurmuzescu Prize for Physics of the Romanian Academy. As a member he assists the European Academy of Sciences and Arts (EASA), Salzburg (1992 to the present), the International Steering Committee of Magnetic Fluids (1993 to the present), the COST expert groups on nanoscience nanotechnologies (NanoSTAG) (1998–2006, national), the National Council for Recognition of Academic Grades, Diplomas and Certificates – Materials Science and Chemical Engineering (2006 to the present), the National University Research Council (NURC)-Engineering Sciences Commission (2003–2004), the Romanian Space Agency-Scientific Council (1998 to the present), and the Romanian Association for Promotion of Magnetic Fluids as chairman (1998 to the present). He is also a member of the editorial board of *Romanian Reports in Physics*, edited by the Romanian Academy, and *Biomagnetic Research and Technology* (BMRT) (BioMedCentral, UK) (2004–2008).

Alexiou C (ed): Nanomedicine – Basic and Clinical Applications in Diagnostics and Therapy.
Else Kröner-Fresenius Symp. Basel, Karger, 2011, vol 2, pp 15–34

Nanocarriers for the Delivery of Drugs, Genes, and Diagnostics

Vladimir P. Torchilin

Department of Pharmaceutical Sciences and Center for Pharmaceutical Biotechnology and Nanomedicine, Northeastern University, Boston, Mass., USA

Abstract

Various pharmaceutical nanocarriers, including liposomes and polymeric micelles, are frequently used for the delivery of a broad variety of both soluble and poorly soluble pharmaceuticals. The use of nanoparticulate pharmaceutical carriers to enhance the in vivo efficiency of many drugs is now well established. Now, within the frame of this concept, it is important to develop multifunctional stimuli-responsive nanocarriers, i.e. nanocarriers that, depending on the particular requirements, can circulate long; target the site of the disease via nonspecific and/or specific mechanisms, such as the enhanced permeability and retention effect and ligand-mediated recognition; respond to local stimuli characteristic of the pathological site by, for example, releasing an entrapped drug or deleting a protective coating under the slightly acidic conditions inside tumors, thus facilitating the contact between drug-loaded nanocarriers and cancer cells, and even provide an enhanced intracellular delivery of an entrapped drug. Additionally, these carriers can be supplied with contrast moieties to follow their real-time biodistribution and target accumulation. Among new developments to be considered in the area of multifunctional pharmaceutical nanocarriers are: drug- or DNA-loaded delivery systems additionally decorated with cell-penetrating peptides for enhanced intracellular delivery, 'smart' multifunctional drug delivery systems which can reveal/expose temporarily hidden functions under the action of certain local stimuli characteristic for the pathological zone, new means for controlled delivery and release of siRNA, and nanocarrier-based new targeted contrast agents for diagnostic imaging.

The use of nanoparticulate pharmaceutical carriers to deliver a broad variety of pharmaceutical agents, including drugs, genes, and diagnostic, established itself well over the past decade both in pharmaceutical research and in the clinical setting. Various pharmaceutical nanocarriers, such as polymeric nanoparticles, nanocapsules, liposomes, micelles, solid lipid nanoparticles, niosomes, carbon nanotubes, dendrimers, and many others, are widely used for experimental and clinical delivery of therapeutic and diagnostic agents [1, 2]. Multiple publications on drug delivery systems (DDS) describe DDS that combine several properties/functions, such as prolonged blood circulation, the ability to specifically accumulate in target areas, contrast (diagnostic) properties, and sensitivity to various local stimuli [3]. Current efforts are directed to preparing scalable and easy-to-use multifunctional DDS, which are expected to bear a sufficient load of a drug or DNA-related material, circulate long, accumulate in the desired area, provide real-time information about their biodistribution and accumulation, deliver their load inside cells, and respond to certain local stimuli. Usually, the desired multifunctionality of pharmaceutical nanocarriers is achieved by proper combination of various functional moieties/modifiers on their surface (fig. 1).

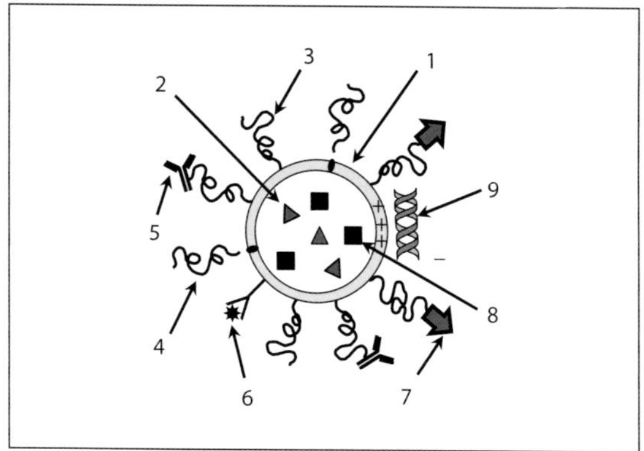

Fig. 1. Schematic structure of a multifunctional nanocarrier. The nanocarrier (1) is loaded with a drug (2) and can be additionally modified (separately or simultaneously in any combination) with a nondetachable (3) or detachable (4) protective polymeric coat for longevity; the targeting ligand attached to the distal tips of protective polymer chains (5), contrast moiety (6) for visualization, and cell-penetrating moiety (7) for intracellular delivery are shown. The system can also carry superparamagnetic particles (8) for visualization or magnetic targeting and positively charged sites for complexing DNA/ODN/siRNA (9).

Longevity of Pharmaceutical Nanocarriers

Long-circulating pharmaceutical nanocarriers represent an important area of biomedical research [4–6]. The longevity of drug carriers allows for maintaining a required level of a pharmaceutical agent in the blood for extended time intervals. Long-circulating drug-containing nanomedicines can slowly accumulate via an impaired filtration mechanism or enhanced permeability and retention (EPR) effect in pathological sites with compromised vasculature (tumors, inflammations, and infarcts) and enhance the delivery of pharmaceutical agents into such areas [7, 8] (fig. 2). The prolonged circulation can also help to achieve a better targeting effect for targeted (specific ligand-modified) drugs and drug carriers allowing for more time for their interaction with the target.

The most usual way to obtain long-circulating nanoparticles is coating them with certain hydrophilic and flexible polymers, such as polyethylene glycol (PEG), as was first suggested for liposomes to sterically hinder interactions of blood components with their surface and reduce the binding of plasma proteins with PEG particles [9–13]. Although quite a few polymers have been tried as steric protectors for nanoparticulate drug carriers [5], the majority of research on long-circulating drugs and drug carriers is still performed with the use of PEG as a sterically protecting polymer because of the excellent solubility of PEG in aqueous solutions, the high flexibility of its polymer chain, its very low toxicity, immunogenicity, and antigenicity, lack of accumulation in the reticuloendothelial system cells, and the minimal influence on specific biological properties of modified pharmaceuticals [14, 15].

To make PEG capable of incorporation into the liposomal membrane for obtaining long-circulating liposomes, the reactive derivative of hydrophilic PEG is a single terminus-modified with a hydrophobic moiety [usually, the residue of phosphatidylethanolamine (PE) or a long-chain fatty acid is attached to a PEG-hydroxysuccinimide ester] [9, 16]. From the clinical point of

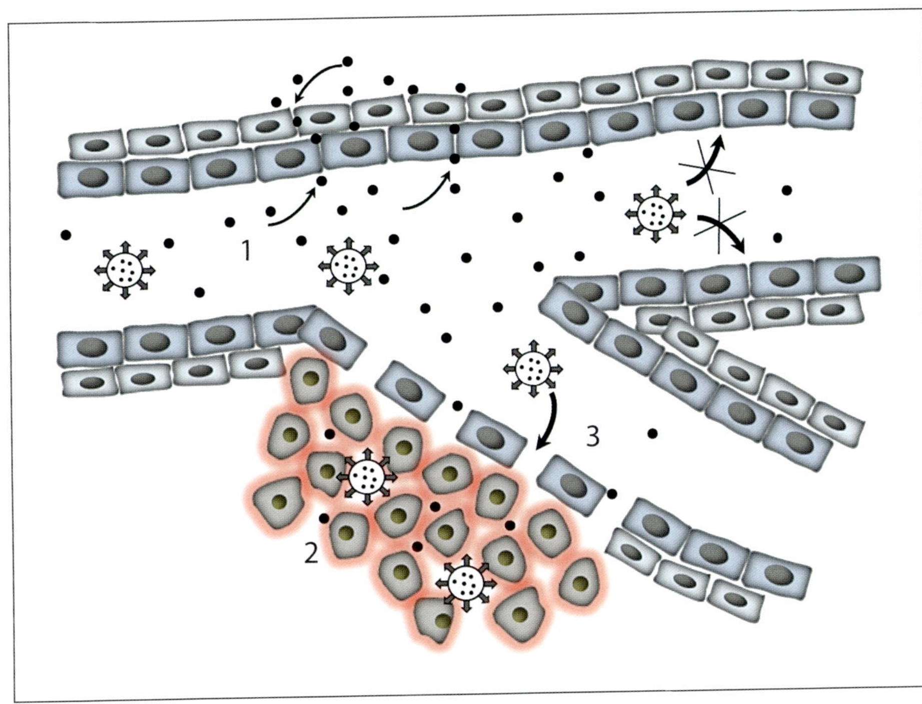

Fig. 2. Schematics of the EPR effect. Normal continuous vasculature does not allow the extravasation of large molecules or small particles (1) in normal tissues; however the extravasation and accumulation in the interstitium become possible in pathological areas (tumors, infarcts, and imflammations) (2) with the compromised ('leaky') vasculature (3).

view, it is important that various long-circulating liposomes of a relatively small size (100–200 nm) were shown to effectively accumulate in many tumors via the EPR mechanism [7, 8, 17, 18]. As a result, PEG-coated and other long-circulating liposomes were prepared containing a variety of anticancer agents, such as doxorubicin, arabinofuranosylcytosine, adriamycin, and vincristine [19–21]. The biggest success was achieved with PEG-liposome-incorporated doxorubicin, which has already demonstrated good clinical results [8, 22].

Since the stability of PEGylated nanocarriers may not always be favorable for drug delivery (drug-containing nanocarriers may become unable to easily release the drug or the presence of the PEG coat on its surface may preclude the contents from escaping the endosome), an additional function can be added to long-circulating PEGylated pharmaceutical carriers, which allows for the detachment of PEG chains under the action of certain local stimuli characteristic of pathological areas, such as a decreased pH value or increased temperature noted for inflamed and neoplastic areas. Polymeric components with pH-sensitive (pH-cleavable) bonds are used to produce stimuli-responsive DDS that are stable in the circulation or in normal tissues. However, they acquire the ability to degrade and release the entrapped drugs in body areas or cell compartments with lowered pH, such as tumors, infarcts, inflammation zones or cell cytoplasm or endosomes [23, 24]. Up to now, a variety of liposomes [25, 26] and micelles [27, 28] have been described which are capable of releasing such drugs as adriamycin [29], paclitaxel (PCT) [30], doxorubicin [31], and DNA [32–34] in acidic cell compartments (endosomes) and pathological body areas under acidosis.

Synthetic amphiphilic polymers have also been used for steric stabilization of polymeric particles with surfaces of a hydrophobic nature in order to prolong their circulation in the blood and to alter their biodistribution. Thus, long-circulating particles have been prepared with insoluble (solid) PLAGA cores and water-soluble PEG shells covalently linked to the core [35]. Similar effects on the longevity and biodistribution of microparticulate drug carriers might be achieved by direct chemical attachment of protective polyethylene oxide chains onto the surface of preformed particles [36]. PEGylation was performed with poly(ε-caprolactone) nanoparticles and significantly suppressed their opsonization and phagocytosis [37]. For a variety of polymeric nanoparticles, the increase in the content of PEG results in increased blood circulation time and a decreased liver uptake [38, 39].

Recently, many studies have been performed with surface modification of superparamagnetic nanoparticles. Similarly to other nanocarriers, PEG-modified magnetite nanospheres demonstrated increased colloidal stability and improved localization in lymph nodes [40]. Grafting PEG onto the surface of gold particles via mercaptosilanes expectedly resulted in decreased protein adsorption onto modified particles and less platelet adhesion [41]. PEGylated gold nanoparticles are suggested, among other applications, for photothermal tumor therapy (ablation) after they accumulate in the tumor via the EPR effect because of their long-circulation property [42]. PEG-coated gold nanoparticles are also suggested as contrast media for computed tomography [43]. Dendrimers, which are currently considered promising drug delivery carriers, have also been modified with PEG to increase their longevity and accumulation, e.g. in the case of PEGylated polyamidoamine (PAMAM) [44]. PEGylated methotrexate-conjugated poly-L-lysine dendrimers were shown to accumulate efficiently in solid tumors via the EPR effect [45]. Importantly, PEGylating reduces the toxicity of positive charge-bearing dendrimers, such as PAMAM [46, 47]. Biomedical application of carbon nanotubes also depends on their PEGylation and this can be achieved by simple sonication of single-wall nanotubes in the presence of PEG-PE [48]. PEGylated nanotubes demonstrate longer circulation in mice upon intravenous administration, which is a must for their future applications [49]. PEGylated carbon nanotubes were also suggested as vehicles for the delivery of antisense oligonucleotides (ODN) [50]. Quantum dots were also modified by PEGylation, and it was demonstrated in mice that PEGylated quantum dots, similar to other nanoparticulates, show a lesser uptake in mononuclear phagocyte system organs (spleen and liver) and prolonged blood circulation [51]. In general, PEGylatin increases the biocompatibility of quantum dots and decreases their toxicity [52, 53].

Polymeric micelles, including micelles made from conjugates of PEG and PE, provide some important advantages as carriers for poorly soluble drugs, including anticancer drugs, such as a small size (10–100 nm), good solubilization efficiency, and high stability [54–56]. Due to their small size they can accumulate efficiently in pathological tissues with leaky vasculature, such as tumors and infarct tissues, via the EPR effect [7, 54]. These micelles are structured in such a way that the outer PEG corona serves as an efficient steric protector in biological media allowing for prolonged circulation of the micelles. On the other hand, the phospholipid residues which represent the micelle core are extremely hydrophobic and can solubilize various poorly soluble drugs such as PCT [57], camptothecin [58], and meso-5,10,15,20-tetraphenyl-21H,23H-porphine [59]. Interestingly, not only can polymeric micelles (such as PEG-PE-based ones) serve as delivery vehicles for drugs but they can also protect drugs from inactivation under the action of the biological surroundings. Thus, camptothecin in

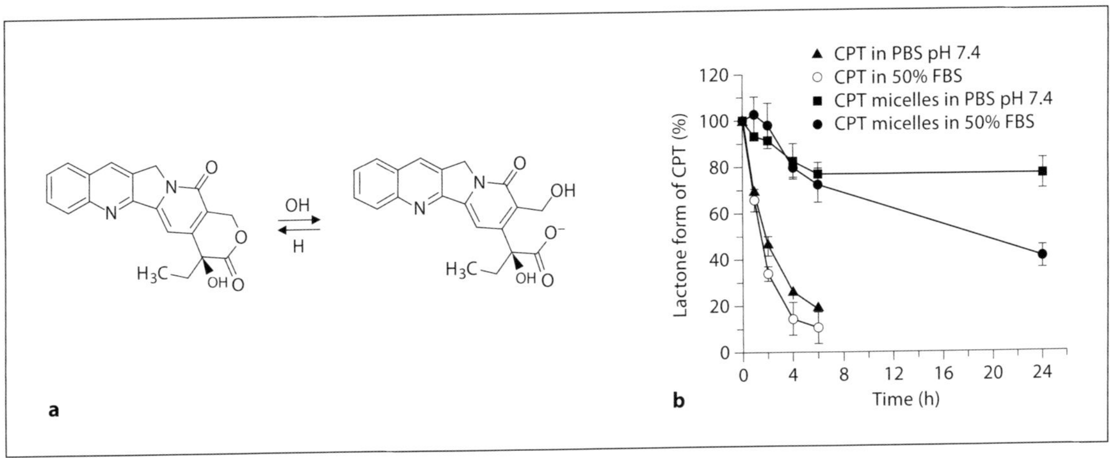

Fig. 3. A micellear carrier stabilizes the active form of the drug. Anticancer drug camptothecin (CPT) in PEG-PE micelles. **a** pH-dependent transformation of the drug from the active lactone form into the inactive carboxylate form. **b** Kinetics of the lactone-carboxylate transformation at different pH values for the free and micelle-incorporated drug. In micelles, drug inactivation is significantly inhibited.

micelles demonstrates a much slower transformation from the active lactone form into the inactive carboxylate form (fig. 3). The critical micelle concentration values of PEG-PE conjugates are very low (in a high-nanomolar-to-low-micromolar range) because of strong hydrophobic interactions between double acyl chains of phospholipids residues which make them extremely stable and prevent their dissociation upon in vivo application [60, 61]. Moreover, PEG-PE micelles can also be made tumor targeted by attaching the specific ligands, e.g. anti-cancer nucleosome-specific monoclonal antibody 2C5 (mAb 2C5), to their outer shell, which allows the modified micelles to recognize a broad variety of tumor cells via tumor cell surface-bound nucleosomes [59, 62–65].

Recently, PEG-PE-based micelles have been suggested as nanocarriers for the stabilization and delivery of siRNA [66]. With this in mind, siRNA was reversibly modified with a lipid moiety via the S-S bond and incorporated into PEG-PE micelles. Such incorporation completely prevented nucleolytic degradation of siRNA because of shielding properties of neighboring PEG chains. Inside cells, however, the S-S bond is reduced by high intracellular glutathione, releases in a free from, and downregulates the target gene (fig. 4).

Combination of Longevity and Targetability

To achieve better selective targeting of PEG-coated nanocarriers, it is advantageous to attach the targeting ligand to the particles not directly but via a PEG spacer arm so that the ligand is extended outside of the dense PEG brush excluding steric hindrances for its binding to the target receptors (fig. 5). Currently, various advanced technologies are used, and the targeting moiety is usually attached above the protecting polymer layer by coupling it with the distal water-exposed terminus of an activated liposome-grafted polymer molecule [67, 68]. The ligand (antibody, another protein, peptide, or carbohydrate) attached to the carrier surface may, however, increase the rate of its uptake by the reticuloendothelial sys-

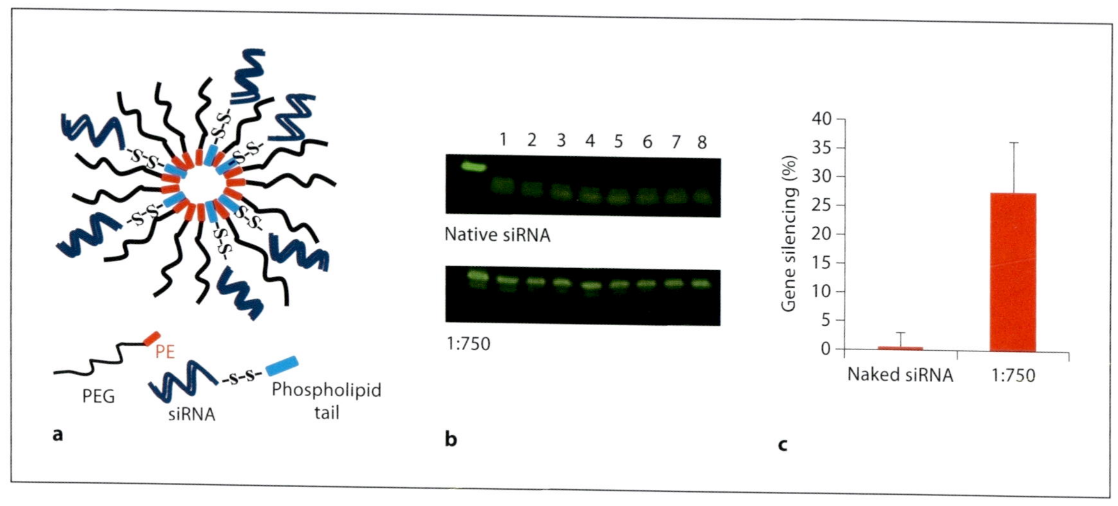

Fig. 4. siRNA in PEG-PE micelles. **a** Schematic structure of the mixed micelles built from PEG-PE and siRNA-S-S-PE conjugate (the S-S bond in the latter can be reduced by high intracellular glutathione, thus releasing free siRNA). **b** Stability of siRNA in PEG-PE micelles against nucleolytic degradation. Free siRNA degrades completely already after a few minutes (band 1), while siRNA in PEG-PE micelles (siRNA; PEG-PE molar ratio 1:750) remains completely intact even after 24 h (band 8). **c** Micelles with siRNA-GFP significantly downregulate GFP production in GFP-transfected cells confirming the liberation of the free active siRNA inside cells.

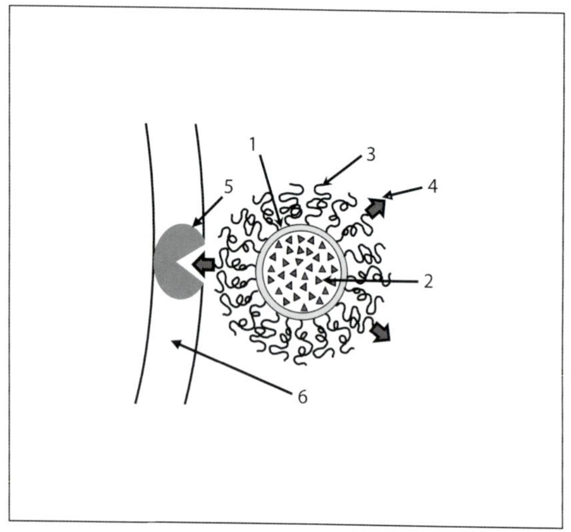

Fig. 5. Schematic structure of the long-circulating targeted nanocarrier. The nanocarrier (1), drug (2), sterically protecting polymer (usually PEG) (3) grafted onto the surface of the nanocarrier, targeting ligand (antibody, folate, transferring, or others) (4) chemically coupled with distant tips of some of the protecting polymer chains, specific receptor (5) targeted with the attached ligand, and cell membrane (6) are shown.

tem despite the presence of sterically protecting grafts [see for example 69]. In addition, ligand-bearing long-circulating nanocarriers could facilitate the development of an unwanted immune response (as was shown with the rise in anti-liposome antibodies), the extent of which depends on the type of ligand (small peptides or Fv fragments are less immunogenic than a complete IgG molecule) and the liposome composition [70, 71]. These issues require certain optimization efforts when preparing multifunctional nanocarriers with longevity and targetability.

Various targeting ligands were attached to PEGylated pharmaceutical nanocarriers to achieve improved delivery [72]. Thus, HER2-overexpressing tumors were targeted using anti-HER2 PEGylated doxorubicin-loaded liposomes [70, 73]. Antibody CC52 against rat colon adenocarcinoma CC531 attached to PEGylated liposomes provided specific accumulation of liposomes in a rat model of metastatic CC531 [74]. A nucleosome-specific monoclonal antibody (mAb

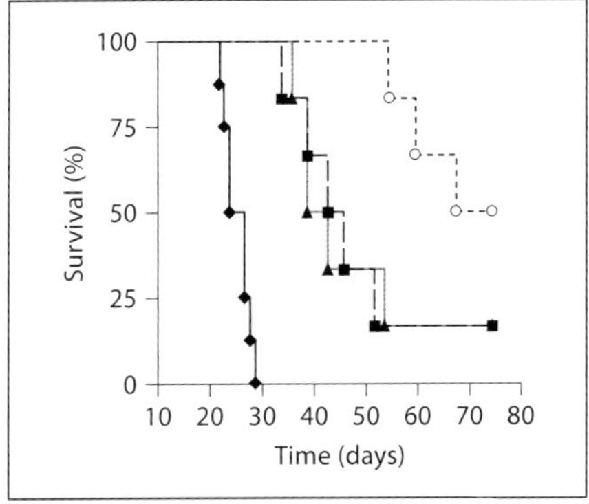

Treatment	Median survival time (days)	p value
Untreated	25.5	–
LS-Doxil	44.5	p < 0.005 vs. untreated
IgG-ILS-Doxil	41.0	p > 0.05 vs. LS-Doxil
2C5-ILS-Doxil	71.5	p < 0.05 vs. LS-Doxil and IgG-ILS-Doxil

Fig. 6. Antibody (mAb 2C5)-targeted Doxil demonstrates significantly higher activity (longer survival of experimental animals) in inhibiting U-87 brain tumors (orthotopic in vivo model in mice) than nontargeted Doxil.

2C5) capable of recognition of various tumor cells via the tumor cell surface-bound nucleosomes significantly improved Doxil® targeting to tumor cells and increased its cytotoxicity [75] both in vitro and in vivo in different test systems [75, 76] including an intracranial human brain U-87 tumor xenograft in nude mice [77] (fig. 6). PEGylated gold nanoparticles were additionally conjugated to monoclonal F19 antibodies as targeted labeling agents for human pancreatic carcinoma tissues [78]. Similarly, PEGylated carbon nanotubes were additionally modified with antibodies specifically targeting cytoplasmic compartment [79].

Since transferrin (Tf) receptors (TfR) are overexpressed on the surface of many tumor cells, antibodies against TfR as well as Tf itself are among popular ligands for targeting various nanoparticulate drug carriers including liposomes to tumors and inside tumor cells [80]. To enhance the interaction of Tf with TfR on TfR-overexpressing tumor cells and increase the efficiency of transfection, Tf could be also attached to the surface of nanoparticles via the PEG spacer [81]. Many studies involve the coupling of Tf to PEG on PEGylated liposomes in order to combine longevity and targetability for drug delivery into solid tumors [82]. Nanoparticles made of poly(lactic acid) were surface modified with PEG and with anti-TfR monoclonal antibody to produce PEGylated immunoparticles with a size of about 120 nm [83]. A similar approach was applied to deliver into tumors agents for photodynamic therapy [84, 85] and for intracellular delivery of cisplatin into gastric cancer [86]. Tf-coupled doxorubicin-loaded PEgylated liposomes demonstrated increased binding and toxicity against tumors in rats [87].

Folate receptors are also frequently overexpressed in many tumor cells. This is why targeting tumors with folate-modified nanocarriers represents a popular approach. Numerous folate-targeted liposomal systems for cancer have been described [see reviews in 88, 89]. Folate was also attached to the surface of cyanoacrylate-based nanoparticles via activated PEG blocks [90]. Similarly, PEG-polycaprolactone-based particles were surface modified with folate and, after being loaded with PCT, demonstrated increased cytotoxicity [91]. Superparamagnetic magnetite nanoparticles were modified with folate (with or without PEG spacer) and demonstrated better uptake by cancer cells, which can be used for both diagnostic [magnetic resonance imaging (MRI) agents] and therapeutic purposes [92, 93]. Folate-modified PEG-grafted hyperbranched polyethyl-

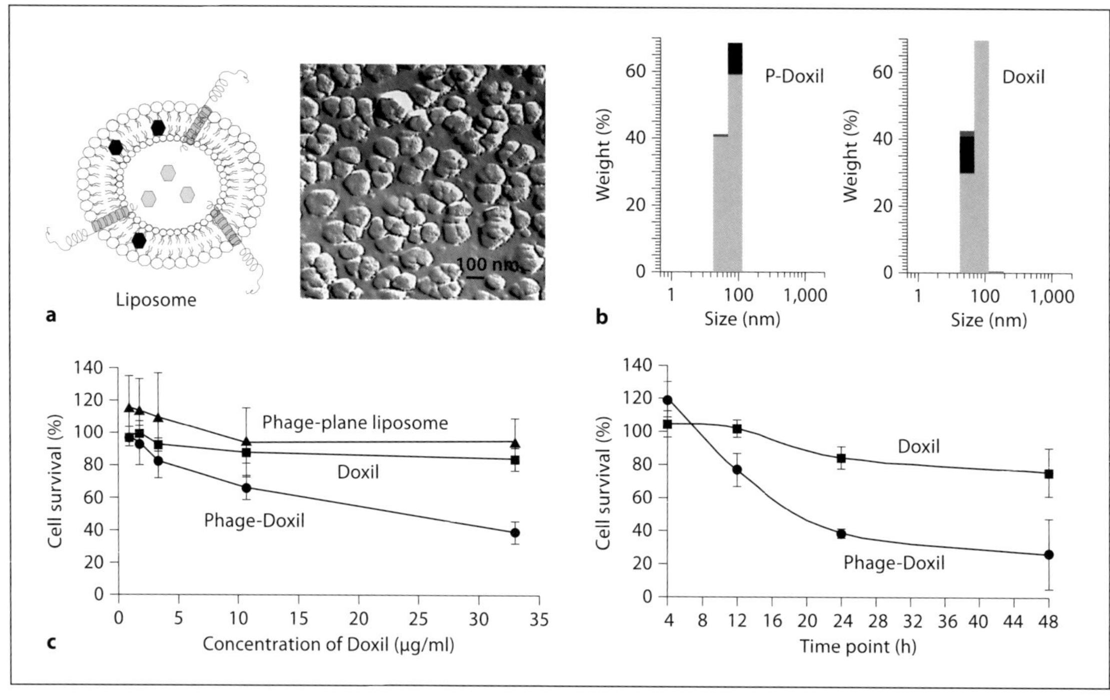

Fig. 7. Phage coat protein-targeted nanopreparation. **a** Amphiphilic phage coat protein can spontaneously incorporate into the membrane of the drug-loaded liposomes resulting in the targeted liposome. **b** Phage coat protein-modified Doxil liposomes demonstrate unchanged morphology and the same size distribution pattern. **c** Doxil modified with phage coat protein obtained from the phage specific towards MCF7 cancer cells demonstrates a significantly enhanced cytotoxicity towards the target cells. P-Doxil = Phage Doxil.

eneimine (PEI) was used for tumor-targeted gene therapy [94]. Folate-modified PEGylated quantum dots were used for the visualization of receptor-mediated endocytosis [95].

Vasoactive intestinal peptide was attached to PEG liposomes with radionuclides to target them to vasoactive intestinal peptide receptors of the tumor, which resulted in enhanced breast cancer inhibition in rats [96]. PEG liposomes were also targeted by RGD peptides to integrins of the tumor vasculature and, being loaded with doxorubicin, demonstrated increased efficiency against C26 colon carcinoma in a murine model [97]. PEGylated liposomes modified with a fibronectin-mimetic peptide were used to target colon cancer cells [98] and prostate cancer [99]. PEGylated particles have also been modified with cell-penetrating peptides (CPP) for better intracellular delivery [100]. Specific attention was paid to the way functions responsible for longevity and targeting should be assembled on the surface of the nanocarriers for optimal performance [101].

Recently, the use of coat proteins of phages specifically selected against various biological targets was suggested for targeted delivery of drug-loaded nanocarriers [102–105]. Since phage coat proteins are amphiphilic, they can be easily incorporated via the hydrophobic blocks into, for example, liposomal membranes or micelle cores leaving their hydrophilic recognition moieties outside. Phage protein-modified PEGylated doxorubicin-loaded liposomes (Doxil) demonstrated effective recognition and killing of target cancer cells (fig. 7).

Table 1. Stimuli utilized to control drug delivery systems

Stimuli	Stimuli origin
pH	Decreased pH in pathological areas, such as tumors, infarcts, and inflammations, because of hypoxia and massive cell death; relatively low pH in endosomes and cytoplasm
Redox conditions	Increased concentration of glutathione inside many pathological cells
Temperature	Inflammation-provoked hyperthermia
	The temperature can be increased in target tissues by locally applied ultrasound or by accumulating magnetic nanoparticles in the target with subsequent action of a high-frequency electromagnetic field
Ultrasound	Sonication can be applied to the body to facilitate DDS penetration into cells and drug/gene release
Magnetic field	A magnetic field can concentrate magneto-sensitive DDS in the required areas

Stimuli-Sensitive Nanocarriers

Multifunctional long-circulating and targeted pharmaceutical nanocarriers can be functionalized still further by imparting them with additional sensitivity to various stimuli – both characteristic for the target pathological zone as well as externally applied (see table 1). Stimuli characteristics for pathological tissues include pH and redox conditions, while temperature can serve as a local stimulus both within the tissue (inflammation and local hyperthermia) and from the outside. Thus, for example, the intratumoral pH value in tumors may drop to 6.5, i.e. approximately 1 pH unit lower than in normal blood, because of hypoxia and massive cell death inside the tumor [106, 107], and it drops still further inside cells, especially inside endosomes (5.5 and even lower) [108]. At the same time, the intracellular concentration of glutathione (i.e. redox potential) in cancer cells is significantly (hundreds-fold) higher than the normal extracellular level of glutathione [109]. Stimuli such as ultrasound and (electro)magnetic fields could be applied only 'artificially' and mainly from the outside.

pH sensitivity was used to modify in a desired way drug/DDS behavior in the pathological areas with a decreased pH value, such as tumors, infarcts, and inflammations. Thus, pH-sensitive liposomes have been made to destabilize and release the incorporated load at lowered pH values. Such liposomes contained phospholipids capable of protonation or formation of nonbilayered structures at a decreased pH and destabilizing liposomal or liposomal and endosomal membranes with the subsequent drug/DNA release from liposomes and endosomes into the cytoplasm [110–113]. In addition to membrane-destabilizing lipid components, there exists a large family of membrane-destabilizing anionic polymers that also can enhance the endosomal escape of various drugs and biomacromolecules [114]. Serum stable, long-circulating PEGylated pH-sensitive liposomes were also prepared using the combination of PEG and pH-sensitive terminally alkylated copolymer of N-isopropylacrylamide and methacrylic [24]. Additional modification of pH-sensitive liposomes with an antibody results in pH-sensitive immunoliposomes. Successful application of pH-sensitive immunoliposomes has been demonstrated in the delivery of a variety of molecules including fluorescent dyes, antitumor drugs, proteins, and DNA [115].

Polymeric micelles can also demonstrate pH sensitivity and the ability to escape from endosomes. Thus, micelles prepared from PEG-poly(aspartate hydrazone adriamycin) easily release an active drug at lowered pH values typical for endosomes and facilitate its cytoplasmic delivery and toxicity against cancer cells [116]. Alternatively, micelles for intracellular delivery of antisense ODN were prepared from ODN-PEG conjugates complexed with a cationic fusogenic peptide, KALA, and provided a much higher intracellular delivery of the ODN that could be achieved with free ODN [117]. PCT was loaded into mixed micelles, which could undergo dissociation into unimers and drug liberation even above the critical micelle concentration value because of the ionization of their components at certain pH values [118].

Polymeric components with pH-sensitive (pH-cleavable) bonds are used to produce stimuli-responsive DDS that are stable in the circulation or in normal tissues; however, they acquire the ability to degrade and release the entrapped drugs in body areas or cell compartments with lowered pH, such as tumors, infarcts, inflammation zones or cell cytoplasm or endosomes [23, 24, 119]. Combination of liposome pH-sensitivity and specific ligand targeting for cytosolic drug delivery utilizing decreased endosomal pH values was described for folate- and Tf-targeted liposomes [120, 121]. The stimuli sensitivity of PEG coats can also allow for the preparation of multifunctional DDS with temporarily 'hidden' functions, which under normal circumstances are 'shielded' by the protective PEG coat; however, they become exposed after the PEG detaches. Such systems require that multiple functions attached to the surface of the nanocarrier function in a certain coordinated way [122].

The idea of using temperature-sensitive nanocarriers naturally came from the fact that many pathological areas demonstrate distinct hyperthermia. Additionally, there exist various means to heat the required area in the body. A significantly greater fraction of the intravenously administered liposomes and other nanocarriers accumulated in the tumor mass upon heating to 42 °C in a human ovarian carcinoma xenograft model and a higher concentration and effectiveness was observed for doxorubicin delivered into tumors in temperature-sensitive liposomes [123, 124]. Temperature-sensitive liposomes frequently include dipalmitoylphosphatidylcholine (DPPC) as the key component since liposomes usually become leaky at the gel-to-liquid crystalline phase transition and this transition for DPPC takes place at 41 °C [125]. Similarly, polymeric micelles can be made temperature sensitive by assembling them from amphiphilic copolymers in which one of the blocks demonstrate properties similar to NIPAM [126].

Drug carriers, such as microcapsules, can be loaded not only with the drug alone but also with magnetic nanoparticles allowing for manipulation of such capsules in a magnetic field, or with metallic nanoparticles which can respond to an external electromagnetic field and control the rate of drug release by oscillating or heating the carrier [127]. In nanomedicine, iron oxide nanoparticles with a particle size of approximately 4–10 nm (superparamagnetic iron oxide nanoparticles; SPION) have drawn special interest [see for example 128–134]. The magnetic drug targeting concept (guiding magnetically susceptible particles towards the intended pathology site under the influence of external magnets) introduced by Widder et al. [135] has recently received increased attention with advances in nanotechnology. Gang et al. [136] have demonstrated targeting of magnetic poly(ε-caprolactone) nanoparticles loaded with gemcitabine in a pancreatic cancer xenograft mouse model using external magnets. Cinteza et al. [137] have also reported coloading polymeric micelles of diacylphospholipid-poly(ethylene glycol) with the photosensitizer drug 2-[1-hexyloxyethyl]-2-devinyl pyropheophorbide-a and magnetic SPION for magnetic drug targeting in vitro. Alexiou et al. [138, 139] have used mitoxantrone-loaded SPION and targeted them to VX2 squamous cell carcinoma in rabbits by using external magnets.

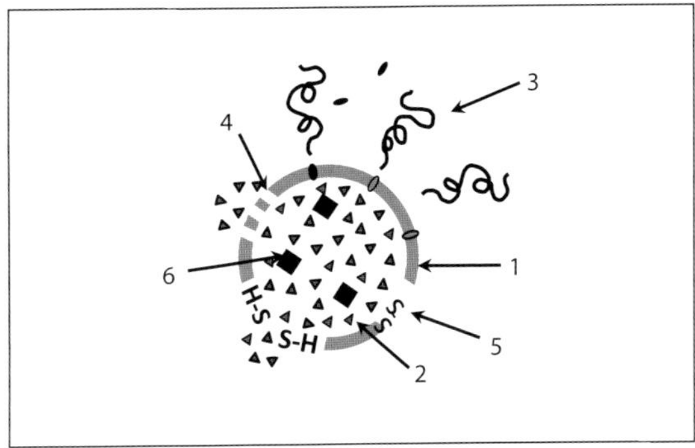

Fig. 8. Schematic picture of different stimuli acting on the stimuli-sensitive nanocarrier and expected responses shows the nanocarrier (1); drug (2); protective polymeric coating (3) attached to the surface of the nanocarrier via pH-sensitive bonds, which could be detached and removed by the action of a lowered pH in certain pathological areas (tumors) or inside individual cellular compartments (cytoplasm and endosome); temperature-sensitive coating or components of the carrier (4), which can be influenced by heat (hyperthermia in certain pathological areas or heat brought upon by an external source) to destabilize the carrier and allow for drug release; redox-sensitive coating or components of the carrier (5), which can be influenced by changing redox conditions (increased glutathione), for example by reducing S-S bonds, and allow for drug release, and particles of magnetic material (SPION) (6), which can allow the whole nanocarrier to be transported to the required site under the action of an external magnetic field.

The application of the external ultrasound to controlled drug delivery and release from nanocarriers is a relatively novel approach, although some publications on this subject go back to 1998 when acoustically active lipospheres were described containing paclitaxel [140]. The whole concept is based on making DDS which upon accumulation in the required areas can be made leaky by the locally applied external ultrasound and liberate incorporated drugs or genes. There already exists a whole set of promising data on drug and gene delivery by ultrasound-sensitive drug carriers, some of which are reviewed in reports by Unger et al. [141] and Liu et al. [142]. Acoustically active liposomes containing a small quantity of a certain gas (air) or perfluorated hydrocarbon and initially developed as ultrasound contrast agents can be loaded with various drugs and release these drugs after being damaged by an applied ultrasound [143, 144]. Polymeric micelles have also been prepared which can incorporate various drugs, such as doxorubicin, and release them after ultrasonication, which can also assist in delivering such DDS inside cells [145, 146]. See the general scheme of stimuli-sensitive DDS in figure 8.

Intracellular Delivery of Pharmaceutical Nanocarriers

Intracellular transport of different biologically active molecules is one of the key problems in drug delivery in general since many pharmaceutical agents, including various large molecules (proteins, enzymes, and antibodies) and even drug-loaded pharmaceutical nanocarriers, need to be delivered intracellularly to exert their therapeutic action inside the cytoplasm or on the nucleus or other specific organelles, such as lysosomes, mitochondria, or endoplasmic reticulum. In addition, the introcytoplasmic drug delivery in cancer treatment may overcome such important obstacles in anticancer chemotherapy as multidrug resistance.

The addition of the positive charge to the nanocarrier can significantly enhance its uptake by cells, and the use of cationic lipids and cationic polymers as transfection vectors for intracellular delivery of DNA was suggested quite a few years ago [147, 148] and was recently reviewed in a report by Elouahabi and Ruysschaert [149]. Complexes between cationic lipids such as Lipofectin®, an equimolar mixture of N-[1-(2,3-dioleyloxy)propyl]-N,N,N-trimethylammonium chloride (DOTMA) and dioleoyl phosphatidylethanolamine (DOPE), and DNA (lipoplexes) and complexes between cationic polymers, such as PEI [150], and DNA (polyplexes) are formed because of strong electrostatic interactions between the positively charged carrier and negatively charged DNA.

Another approach is based on the modification of drugs and drug carriers with certain proteins and peptides demonstrating a unique ability to penetrate into cells ('transduction' phenomenon). CPP-mediated intracellular delivery of large molecules and nanoparticles was proved to proceed via energy-dependent macropinocytosis with subsequent enhanced escape from the endosome into the cell cytoplasm [151], while individual CPPs or CPP-conjugated small molecules penetrate cells via electrostatic interactions and hydrogen bonding and do not seem to depend on the energy [152]. Clearly, this important function can be added on top of the longevity, targetability, and stimuli-sensitivity of the pharmaceutical drug-loaded nanocarriers.

It has been shown that CPPs can internalize nanosized particles into the cells [153, 154]. Dextran-coated SPION derivatized with TATp were internalized into lymphocytes over 100-fold more efficiently than nonmodified particles [155]. Even relatively large particles, such as liposomes, could be delivered into various cells by multiple TATp or other CPP molecules attached to their surface [156–158]. Complexes of TATp-liposomes with a plasmid encoding for the green fluorescence protein (GFP) were used for successful in vitro transfection of various tumor and normal cells as well as for in vivo transfection of tumor cells in mice bearing Lewis lung carcinoma [159]. Antp and TATp coupled to small unilamellar liposomes were accumulated within tumor cells and dendritic cells more effectively than unmodified control liposomes [160]. An octamer of arginine (R8) attached to the surface of siRNA-loaded liposomes provided their effective intracellular delivery and silencing of the targeted gene [161].

The cell-penetrating function could be beneficially combined with the stimuli sensitivity discussed earlier allowing nanocarriers to: (1) accumulate in the required organ or tissue, and then (2) penetrate inside target cells delivering there their load (drug or DNA). Organ or tissue (tumor or infarct) accumulation could be achieved by passive targeting via the EPR effect or by antibody-mediated active targeting, while intracellular delivery could be mediated by certain internalizable ligands (folate and Tf) or by CPPs (such as TAT or polyArg) [162, 163]. When in the blood, the cell-penetrating function should be temporarily inactivated (shielded) to prevent nonspecific drug delivery into nontarget cells; however once inside the target, the delivery system should lose its protective coat, expose the cell-penetrating function, and provide intracellular drug delivery [164]. Thus, the administration of pGFP-TATp-liposomes with a non-pH-sensitive PEG coating has resulted in only minimal transfection of tumor cells because of steric hindrances to the liposome-to-cell interaction created by the PEG coat. Contrarily, the administration of pGFP-TATp-liposomes with the low pH-detachable PEG resulted in a highly efficient transfection since the removal of PEG under the action of the decreased intratumoral pH leads to exposure of the liposome-attached TATp residues, enhanced penetration of the liposomes inside tumor cells, and effective intracellular delivery of the pGFP [165].

Recently, targeted long-circulating PEGylated liposomes and PEG-PE-based micelles possessing several functionalities [122, 166] have been described. Such systems are capable of targeting a specific cell or organ by attaching the monoclonal antibody (infarct-specific antimyosin antibody 2G4 or cancer-specific anti-nucleosome antibody 2C5) to their surface (see fig. 9 on infarct targeting) [167].

Theranostic Nanopreparations

A recent challenge in the area of multifunctional pharmaceutical nanocarriers is the use of pharmaceutical nanocarriers for diagnostic/imaging

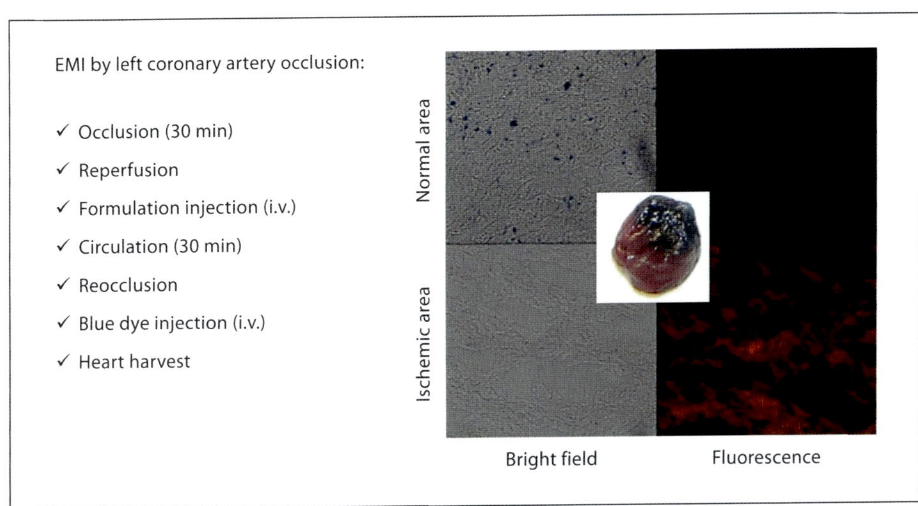

Fig. 9. Liposomes labeled with rhodamine and modified with the infarct-specific monoclonal antibody 2G4 and cell-penetrating TAT-peptide demonstrate a significantly enhanced accumulation (rhodamine fluorescence) in the area of experimental myocardial infarction (EMI).

purposes simultaneously with their therapeutic use. With this in mind, the contrast reporter moieties can be added to multifunctionalized nanocarriers. Liposomes, polymeric micelles, dendrimers, and iron oxide nanoparticles are frequently used as templates for engineering theranostic carriers. In the case of liposomes for gamma imaging and MRI, when heavy metal atoms are used as contrast moieties, the reporter metal could be chelated into a soluble chelator (such as diethylene triamine pentaacetic acid; DTPA) and then incorporated into the interior of a liposome [168], or a chelating compound could be chemically modified with a hydrophobic group, which can anchor it onto the liposome surface [169]. Polychelating amphiphilic polymers (PAP) were synthesized consisting of a main chain with multiple side chelating groups capable of firmly binding many reporter metal atoms and a hydrophobic terminal group allowing for polymer adsorption onto hydrophobic nanoparticles or incorporation into hydrophobic domains of liposomes or micelles and for a sharp increase in the number of bound reporter metal atoms per particle and image signal intensity [170]. Such PAP nanoparticles were used for in vivo MRI of lymphatic system components with Gd-loaded nanocarriers. The performance of Gd-PAP liposomes or Gd-PAP micelles could be further improved in case of the coincorporation of amphiphilic PEG onto the liposome membrane or micelle surface, which can be explained by the increased relaxivity of PEG-Gd liposomes because of the presence of increased amounts of PEG-associated water protons in the close vicinity of chelated Gd ions [171, 172]. Gd-PAP-PEG liposomes additionally modified with the cancer-specific monoclonal antibody demonstrated fast and specific tumor accumulation and could serve as effective contrast agents for tumor MRI [173, 174].

The combination of drug loading, longevity, targetability, and contrast properties results in multifunctional nanopharmaceuticals of a new generation. Thus, long-circulating PEGylated liposomes loaded with doxorubicin and additionally decorated with a tumor-specific antibody and contrast moieties [175, 176] demonstrated in-

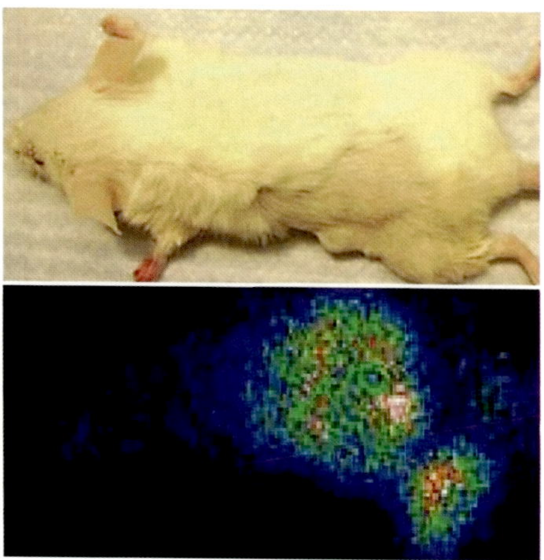

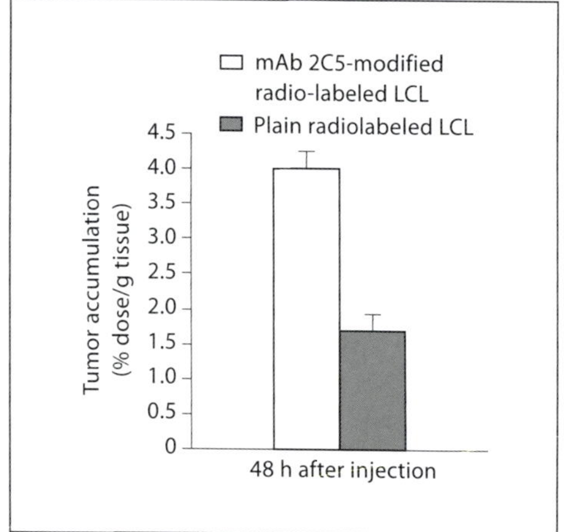

Fig. 10. Combination of longevity, targetability, and contrast function. Radiolabeled (^{111}In) long-circulating PEGylated liposomes (LCL) modified with the cancer-specific monoclonal antibody 2C5 demonstrate an enhanced tumor accumulation in experimental mice. This can be used for both drug delivery and fast and specific tumor visualization by gamma scintigraphy.

creased therapeutic activity in vivo, and their target accumulation could be easily followed by gamma scintigraphy (see fig. 10) or MRI. The combination of drug (chemotherapeutics) and contrast agent load on the same nanocarrier allows for the development of personalized nanotherapeutic approaches since the efficiency of nanocarrier accumulation in the tumor (determined by the leakiness of the blood vessels and traced by the contrast agent, e.g. by mammography) can assist in determining the proper choice of the dose of therapeutic agent for the treatment of each individual tumor [177, 178].

A variety of polymeric nanoparticles has been described designed to serve as imaging agents for two different imaging modalities. Thus, multifunctional chitosan nanoparticles have been prepared encapsulating quantum dots and Gd-DTPA and have been suitable for optical and MRI [179]. Nanoparticles which can serve for magnetic and fluorescent or magnetic, fluorescent, and PET imaging have been reviewed in several reports [180–182]. Such nanoprobes for dual modality magnetofluorescent imaging have been additionally modified with PEG for longevity and with folate for tumor targeting [183].

Disclosure Statement

The authors have nothing to disclose.

References

1 Torchilin VP (ed): Nanoparticulates as Drug Carriers. London, Imperial College Press, 2006.

2 Thassu D, Deleers M, Pathak Y (eds): Nanoparticulate Drug Delivery Systems. New York, Informa Healthcare, 2007.

3 Torchilin V: Multifunctional and stimuli-sensitive pharmaceutical nanocarriers. Eur J Pharm Biopharm 2009;71: 431–434.

4 Lasic DD, Martin FJ (eds): Stealth Liposomes. Boca Raton, CRC Press, 1995.

5 Torchilin VP, Trubetskoy VS: Which polymers can make nanoparticulate drug carriers long-circulating? Adv Drug Deliv Rev 1995;16:141–155.
6 Moghimi SM, Szebeni J: Stealth liposomes and long circulating nanoparticles: critical issues in pharmacokinetics, opsonization and protein-binding properties. Prog Lipid Res 2003;42: 463–478.
7 Maeda H, Wu J, Sawa T, Matsumura Y, Hori K: Tumor vascular permeability and the EPR effect in macromolecular therapeutics: a review. J Control Release 2000;65:271–284.
8 Gabizon AA: Liposome circulation time and tumor targeting: implications for cancer chemotherapy. Adv Drug Deliv Rev 1995;16:285–294.
9 Klibanov AL, Maruyama K, Torchilin VP, Huang L: Amphipathic polyethyleneglycols effectively prolong the circulation time of liposomes. FEBS Lett 1990;268:235–237.
10 Maruyama K, Yuda T, Okamoto A, Ishikura C, Kojima S, Iwatsuru M: Effect of molecular weight in amphipathic polyethyleneglycol on prolonging the circulation time of large unilamellar liposomes. Chem Pharm Bull (Tokyo) 1991;39:1620–1622.
11 Papahadjopoulos D, Allen TM, Gabizon A, et al: Sterically stabilized liposomes: improvements in pharmacokinetics and antitumor therapeutic efficacy. Proc Natl Acad Sci USA 1991;88:11460–11464.
12 Allen TM: The use of glycolipids and hydrophilic polymers in avoiding rapid uptake of liposomes by the mononuclear phagocyte system. Adv Drug Deliv Rev 1994;13:285–309.
13 Chonn A, Semple SC, Cullis PR: Association of blood proteins with large unilamellar liposomes in vivo: relation to circulation lifetimes. J Biol Chem 1992;267:18759–18765.
14 Zalipsky S: Chemistry of polyethylene glycol conjugates with biologically active molecules. Adv Drug Deliv Rev 1995;16:157–182.
15 Yamaoka T, Tabata Y, Ikada Y: Distribution and tissue uptake of poly(ethylene glycol) with different molecular weights after intravenous administration to mice. J Pharm Sci 1994;83:601–606.
16 Klibanov AL, Maruyama K, Beckerleg AM, Torchilin VP, Huang L: Activity of amphipathic poly(ethylene glycol) 5000 to prolong the circulation time of liposomes depends on the liposome size and is unfavorable for immunoliposome binding to target. Biochim Biophys Acta 1991;1062:142–148.
17 Maeda H: The enhanced permeability and retention (EPR) effect in tumor vasculature: the key role of tumor-selective macromolecular drug targeting. Adv Enzyme Regul 2001;41:189–207.
18 Gabizon A, Papahadjopoulos D: Liposome formulations with prolonged circulation time in blood and enhanced uptake by tumors. Proc Natl Acad Sci USA 1988;85:6949–6953.
19 Gabizon A, Catane R, Uziely B, et al: Prolonged circulation time and enhanced accumulation in malignant exudates of doxorubicin encapsulated in polyethylene-glycol coated liposomes. Cancer Res 1994;54:987–992.
20 Boman NL, Masin D, Mayer LD, Cullis PR, Bally MB: Liposomal vincristine which exhibits increased drug retention and increased circulation longevity cures mice bearing P388 tumors. Cancer Res 1994;54:2830–2833.
21 Allen TM, Mehra T, Hansen C, Chin YC: Stealth liposomes: an improved sustained release system for 1-beta-D-arabinofuranosylcytosine. Cancer Res 1992;52:2431–2439.
22 Rose PG: PEGylated liposomal doxorubicin: optimizing the dosing schedule in ovarian cancer. Oncologist 2005;10: 205–214.
23 Simoes S, Moreira JN, Fonseca C, Duzgunes N, de Lima MC: On the formulation of pH-sensitive liposomes with long circulation times. Adv Drug Deliv Rev 2004;56:947–965.
24 Roux E, Passirani C, Scheffold S, Benoit JP, Leroux JC: Serum-stable and long-circulating, PEGylated, pH-sensitive liposomes. J Control Release 2004;94: 447–451.
25 Leroux J, Roux E, Le Garrec D, Hong K, Drummond DC: N-isopropylacrylamide copolymers for the preparation of pH-sensitive liposomes and polymeric micelles. J Control Release 2001; 72:71–84.
26 Roux E, Stomp R, Giasson S, Pezolet M, Moreau P, Leroux JC: Steric stabilization of liposomes by pH-responsive N-isopropylacrylamide copolymer. J Pharm Sci 2002;91:1795–1802.
27 Lee ES, Na K, Bae YH: Polymeric micelle for tumor pH and folate-mediated targeting. J Control Release 2003;91: 103–113.
28 Lee ES, Shin HJ, Na K, Bae YH: Poly(L-histidine)-PEG block copolymer micelles and pH-induced destabilization. J Control Release 2003;90:363–374.
29 Jones MC, Ranger M, Leroux JC: pH-sensitive unimolecular polymeric micelles: synthesis of a novel drug carrier. Bioconjug Chem 2003;14:774–781.
30 Suzawa T, Nagamura S, Saito H, et al: Enhanced tumor cell selectivity of adriamycin-monoclonal antibody conjugate via a poly(ethylene glycol)-based cleavable linker. J Control Release 2002; 79:229–242.
31 Potineni A, Lynn DM, Langer R, Amiji MM: Poly(ethylene oxide)-modified poly(beta-amino ester) nanoparticles as a pH-sensitive biodegradable system for paclitaxel delivery. J Control Release 2003;86:223–234.
32 Yoo HS, Lee EA, Park TG: Doxorubicin-conjugated biodegradable polymeric micelles having acid-cleavable linkages. J Control Release 2002;82:17–27.
33 Cheung CY, Murthy N, Stayton PS, Hoffman AS: A pH-sensitive polymer that enhances cationic lipid-mediated gene transfer. Bioconjug Chem 2001;12: 906–910.
34 Venugopalan P, Jain S, Sankar S, Singh P, Rawat A, Vyas SP: pH-sensitive liposomes: mechanism of triggered release to drug and gene delivery prospects. Pharmazie 2002;57:659–671.
35 Gref R, Minamitake Y, Peracchia MT, Trubetskoy V, Torchilin V, Langer R: Biodegradable long-circulating polymeric nanospheres. Science 1994;263: 1600–1603.
36 Harper GR, Davies MC, Davis SS, et al: Steric stabilization of microspheres with grafted polyethylene oxide reduces phagocytosis by rat Kupffer cells in vitro. Biomaterials 1991;12: 695–700.
37 Shan X, Yuan Y, Liu C, Tao X, Sheng Y, Xu F: Influence of PEG chain on the complement activation suppression and longevity in vivo prolongation of the PCL biomedical nanoparticles. Biomed Microdevices 2009;11:1187–1194.

38 Gindy ME, Ji S, Hoye TR, Panagiotopoulos AZ, Prud'homme RK: Preparation of poly(ethylene glycol) protected nanoparticles with variable bioconjugate ligand density. Biomacromolecules 2008;9:2705–2711.
39 Shan X, Liu C, Yuan Y, et al: In vitro macrophage uptake and in vivo biodistribution of long-circulation nanoparticles with poly(ethylene-glycol)-modified PLA (BAB type) triblock copolymer. Colloids Surf B Biointerfaces 2009;72:303–311.
40 Illum L, Church AE, Butterworth MD, Arien A, Whetstone J, Davis SS: Development of systems for targeting the regional lymph nodes for diagnostic imaging: in vivo behaviour of colloidal PEG-coated magnetite nanospheres in the rat following interstitial administration. Pharm Res 2001;18:640–645.
41 Zhang F, Kang ET, Neoh KG, Huang W: Modification of gold surface by grafting of poly(ethylene glycol) for reduction in protein adsorption and platelet adhesion. J Biomater Sci Polym Ed 2001;12:515–531.
42 von Maltzahn G, Park JH, Agrawal A, et al: Computationally guided photothermal tumor therapy using long-circulating gold nanorod antennas. Cancer Res 2009;69:3892–3900.
43 Kim D, Park S, Lee JH, Jeong YY, Jon S: Antibiofouling polymer-coated gold nanoparticles as a contrast agent for in vivo X-ray computed tomography imaging. J Am Chem Soc 2007;129:7661–7665.
44 Lee H, Larson RG: Molecular dynamics study of the structure and interparticle interactions of polyethylene glycol-conjugated PAMAM dendrimers. J Phys Chem B 2009;113:13202–13207.
45 Kaminskas LM, Kelly BD, McLeod VM, et al: Pharmacokinetics and tumor disposition of PEGylated, methotrexate conjugated poly-L-lysine dendrimers. Mol Pharm 2009;6:1190–1204.
46 Wang W, Xiong W, Wan J, Sun X, Xu H, Yang X: The decrease of PAMAM dendrimer-induced cytotoxicity by PEGylation via attenuation of oxidative stress. Nanotechnology 2009;20:105103.
47 Lopez AI, Reins RY, McDermott AM, Trautner BW, Cai C: Antibacterial activity and cytotoxicity of PEGylated poly(amidoamine) dendrimers. Mol Biosyst 2009;5:1148–1156.
48 Liu Z, Tabakman SM, Chen Z, Dai H: Preparation of carbon nanotube bioconjugates for biomedical applications. Nat Protoc 2009;4:1372–1382.
49 Murray AR, Kisin E, Leonard SS, et al: Oxidative stress and inflammatory response in dermal toxicity of single-walled carbon nanotubes. Toxicology 2009;257:161–171.
50 Delogu LG, Magrini A, Bergamaschi A, et al: Conjugation of antisense oligonucleotides to PEGylated carbon nanotubes enables efficient knockdown of PTPN22 in T lymphocytes. Bioconjug Chem 2009;20:427–431.
51 Schipper ML, Iyer G, Koh AL, et al: Particle size, surface coating, and PEGylation influence the biodistribution of quantum dots in living mice. Small 2009;5:126–134.
52 Susumu K, Mei BC, Mattoussi H: Multifunctional ligands based on dihydrolipoic acid and polyethylene glycol to promote biocompatibility of quantum dots. Nat Protoc 2009;4:424–436.
53 Warnement MR, Tomlinson ID, Chang JC, Schreuder MA, Luckabaugh CM, Rosenthal SJ: Controlling the reactivity of ampiphilic quantum dots in biological assays through hydrophobic assembly of custom PEG derivatives. Bioconjug Chem 2008;19:1404–1413.
54 Torchilin VP: Structure and design of polymeric surfactant-based drug delivery systems. J Control Release 2001;73:137–172.
55 Torchilin VP: Micellar nanocarriers: pharmaceutical perspectives. Pharm Res 2007;24:1–16.
56 Lukyanov AN, Torchilin VP: Micelles from lipid derivatives of water-soluble polymers as delivery systems for poorly soluble drugs. Adv Drug Deliv Rev 2004;56:1273–1289.
57 Gao Z, Lukyanov AN, Chakilam AR, Torchilin VP: PEG-PE/phosphatidylcholine mixed immunomicelles specifically deliver encapsulated taxol to tumor cells of different origin and promote their efficient killing. J Drug Target 2003;11:87–92.
58 Mu L, Chrastina A, Levchenko T, Torchilin VP: Micelles from poly(ethylene glycol)-phosphatidyl ethanolamine conjugate (PEG-PE) as pharmaceutical nanocarriers for poorly soluble drug camptothecin. J Biomed Nanotechnol 2005;1:190–196.
59 Skidan I, Dholakia P, Torchilin V: Photodynamic therapy of experimental B-16 melanoma in mice with tumor-targeted 5,10,15,20-tetraphenylporphin-loaded PEG-PE micelles. J Drug Target 2008;16:486–493.
60 Gao Z, Lukyanov A, Singhal A, Torchilin V: Diacylipid-polymer micelles as nanocarriers for poorly soluble anticancer drugs. Nano Lett 2002;2:979–982.
61 Lukyanov AN, Gao Z, Mazzola L, Torchilin VP: Polyethylene glycol-diacyllipid micelles demonstrate increased acculumation in subcutaneous tumors in mice. Pharm Res 2002;19:1424–1429.
62 Sawant RR, Sawant RM, Torchilin VP: Mixed PEG-PE/vitamin E tumor-targeted immunomicelles as carriers for poorly soluble anti-cancer drugs: improved drug solubilization and enhanced in vitro cytotoxicity. Eur J Pharm Biopharm 2008;70:51–57.
63 Roby A, Erdogan S, Torchilin VP: Solubilization of poorly soluble PDT agent, meso-tetraphenylporphin, in plain or immunotargeted PEG-PE micelles results in dramatically improved cancer cell killing in vitro. Eur J Pharm Biopharm 2006;62:235–240.
64 Lukyanov AN, Gao Z, Torchilin VP: Micelles from polyethylene glycol/phosphatidylethanolamine conjugates for tumor drug delivery. J Control Release 2003;91:97–102.
65 Torchilin VP, Lukyanov AN, Gao Z, Papahadjopoulos-Sternberg B: Immunomicelles: targeted pharmaceutical carriers for poorly soluble drugs. Proc Natl Acad Sci USA 2003;100:6039–6044.
66 Musacchio T, Laquintana V, Latrofa A, Trapani G, Torchilin VP: PEG-PE micelles loaded with paclitaxel and surface-modified by a PBR-ligand: synergistic anticancer effect. Mol Pharm 2009;6:468–479.
67 Blume G, Cevc G, Crommelin MD, Bakker-Woudenberg IA, Kluft C, Storm G: Specific targeting with poly(ethylene glycol)-modified liposomes: coupling of homing devices to the ends of the polymeric chains combines effective target binding with long circulation times. Biochim Biophys Acta 1993;1149:180–184.

68 Torchilin VP, Levchenko TS, Lukyanov AN, et al: p-Nitrophenylcarbonyl-PEG-PE-liposomes: fast and simple attachment of specific ligands, including monoclonal antibodies, to distal ends of PEG chains via p-nitrophenylcarbonyl groups. Biochim Biophys Acta 2001;1511:397–411.
69 Klibanov AL: Antibody-mediated targeting of PEG-coated liposomes; in Woodle MC, Storm G (eds): Long circulating liposomes: old drugs, new therapeutics. Berlin, Springer, 1998, p 269.
70 Park JW, Kirpotin DB, Hong K, et al: Tumor targeting using anti-her2 immunoliposomes. J Control Release 2001;74:95–113.
71 Benhar I, Padlan EA, Jung SH, Lee B, Pastan I: Rapid humanization of the Fv of monoclonal antibody B3 by using framework exchange of the recombinant immunotoxin B3(Fv)-PE38. Proc Natl Acad Sci USA 1994;91:12051–12055.
72 Brannon-Peppas L, Blanchette JO: Nanoparticle and targeted systems for cancer therapy. Adv Drug Deliv Rev 2004;56:1649–1659.
73 Gao J, Zhong W, He J, et al: Tumor-targeted PE38KDEL delivery via PEGylated anti-HER2 immunoliposomes. Int J Pharm 2009;374:145–152.
74 Kamps JA, Koning GA, Velinova MJ, et al: Uptake of long-circulating immunoliposomes, directed against colon adenocarcinoma cells, by liver metastases of colon cancer. J Drug Target 2000;8: 235–245.
75 Lukyanov AN, Elbayoumi TA, Chakilam AR, Torchilin VP: Tumor-targeted liposomes: doxorubicin-loaded long-circulating liposomes modified with anti-cancer antibody. J Control Release 2004;100:135–144.
76 ElBayoumi TA, Torchilin VP: Tumor-targeted nanomedicines: enhanced antitumor efficacy in vivo of doxorubicin-loaded, long-circulating liposomes modified with cancer-specific monoclonal antibody. Clin Cancer Res 2009; 15:1973–1980.
77 Gupta B, Torchilin VP: Monoclonal antibody 2C5-modified doxorubicin-loaded liposomes with significantly enhanced therapeutic activity against intracranial human brain U-87 MG tumor xenografts in nude mice. Cancer Immunol Immunother 2007;56:1215–1223.
78 Eck W, Craig G, Sigdel A, et al: PEGylated gold nanoparticles conjugated to monoclonal F19 antibodies as targeted labeling agents for human pancreatic carcinoma tissue. ACS Nano 2008;2: 2263–2272.
79 Cato MH, D'Annibale F, Mills DM, et al: Cell-type specific and cytoplasmic targeting of PEGylated carbon nanotube-based nanoassemblies. J Nanosci Nanotechnol 2008;8:2259–2269.
80 Hatakeyama H, Akita H, Maruyama K, Suhara T, Harashima H: Factors governing the in vivo tissue uptake of transferrin-coupled polyethylene glycol liposomes in vivo. Int J Pharm 2004; 281:25–33.
81 Li Y, Ogris M, Wagner E, Pelisek J, Ruffer M: Nanoparticles bearing polyethyleneglycol-coupled transferrin as gene carriers: preparation and in vitro evaluation. Int J Pharm 2003;259:93–101.
82 Ishida O, Maruyama K, Tanahashi H, et al: Liposomes bearing polyethyleneglycol-coupled transferrin with intracellular targeting property to the solid tumors in vivo. Pharm Res 2001;18: 1042–1048.
83 Olivier JC, Huertas R, Lee HJ, Calon F, Pardridge WM: Synthesis of PEGylated immunonanoparticles. Pharm Res 2002;19:1137–1143.
84 Derycke AS, De Witte PA: Transferrin-mediated targeting of hypericin embedded in sterically stabilized PEG-liposomes. Int J Oncol 2002;20:181–187.
85 Gijsens A, Derycke A, Missiaen L, et al: Targeting of the photocytotoxic compound AlPcS4 to Hela cells by transferrin conjugated PEG-liposomes. Int J Cancer 2002;101:78–85.
86 Iinuma H, Maruyama K, Okinaga K, et al: Intracellular targeting therapy of cisplatin-encapsulated transferrin-polyethylene glycol liposome on peritoneal dissemination of gastric cancer. Int J Cancer 2002;99:130–137.
87 Li X, Ding L, Xu Y, Wang Y, Ping Q: Targeted delivery of doxorubicin using stealth liposomes modified with transferrin. Int J Pharm 2009;373:116–123.
88 Lu Y, Low PS: Folate-mediated delivery of macromolecular anticancer therapeutic agents. Adv Drug Deliv Rev 2002;54:675–693.
89 Gabizon A, Shmeeda H, Horowitz AT, Zalipsky S: Tumor cell targeting of liposome-entrapped drugs with phospholipid-anchored folic acid-PEG conjugates. Adv Drug Deliv Rev 2004;56: 1177–1192.
90 Stella B, Arpicco S, Peracchia MT, et al: Design of folic acid-conjugated nanoparticles for drug targeting. J Pharm Sci 2000;89:1452–1464.
91 Park EK, Lee SB, Lee YM: Preparation and characterization of methoxy poly(ethylene glycol)/poly(ε-caprolactone) amphiphilic block copolymeric nanospheres for tumor-specific folate-mediated targeting of anticancer drugs. Biomaterials 2005;26:1053–1061.
92 Zhang Y, Kohler N, Zhang M: Surface modification of superparamagnetic magnetite nanoparticles and their intracellular uptake. Biomaterials 2002; 23:1553–1561.
93 Choi H, Choi SR, Zhou R, Kung HF, Chen IW: Iron oxide nanoparticles as magnetic resonance contrast agent for tumor imaging via folate receptor-targeted delivery. Acad Radiol 2004;11: 996–1004.
94 Liang B, He ML, Xiao ZP, et al: Synthesis and characterization of folate-PEG-grafted-hyperbranched-PEI for tumor-targeted gene delivery. Biochem Biophys Res Commun 2008;367:874–880.
95 Song EQ, Zhang ZL, Luo QY, Lu W, Shi YB, Pang DW: Tumor cell targeting using folate-conjugated fluorescent quantum dots and receptor-mediated endocytosis. Clin Chem 2009;55:955–963.
96 Dagar S, Krishnadas A, Rubinstein I, Blend MJ, Onyuksel H: VIP grafted sterically stabilized liposomes for targeted imaging of breast cancer: in vivo studies. J Control Release 2003;91:123–133.
97 Schiffelers RM, Koning GA, ten Hagen TL, et al: Anti-tumor efficacy of tumor vasculature-targeted liposomal doxorubicin. J Control Release 2003;91:115–122.
98 Garg A, Tisdale AW, Haidari E, Kokkoli E: Targeting colon cancer cells using PEGylated liposomes modified with a fibronectin-mimetic peptide. Int J Pharm 2009;366:201–210.
99 Demirgoz D, Garg A, Kokkoli E: PR_b-targeted PEGylated liposomes for prostate cancer therapy. Langmuir 2008;24: 13518–13524.

100 Maitani Y, Hattori Y: Oligoarginine-PEG-lipid particles for gene delivery. Expert Opin Drug Deliv 2009;6:1065–1077.
101 Sawant RR, Sawant RM, Kale AA, Torchilin VP: The architecture of ligand attachment to nanocarriers controls their specific interaction with target cells. J Drug Target 2008;16:596–600.
102 Wang T, D'Souza GG, Bedi D, et al: Enhanced binding and killing of target tumor cells by drug-loaded liposomes modified with tumor-specific phage fusion coat protein. Nanomedicine (Lond) 2010;5:563–574.
103 Wang T, Petrenko VA, Torchilin VP: Paclitaxel-loaded polymeric micelles modified with MCF-7 cell-specific phage protein: enhanced binding to target cancer cells and increased cytotoxicity. Mol Pharm 2010;7:1007–1014.
104 Wang T, Yang S, Petrenko VA, Torchilin VP: Cytoplasmic delivery of liposomes into MCF-7 breast cancer cells mediated by cell-specific phage fusion coat protein. Mol Pharm 2010;7:1149–1158.
105 Jayanna PK, Bedi D, Gillespie JW, et al: Landscape phage fusion protein-mediated targeting of nanomedicines enhances their prostate tumor cell association and cytotoxic efficiency. Nanomedicine 2010;6:538–546.
106 Vaupel P, Kallinowski F, Okunieff P: Blood flow, oxygen and nutrient supply, and metabolic microenvironment of human tumors: a review. Cancer Res 1989;49:6449–6465.
107 Wike-Hooley JL, Haveman J, Reinhold HS: The relevance of tumour pH to the treatment of malignant disease. Radiother Oncol 1984;2:343–366.
108 Gerweck LE, Seetharaman K: Cellular pH gradient in tumor versus normal tissue: potential exploitation for the treatment of cancer. Cancer Res 1996; 56:1194–1198.
109 Saito G, Swanson JA, Lee KD: Drug delivery strategy utilizing conjugation via reversible disulfide linkages: role and site of cellular reducing activities. Adv Drug Deliv Rev 2003;55:199–215.
110 Litzinger DC, Huang L: Phosphatidylethanolamine liposomes: drug delivery, gene transfer and immunodiagnostic applications. Biochim Biophys Acta 1992;1113:201–227.
111 Connor J, Huang L: pH-sensitive immunoliposomes as an efficient and target-specific carrier for antitumor drugs. Cancer Res 1986;46:3431–3435.
112 Torchilin VP, Zhou F, Huang L: pH-sensitive liposomes. J Liposome Res 1993;3:201–255.
113 Sheff D: Endosomes as a route for drug delivery in the real world. Adv Drug Deliv Rev 2004;56:927–930.
114 Yessine MA, Leroux JC: Membrane-destabilizing polyanions: interaction with lipid bilayers and endosomal escape of biomacromolecules. Adv Drug Deliv Rev 2004;56:999–1021.
115 Geisert EE Jr, Del Mar NA, Owens JL, Holmberg EG: Transfecting neurons and glia in the rat using pH-sensitive immunoliposomes. Neurosci Lett 1995; 184:40–43.
116 Bae Y, Nishiyama N, Fukushima S, Koyama H, Yasuhiro M, Kataoka K: Preparation and biological characterization of polymeric micelle drug carriers with intracellular pH-triggered drug release property: tumor permeability, controlled subcellular drug distribution, and enhanced in vivo antitumor efficacy. Bioconjug Chem 2005;16: 122–130.
117 Jeong JH, Kim SW, Park TG: Novel intracellular delivery system of antisense oligonucleotide by self-assembled hybrid micelles composed of DNA/PEG conjugate and cationic fusogenic peptide. Bioconjug Chem 2003;14:473–479.
118 Shim WS, Kim SW, Choi EK, Park HJ, Kim JS, Lee DS: Novel pH sensitive block copolymer micelles for solvent free drug loading. Macromol Biosci 2006;6:179–186.
119 Roux E, Francis M, Winnik FM, Leroux JC: Polymer based pH-sensitive carriers as a means to improve the cytoplasmic delivery of drugs. Int J Pharm 2002; 242:25–36.
120 Turk MJ, Reddy JA, Chmielewski JA, Low PS: Characterization of a novel pH-sensitive peptide that enhances drug release from folate-targeted liposomes at endosomal pHs. Biochim Biophys Acta 2002;1559:56–68.
121 Shi G, Guo W, Stephenson SM, Lee RJ: Efficient intracellular drug and gene delivery using folate receptor-targeted pH-sensitive liposomes composed of cationic/anionic lipid combinations. J Control Release 2002;80:309–319.
122 Sawant RM, Hurley JP, Salmaso S, et al: 'SMART' drug delivery systems: double-targeted pH-responsive pharmaceutical nanocarriers. Bioconjug Chem 2006;17:943–949.
123 Meyer DE, Shin BC, Kong GA, Dewhirst MW, Chilkoti A: Drug targeting using thermally responsive polymers and local hyperthermia. J Control Release 2001;74:213–224.
124 Ponce AM, Vujaskovic Z, Yuan F, Needham D, Dewhirst MW: Hyperthermia mediated liposomal drug delivery. Int J Hyperthermia 2006;22:205–213.
125 Yatvin MB, Weinstein JN, Dennis WH, Blumenthal R: Design of liposomes for enhanced local release of drugs by hyperthermia. Science 1978;202:1290–1293.
126 Yoshida R, Uchida K, Kaneko Y, et al: Comb-type grafted hydrogels with rapid deswelling response to temperature changes. Nature 1995;374:240–242.
127 Sukhorukov GB, Rogach AL, Garstka M, et al: Multifunctionalized polymer microcapsules: novel tools for biological and pharmacological applications. Small 2007;3:944–955.
128 Gupta AK, Naregalkar RR, Vaidya VD, Gupta M: Recent advances on surface engineering of magnetic iron oxide nanoparticles and their biomedical applications. Nanomedicine (Lond) 2007;2:23–39.
129 Gupta AK, Gupta M: Synthesis and surface engineering of iron oxide nanoparticles for biomedical applications. Biomaterials 2005;26:3995–4021.
130 Ito A, Shinkai M, Honda H, Kobayashi T: Medical application of functionalized magnetic nanoparticles. J Biosci Bioeng 2005;100:1–11.
131 Shinkai M, Ito A: Functional magnetic particles for medical application. Adv Biochem Eng Biotechnol 2004;91:191–220.
132 Wagner S, Schnorr J, Pilgrimm H, Hamm B, Taupitz M: Monomer-coated very small superparamagnetic iron oxide particles as contrast medium for magnetic resonance imaging: preclinical in vivo characterization. Invest Radiol 2002;37:167–177.
133 Anzai Y, Prince MR: Iron oxide-enhanced MR lymphography: the evaluation of cervical lymph node metastases in head and neck cancer. J Magn Reson Imaging 1997;7:75–81.

134 Ngaboni Okassa L, Marchais H, Douziech-Eyrolles L, et al: Development and characterization of sub-micron poly(D,L-lactide-co-glycolide) particles loaded with magnetite/maghemite nanoparticles. Int J Pharm 2005;302:187–196.
135 Widder KJ, Senyel AE, Scarpelli GD: Magnetic microspheres: a model system of site specific drug delivery in vivo. Proc Soc Exp Biol Med 1978;158:141–146.
136 Gang J, Park SB, Hyung W, et al: Magnetic poly epsilon-caprolactone nanoparticles containing Fe3O4 and gemcitabine enhance anti-tumor effect in pancreatic cancer xenograft mouse model. J Drug Target 2007;15:445–453.
137 Cinteza LO, Ohulchanskyy TY, Sahoo Y, Bergey EJ, Pandey RK, Prasad PN: Diacyllipid micelle-based nanocarrier for magnetically guided delivery of drugs in photodynamic therapy. Mol Pharm 2006;4:415–423.
138 Alexiou C, Arnold W, Klein RJ, et al: Locoregional cancer treatment with magnetic drug targeting. Cancer Res 2000;60:6641–6648.
139 Alexiou C, Jurgons R, Schmid RJ, et al: Magnetic drug targeting – biodistribution of the magnetic carrier and the chemotherapeutic agent mitoxantrone after locoregional cancer treatment. J Drug Target 2003;11:139–149.
140 Unger EC, McCreery TP, Sweitzer RH, Caldwell VE, Wu Y: Acoustically active lipospheres containing paclitaxel: a new therapeutic ultrasound contrast agent. Invest Radiol 1998;33:886–892.
141 Unger EC, Porter T, Culp W, Labell R, Matsunaga T, Zutshi R: Therapeutic applications of lipid-coated microbubbles. Adv Drug Deliv Rev 2004;56:1291–1314.
142 Liu Y, Miyoshi H, Nakamura M: Encapsulated ultrasound microbubbles: therapeutic application in drug/gene delivery. J Control Release 2006;114:89–99.
143 Huang SL, MacDonald RC: Acoustically active liposomes for drug encapsulation and ultrasound-triggered release. Biochim Biophys Acta 2004;1665:134–141.
144 Tartis MS, McCallan J, Lum AF, et al: Therapeutic effects of paclitaxel-containing ultrasound contrast agents. Ultrasound Med Biol 2006;32:1771–1780.
145 Rapoport N, Gao Z, Kennedy A: Multifunctional nanoparticles for combining ultrasonic tumor imaging and targeted chemotherapy. J Natl Cancer Inst 2007;99:1095–1106.
146 Husseini GA, Diaz de la Rosa MA, Gabuji T, Zeng Y, Christensen DA, Pitt WG: Release of doxorubicin from unstabilized and stabilized micelles under the action of ultrasound. J Nanosci Nanotechnol 2007;7:1028–1033.
147 Xu Y, Szoka FC, Jr: Mechanism of DNA release from cationic liposome/DNA complexes used in cell transfection. Biochemistry 1996;35:5616–5623.
148 Wu GY, Wu CH: Receptor-mediated in vitro gene transformation by a soluble DNA carrier system. J Biol Chem 1987;262:4429–4432.
149 Elouahabi A, Ruysschaert JM: Formation and intracellular trafficking of lipoplexes and polyplexes. Mol Ther 2005;11:336–347.
150 Kunath K, von Harpe A, Fischer D, et al: Low-molecular-weight polyethylenimine as a non-viral vector for DNA delivery: comparison of physicochemical properties, transfection efficiency and in vivo distribution with high-molecular-weight polyethylenimine. J Control Release 2003;89:113–125.
151 Wadia JS, Stan RV, Dowdy SF: Transducible TAT-HA fusogenic peptide enhances escape of TAT-fusion proteins after lipid raft macropinocytosis. Nat Med 2004;10:310–315.
152 Rothbard JB, Jessop TC, Lewis RS, Murray BA, Wender PA: Role of membrane potential and hydrogen bonding in the mechanism of translocation of guanidinium-rich peptides into cells. J Am Chem Soc 2004;126:9506–9507.
153 Josephson L, Tung CH, Moore A, Weissleder R: High-efficiency intracellular magnetic labeling with novel superparamagnetic-Tat peptide conjugates. Bioconjug Chem 1999;10:186–191.
154 Torchilin VP: Tat peptide-mediated intracellular delivery of pharmaceutical nanocarriers. Adv Drug Deliv Rev 2008;60:548–558.
155 Zhao M, Kircher MF, Josephson L, Weissleder R: Differential conjugation of tat peptide to superparamagnetic nanoparticles and its effect on cellular uptake. Bioconjug Chem 2002;13:840–844.
156 Tseng YL, Liu JJ, Hong RL: Translocation of liposomes into cancer cells by cell-penetrating peptides penetratin and TAT: a kinetic and efficacy study. Mol Pharmacol 2002;62:864–872.
157 Gorodetsky R, Levdansky L, Vexler A, et al: Liposome transduction into cells enhanced by haptotactic peptides (Haptides) homologous to fibrinogen C-termini. J Control Release 2004;95:477–488.
158 Levchenko TS, Rammohan R, Volodina N, Torchilin VP: Tat peptide-mediated intracellular delivery of liposomes. Methods Enzymol 2003;372:339–349.
159 Torchilin VP, Levchenko TS, Rammohan R, Volodina N, Papahadjopoulos-Sternberg B, D'Souza GG: Cell transfection in vitro and in vivo with nontoxic TAT peptide-liposome-DNA complexes. Proc Natl Acad Sci USA 2003;100:1972–1977.
160 Marty C, Meylan C, Schott H, Ballmer-Hofer K, Schwendener RA: Enhanced heparan sulfate proteoglycan-mediated uptake of cell-penetrating peptide-modified liposomes. Cell Mol Life Sci 2004;61:1785–1794.
161 Zhang C, Tang N, Liu X, Liang W, Xu W, Torchilin VP: siRNA-containing liposomes modified with polyarginine effectively silence the targeted gene. J Control Release 2006;112:229–239.
162 Gupta B, Levchenko TS, Torchilin VP: Intracellular delivery of large molecules and small particles by cell-penetrating proteins and peptides. Adv Drug Deliv Rev 2005;57:637–651.
163 Lochmann D, Jauk E, Zimmer A: Drug delivery of oligonucleotides by peptides. Eur J Pharm Biopharm 2004;58:237–251.
164 Sawant RM, Hurley JP, Huang Z, Szoka FC, Torchilin VP: Creating multifunctional drug delivery systems: micelles made of pH-responsive amohiphilic polymers with hidden biotin moiety. Proceedings 32nd Int Symp on Control Release Bioact Mater, Miami, 2005, p 406.
165 Kale AA, Torchilin VP: Enhanced transfection of tumor cells in vivo using 'Smart' pH-sensitive TAT-modified PEGylated liposomes. J Drug Target 2007;15:538–545.

166 Kale AA, Torchilin VP: Design, synthesis, and characterization of pH-sensitive PEG-PE conjugates for stimuli-sensitive pharmaceutical nanocarriers: the effect of substitutes at the hydrazone linkage on the ph stability of PEG-PE conjugates. Bioconjug Chem 2007;18: 363–370.
167 Ko YT, Hartner WC, Kale A, Torchilin VP: Gene delivery into ischemic myocardium by double-targeted lipoplexes with anti-myosin antibody and TAT peptide. Gene Ther 2009;16:52–59.
168 Tilcock C, Unger E, Cullis P, MacDougall P: Liposomal Gd-DTPA: preparation and characterization of relaxivity. Radiology 1989;171:77–80.
169 Kabalka GW, Davis MA, Holmberg E, Maruyama K, Huang L: Gadolinium-labeled liposomes containing amphiphilic Gd-DTPA derivatives of varying chain length: targeted MRI contrast enhancement agents for the liver. Magn Reson Imaging 1991;9:373–377.
170 Torchilin VP: Polymeric contrast agents for medical imaging. Curr Pharm Biotechnol 2000;1:183–215.
171 Torchilin VP, Trubetskoy VS, Narula J, Khaw BA: PEG-modified liposomes for gamma- and magnetic resonance imaging; in Lasic DD, Martin FJ (eds): Stealth Liposomes. Boca Raton, CRC Press, 1995, pp 225–231.
172 Trubetskoy VS, Cannillo JA, Milshtein A, Wolf GL, Torchilin VP: Controlled delivery of Gd-containing liposomes to lymph nodes: surface modification may enhance MRI contrast properties. Magn Reson Imaging 1995;13:31–37.
173 Erdogan S, Roby A, Torchilin VP: Enhanced tumor visualization by gamma-scintigraphy with ^{111}In-labeled polychelating-polymer-containing immunoliposomes. Mol Pharm 2006;3: 525–530.
174 Erdogan S, Medarova ZO, Roby A, Moore A, Torchilin VP: Enhanced tumor MR imaging with gadolinium-loaded polychelating polymer-containing tumor-targeted liposomes. J Magn Reson Imaging 2008;27:574–580.
175 Elbayoumi TA, Torchilin VP: Enhanced accumulation of long-circulating liposomes modified with the nucleosome-specific monoclonal antibody 2C5 in various tumours in mice: gamma-imaging studies. Eur J Nucl Med Mol Imaging 2006;33:1196–1205.
176 Elbayoumi TA, Pabba S, Roby A, Torchilin VP: Antinucleosome antibody-modified liposomes and lipid-core micelles for tumor-targeted delivery of therapeutic and diagnostic agents. J Liposome Res 2007;17:1–14.
177 Koning GA, Krijger GC: Targeted multifunctional lipid-based nanocarriers for image-guided drug delivery. Anticancer Agents Med Chem 2007;7:425–440.
178 Sajja HK, East MP, Mao H, Wang YA, Nie S, Yang L: Development of multifunctional nanoparticles for targeted drug delivery and noninvasive imaging of therapeutic effect. Curr Drug Discov Technol 2009;6:43–51.
179 Tan WB, Zhang Y: Multi-functional chitosan nanoparticles encapsulating quantum dots and Gd-DTPA as imaging probes for bio-applications. J Nanosci Nanotechnol 2007;7:2389–2393.
180 Mulder WJ, Griffioen AW, Strijkers GJ, Cormode DP, Nicolay K, Fayad ZA: Magnetic and fluorescent nanoparticles for multimodality imaging. Nanomedicine (Lond) 2007;2:307–324.
181 Cheon J, Lee JH: Synergistically integrated nanoparticles as multimodal probes for nanobiotechnology. Acc Chem Res 2008;41:1630–1640.
182 Wehrl HF, Judenhofer MS, Wiehr S, Pichler BJ: Pre-clinical PET/MR: technological advances and new perspectives in biomedical research. Eur J Nucl Med Mol Imaging 2009;36(suppl 1): S56–S68.
183 Ke JH, Lin JJ, Carey JR, Chen JS, Chen CY, Wang LF: A specific tumor-targeting magnetofluorescent nanoprobe for dual-modality molecular imaging. Biomaterials 2010;31:1707–1715.

Vladimir P. Torchilin
Department of Pharmaceutical Sciences
Center for Pharmaceutical Biotechnology and Nanomedicine, Northeastern University
360 Huntington Ave., Mugar Building, Room 312, Boston, MA 02115 (USA)
Tel. +1 617 373 3206, E-Mail v.torchilin@neu.edu

Alexiou C (ed): Nanomedicine – Basic and Clinical Applications in Diagnostics and Therapy.
Else Kröner-Fresenius Symp. Basel, Karger, 2011, vol 2, pp 35–52

Synthesis of Magnetic Nanoparticles and Magnetic Fluids for Biomedical Applications

L. Vékás[a, b] · Etelka Tombácz[e] · Rodica Turcu[d] · I. Morjan[c] · M.V. Avdeev[f] · Theodora Krasia-Christoforou[g] · V. Socoliuc[a]

[a]Laboratory of Magnetic Fluids, Romanian Academy-Timisoara Branch, CFATR, and [b]Politehnica University of Timisoara, NCESCF, Timisoara, [c]National Institute for Laser, Plasma and Radiation Physics, Bucharest, and [d]National Institute for R&D of Isotopic and Molecular Technologies, Cluj-Napoca, Romania; [e]Department of Physical Chemistry and Material Science, University of Szeged, Szeged, Hungary; [f]Joint Institute for Nuclear Research, FLNP, Dubna, Russia; [g]Department of Mechanical and Manufacturing Engineering, University of Cyprus, Nicosia, Cyprus

Abstract

Chemical coprecipitation and gas-phase laser pyrolysis procedures were applied to obtain various iron-based magnetic nanoparticles (magnetite, maghemite, and carbon layer-coated iron) in the size range of 3–15 nm used as basic building blocks for functionalized core-shell particles, magnetic nanofluids, as well as multifunctional hybrid nanostructures based on stimuli-responsive biocompatible polymers and block copolymers. The particle size distribution, magnetostatic properties, surface coating efficiency, and embedding/encapsulation mechanisms of magnetic nanoparticles and particle clusters in various biocompatible polymer matrices (core-shell nanostructures, microgels, and micelles) were examined by TEM/HRTEM, vibrational sample magnetometry, dynamic light scattering, Fourier transform infrared spectroscopy, and small-angle neutron scattering. The novel magnetic hybrid nanostructured materials envisaged for MRI contrast agents, magnetic carriers in bioseparation equipment, or magnetothermally triggered drug delivery systems have superparamagnetic behavior and exhibit magneto- and thermoresponsive properties, high stability, and in vitro biocompatibility.

Various pharmaceutical nanocarriers, such as liposomes, micelles, nanoemulsions, polymeric nanoparticles, and many others, demonstrate a broad variety of useful properties for biomedical applications [1].

Magnetic nanoparticles (MNPs), well separated or in clusters, with proper surface functionalization usually dispersed in liquid carriers [magnetic nanofluids (ferrofluids)] or embedded/encapsulated in polymeric networks (magnetic nanocomposites), are the basic building blocks of a large variety of multifunctional carriers. Such multifunctional nanostructured particles are magnetically susceptible and are specially designed for applications in the fields of biotechnology and biomedicine (see, for example, recent reviews [2, 3]), such as multimodal imaging, analyte monitoring, nanotherapeutics [4], enzyme and protein separations [5, 6], targeted drug delivery, and magnetic ferrofluid hyperthermia [7, 8].

The envisaged applications require the MNPs to be nontoxic, chemically stable, uniform in size, and well dispersed under physiological conditions. MNPs are mostly of magnetite and maghemite because iron oxides are excreted via the liver after the treatment, but recently surface passivated iron nanoparticles are also frequently applied due to their high specific magnetic moment. The magnetic characteristics and colloidal stability are overemphasized since particle aggregation must be excluded in the magnetic field during application with reference to the risk of blood clots in blood vessels. To fulfill the stability criterion, different coatings on the surface of particles are developed to prevent their aggregation and to improve their colloidal and chemical stability. In general, colloidal stabilization of MNPs in aqueous medium is assigned to the surface accumulation (adsorption) of appropriate dissolved species forming the innermost layer on the particle surface due to either: (1) physical interaction [e.g. ions by Coulombic attraction, nonionic polymers [polyethylene glycol (PEG), PVA, and dextran] and surfactants (Pluronics) by van der Waals forces] or (2) chemical interaction, i.e. chemical bond formation on active sites of the surface such as surface complexation, e.g. –COOH groups [of fatty acids, citric acid (CA), or polyacrylic acid (PAA)] on ≡FeOH sites [9]. In several cases a second layer has to be built to enhance colloidal stability in general, but especially in biomedical applications, to provide biocompatibility by inert coating (e.g. PEG or PEO) or to functionalize magnetic particles for specific interactions with antibodies. The second layer formation may take place due to either: (1) physical causes such as hydrophobic interactions (e.g. fatty acid double layers [10]) or (2) chemical, profoundly covalent binding (e.g. streptavidin [2]).

While the colloidal stability of magnetic fluids (MFs), especially that focusing on particle aggregation in strong magnetic fields, has been studied extensively, less attention has been paid to the adsorption of different stabilizers, the charge neutralization, and re- or overcharging of the surface of magnetite nanoparticles. The effect of MF dilution and the common parameters in aqueous solutions such as pH and salt concentration have also remained in the background, although these are the most important factors in biomedical applications. In this contribution recent results are summarized concerning the synthesis and manifold characterization of MNPs, magnetic nanofluids, and multifunctional nanocomposites designed for biotechnological and biomedical applications.

Synthesis of Biocompatible Magnetic Nanoparticles and Preparation of Magnetic Nanofluids

Any biomedical use of MNPs is basically related to their physicochemical and colloidal properties such as composition (chemical, crystalline, and magnetic), morphology (shape), size (magnetic core, coating layer, and hydrodynamic), charge (pH-dependent), and other biorelevant surface properties (e.g. salt tolerance, hydrophilicity/hydrophobicity, targeting ligands, and other anchored functionalities). These properties have to be considered during the design and fabrication of MNPs. In general, the magnetic nanocrystallites (mostly iron oxide but also iron or ferrite nanoparticles) provide a magnetically active core which is coated with a stabilizing shell. Although several routes of synthesis have been reported, two ways of preparation are basically followed:

- The 2-step route, where the magnetic core (e.g. naked iron or Massart type magnetite [11, 12]) is prepared first and then the designed coating on the MNP surface is built in separate chemical processes.
- The 1-step route, where the magnetic core is in situ coated with a chemisorbed layer of surfactants, e.g. C12 to C18 carboxylic acids, or complexants such as CA to prevent the adhesion of MNPs formed in a coprecipitation process.

Preparation of magnetic nanofluids always needs more or less adjustment of the stabilizing layers, e.g. the second layer formation in water-based MFs.

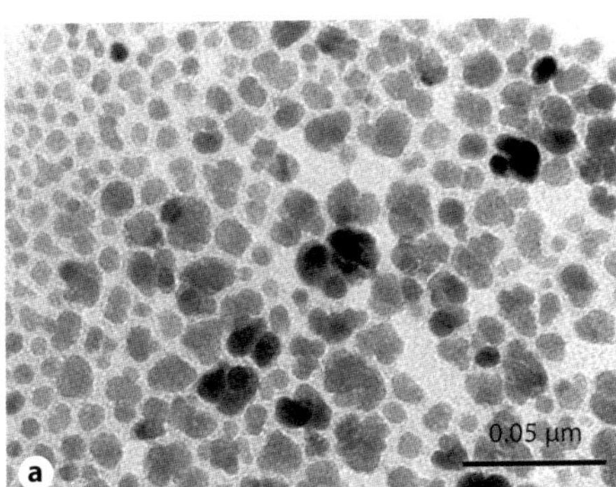

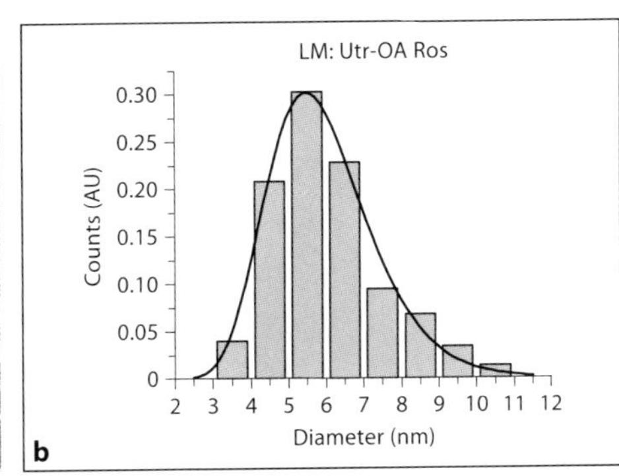

Fig. 1. TEM (**a**) and solid particle size distribution (**b**) of OA-Fe_3O_4 nanoparticles prepared on a micropilot scale (ROSEAL Co., Romania). Mean size 5.9 nm (single particles); standard deviation 1.4 nm. LM = MF (magnetic fluid): Utr (transformer oil carrier)-OA (oleic acid) Ros (ROSEAL Co.).

The *coprecipitation method* proved to be versatile for the preparation of magnetite, maghemite, and substituted ferrite nanoparticles suitable for MF preparation. In the case of magnetite, nanoparticles can be prepared from an aqueous solution of Fe^{3+} and Fe^{2+} in the mole ratio of 2:1 (under atmospheric conditions, with $Fe^{3+}/Fe^{2+} \approx 1.7$) at 80–82°C using a base solution in excess [13, 14]. The resulting subdomain magnetite particles are often surfacted in situ by chemisorption of C12 to C18 carboxylic acids and mostly oleic acid at 80–82 °C. The oleic acid monolayer-coated magnetite nanoparticles obtained along the 1-step route (see an example of their TEM image in fig. 1) are *hydrophobic* and disperse well in organic nonpolar liquid carriers. Magnetite nanoparticles with *hydrophilic* surface properties need surfactant double layer coating, e.g. the first chemisorbed layer should be covered by a second layer of the selected carboxylic acids adsorbed physically [10]. The solid particle (physical) mean size is slightly greater, usually in the interval of 6–9 nm, depending on the surfactant bilayer applied, as illustrated in figure 2 for the OA+OA coating.

Infrared laser pyrolysis is recommended as a versatile technique for producing iron-based nanoparticles [15]. Nanosized iron oxide-based materials, such as γ-Fe_2O_3/Fe_3O_4 or Fe-Fe_3O_4, are obtained via two procedures [16]: (1) a standard experimental procedure in which the oxidation process initiates and develops inside the laser-induced reaction zone, by the simultaneous addition of Fe precursors and an oxidizing agent (air), resulting in MNPs denoted by SF, and (2) a more complex experimental procedure in which the iron precursor is allowed to dissociate alone in the flame but in a surrounding oxidizing atmosphere, resulting in magnetic nanopowder denoted by F. The procedures are based on the resonance between the emission of a CW CO_2 laser line and the infrared absorption band of a gas precursor and upon subsequent heating of precursors by collisional energy transfer. A sensitizer (usually ethylene or sulfur hexafluoride) serves as the energy transfer agent.

As determined by TEM (fig. 3a, b), mean particle sizes depend on the laser power density and vary between about 4 and 6 nm and between 9 and 11 nm in the cases of samples SF and F, respectively. From the hysteresis loops at room temperature [16], the saturation magnetization (M_S) values of sample F are much higher (about 80 Am^2/kg) as compared to SF samples (between 16 and 46 Am^2/kg).

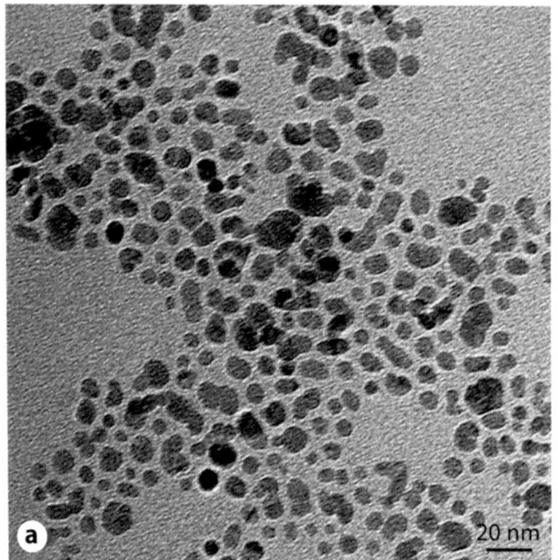

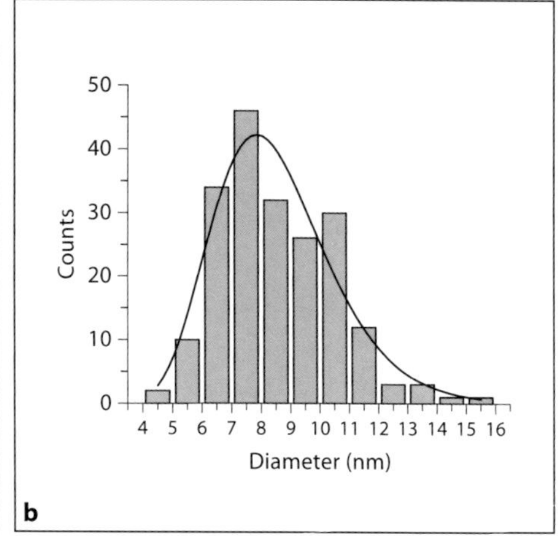

Fig. 2. TEM (**a**) and solid (physical) size distribution (**b**) for hydrophilic (OA+OA)-Fe_3O_4 nanoparticles. Mean size 8.6 nm; standard deviation 1.9 nm.

Carbon-encapsulated iron nanoparticles represent a core-shell type of nanostructure which is rather attractive for biomedical uses. Recent advances concerning a 1-step laser pyrolysis technique refer to the optimization of the relative carbon shell thickness in order to improve the overall magnetic properties of Fe-C nanoparticles. Acetylene/ethylene mixtures were used as carbon precursors. TEM images of the Fe-C particles and the histograms of the particle size distributions suggest an increase in particle diameter (from about 3.5 to 10.5 nm; fig. 3c) at an increase in nozzle diameter. HRTEM images show that onion-like surrounding graphenic layers often cover the buried iron-based (Fe/Fe_3C) cores (fig. 3c). The magnetic measurements of the composite nanoparticles indicate an almost direct correlation between the residence time in the laser beam and the M_S, i.e. from 30 to 70 Am^2/kg.

Water-Based Magnetic Nanofluids: Hydrophilic Magnetite Nanoparticles Dispersed in Water
In order to prepare water-based MFs, the hydrophilic MNPs must be dispersed in water. Superparamagnetic iron oxide nanoparticles (SPIONs) can be prepared according to the coprecipitation of Fe(II) and Fe(III) salts. The process described above is the simplest and most widely used route for the synthesis of SPIONs, because it produces a large amount of material with a particle size of around 10 nm. The surface of nanoparticles can be coated either in situ using additives such as organic compounds (e.g. sodium citric) or polymers (e.g. dextran and PAA), or in separate chemical processes using surface active or complexing agents in order to disperse them in water. The most common surfactant is sodium oleate (NaO) and well-known complexants are CA and PAA. The details of the preparation and characterization of MNPs themselves and the carboxylated and surfacted nanofluids can be found in previously published papers [9, 10, 17–19].

The study of the adsorption of different stabilizers on MNPs has not been widespread. Oleic acid or oleate is the most often used surfactant in MF stabilization; however, to the best of our knowledge its adsorption on magnetite nanoparticles has not been quantitatively studied yet. It is used in a general way, i.e. empirical doses are simply added to magnetite just after precipitating nanoparticles and its excess related to the mono-

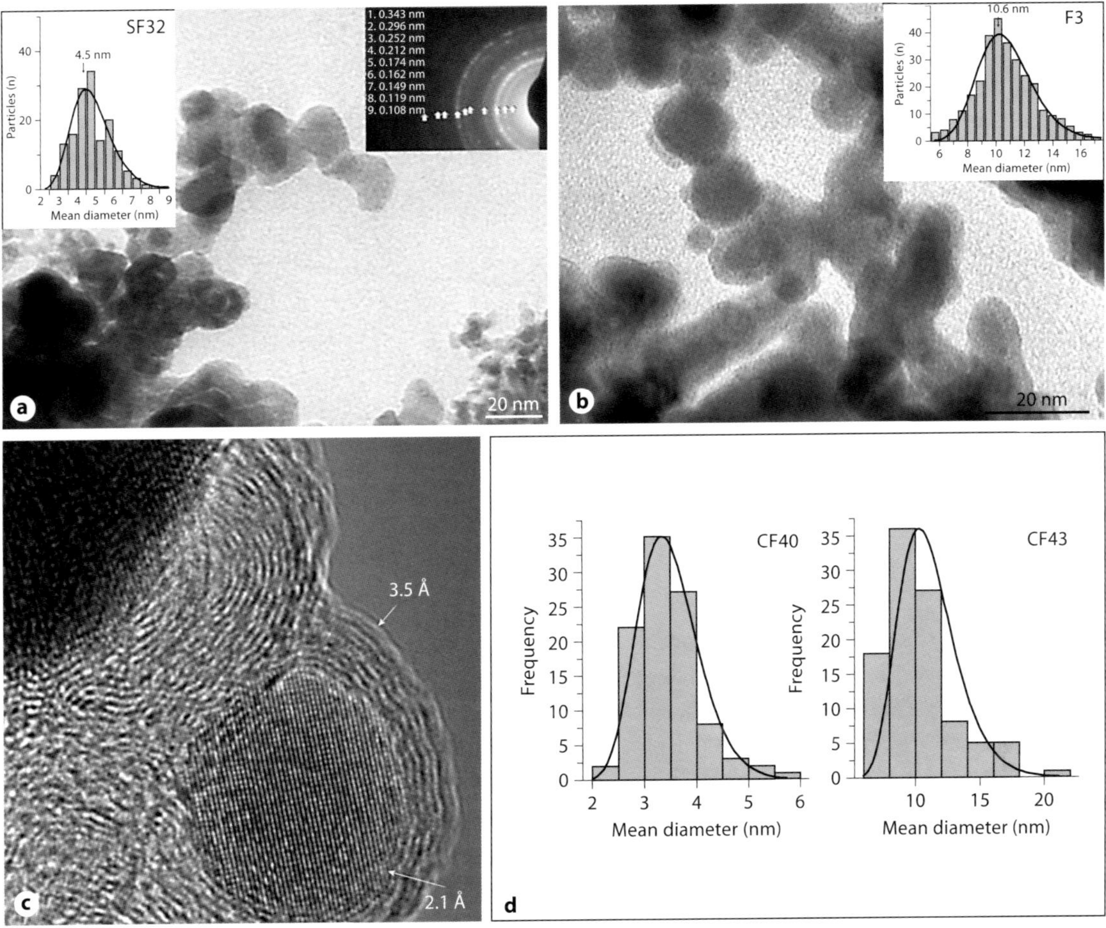

Fig. 3. TEM/HRTEM analysis of maghemite/magnetite and Fe-C nanoparticles obtained by laser pyrolysis applying different experimental conditions: SF type nanopowder (**a**) and F type nanopowder (**b**). **c** HRTEM image of the Fe-C nanoparticle; there is a well-distinguishable quasispherical iron core of about 11 nm in diameter (revealing the identified interplanar distance of d = 2.1 Å for Fe/Fe_3C) surrounded by about 8 graphene layers (with an identified interplanar distance of d = 3.5 Å). **d** Size histograms of Fe-C nanoparticle samples.

layer coverage is removed by washing solids with ethanol or acetone [20], and the organic coated particles are ready to disperse spontaneously in organic carriers. If water-based MFs are prepared, the second layer development is followed by visual observation of dispersing solid particles in aqueous medium. Unfortunately the empirical way is preferred not only in the preparation of the surfacted MFs but also in the case of other stabilizers like citrate, dextran, or PEG/PEO. Oleate and citrate are the most often used stabilizers and their adsorption mechanism is very different.

The synthesized SPIONs were stabilized with different carboxylated compounds, each of which can bind chemically to ≡FeOH sites of the MNPs; however, they present systematic structural differences since CA is a small molecular complexant, while PAA is a macromolecular compound, and NaO is a surfactant with a hydrophobic alkyl chain. In order to prepare well-characterized car-

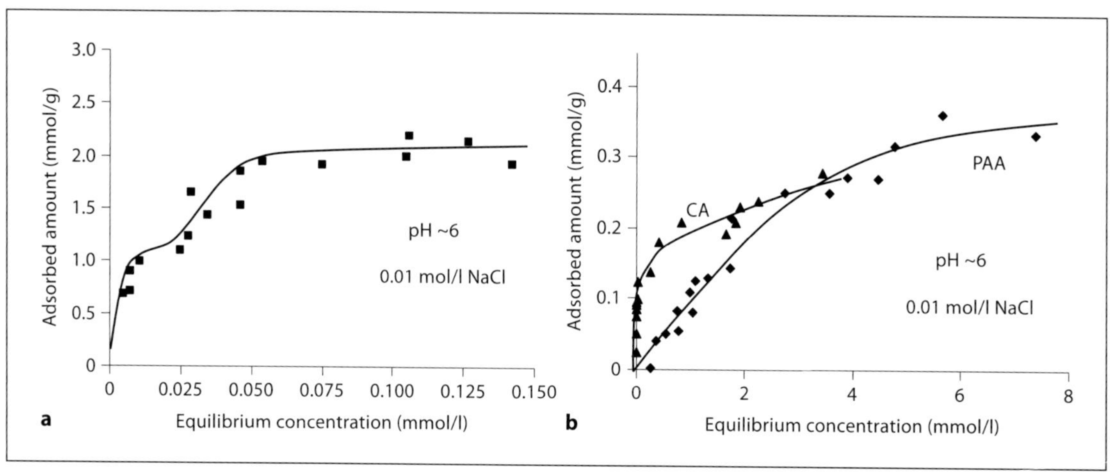

Fig. 4. Adsorption isotherms of surfactant NaO (**a**) and small and macromolecular organic acids (CA and PAA, **b**) on magnetite at room temperature.

boxylated MFs, the adsorption of oleate and small and macromolecular organic acids on magnetite nanoparticles was studied at pH ~6 in the presence of 0.01 mol/l NaCl. Adsorption isotherms are shown in figure 4. Double layers of oleate are adsorbed on magnetite as explained in detail [16, 21]. Monolayer coverage is completed below a specific amount of ~1 mmol/g, while the saturation value of ~2 mmol/g indicates the completion of double layer formation as seen in the 2-step isotherm shown in figure 4a. Oleate is chemisorbed in the first layer due to the surface complex formation between the $-COO^-$ groups of oleate and the ≡FeOH sites on the surface of magnetite as assumed. A second layer of oleate anions can be adsorbed on the hydrophobic shell of oriented surfactant molecules via hydrophobic interaction.

CA is also applied in MF stabilization without systematically studying the adsorption process. In general, empirical doses, sometimes in huge amounts [22], are simply added to magnetite just after precipitating nanoparticles, and the products are interpreted as CA surface coating, which provides good/excellent stability of water-based MFs. Our adsorption study shows the high-affinity character of CA adsorption up to the amount of ~0.1 mmol CA on 1 g of magnetite (fig. 4b) due to complex formation on the surface sites of magnetite. This adsorbed amount is far below the amount of the available ≡FeOH surface sites, i.e. 0.8 mmol/g, even supposing that all 3 carboxylic groups of CA molecules are bound on ≡FeOH sites. The feature of adsorption isotherm of PAA and its adsorbed amount related to its –COOH groups are different from those of CA as seen in figure 4b. The chemical interactions between the ≡FeOH sites and the adsorbed carboxylic groups are similar, but their bond strengths differ significantly and their stabilizing efficiency changes definitely due to the different structure and thickness of the coating layer on the MNPs, i.e. the composition of the aqueous interface at the particle surface. It should be noted that dissolution of iron oxides is enhanced in the presence of complexants like CA. Although each carboxylated MNP holds excess negative charges around pH ~7, the CA coating is thin and its density is low (less than 1 CA/nm^2), the PAA is adsorbed in a thicker layer, while oleate forms a really thick double layer with a quite high packing density (~2.5 $oleate/nm^2$ in the first layer and ~5 $oleate/nm^2$ in

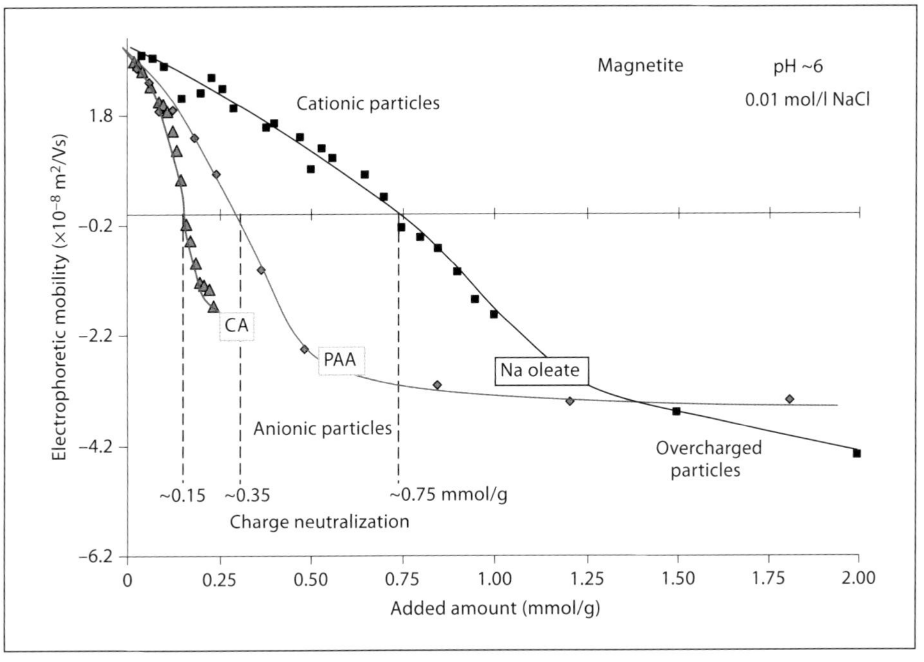

Fig. 5. Effect of different stabilizers on the charge state of magnetite nanoparticles.

the second layer) on MNPs. Well-stabilized MFs can be prepared by using these compounds.

In the case of magnetic nanopowders obtained by laser pyrolysis, the main task is the surfactant coating of individual MNPs and their stable dispersion in carrier liquids.

Magnetic core-shell iron-carbon nanoparticles with a high specific magnetic moment are among the most interesting magnetic nanofluids designed for biomedical applications. The dispersion of Fe-C nanopowder in polar liquid matrices, especially in water, implies the hydrophilization and surfactant coating of individual particles following the procedures described in reports by Hajdu et al. [9] and Bica et al. [10]. Due to the existence of rather tight particle agglomerates, as revealed by TEM analysis, the volume fractions of the dispersed nanoparticles do not exceed 1% and the values of the hydrodynamic diameter (D_H) of dispersed particle clusters exceed 100 nm.

Particle Charge, Surface Coating, and Aggregation under Different Conditions: Colloidal Stability Requirements

Tuning pH- and salt concentration-dependent properties is a recent challenge in the preparation of water-based MFs for biomedical use. The effect of different stabilizers on particle charge is worth studying. Electrophoretic mobilities determined in the presence of increasing concentrations of additives at pH ~6 can be seen in figure 5. Each curve starts at a fairly well-defined positive value indicating the dominance of positive charges on magnetite under acidic conditions. The cationic particles exist up to the sign reversal of measured electrophoretic mobility. The latter indicates reversal of the charge sign at a shear plane of magnetite particles due to the chemical adsorption of anionic compounds. The neutralization of positive surface charges of iron oxide particles occurs

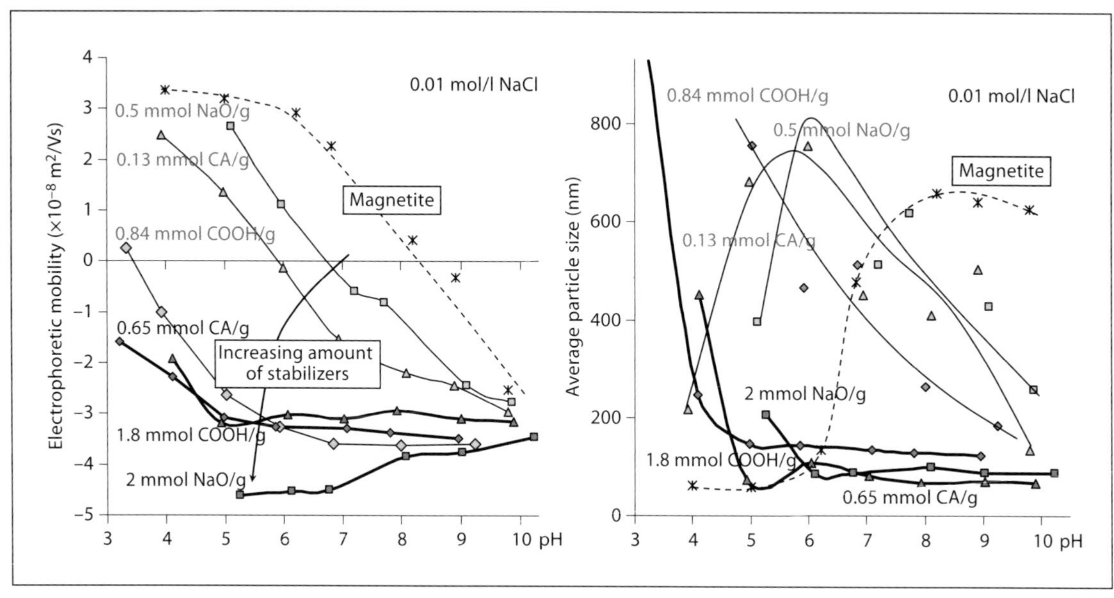

Fig. 6. Effect of NaO (square), CA (triangle), and PAA (diamond) loading on the pH-dependent charge state (left) and aggregation (right) of magnetite nanoparticles at 25 ± 0.1 °C. The molar amount of PAA is related to its COOH groups. The reproducibility of electrophoretic mobility was not better than ±0.4 10^{-8} m^2/Vs. Reproducible size values below ~150 nm were measured in stable sols. The larger values even measured under the same kinetic conditions show only the presence of larger aggregates in unstable systems.

at different amounts of added anions as shown in figure 5, but it never reaches the amount of the available ≡FeOH surface sites, i.e. 0.8 mmol/g. The charge neutralization points for CA and PAA at ~0.15 and ~0.35 mmol/g, respectively, are comparable with their adsorption coverage (fig. 4). The specific amount of the added oleate is ~0.75 mmol/g at this charge neutralization point, which is in fairly good agreement with the amount of oleate adsorbed in the first layer (fig. 4). Anionic particles are formed above the charge neutralization points due to the excess charges accumulated in the adsorption layers of MNPs. In the case of oleate, the amount of negative charge carried by MNPs increases significantly, and particles become overcharged similarly to the PAA-stabilized particles where a thick layer of adsorbed macroions holds excess charges due to the dissociation of unbound carboxylic groups at pH ~6.

The effect of stabilizer loading on the pH-dependent charge state and aggregation of MNPs can be observed over a broad range of pH in figure 6. It can be seen that the naked magnetite particles (broken line) are positively charged below pH ~8 and have negative charges above it. This pH may be identified as the isoelectric point (IEP) of magnetite particles. The effect of increasing amounts of stabilizers on the surface charge properties of magnetite was examined at several concentrations. The measured data at 2 characteristic loadings, one not far from the charge neutralization points (fig. 5) and the other in the high excess of stabilizers, are shown in figure 6. It is obvious that electrophoretic mobility decreases gradually with increased loading of additives over the whole range of pH. The IEP values shift by 1–2 units to the lower pHs. At the higher loadings, however, MNPs become negatively charged even at low pHs since particles are covered completely in the high excess of stabilizers. In the case of oleate, the second adsorption layer is built up and magnetite nanoparticles become overcharged at 2 mmol NaO/g loading. Due to the close packed structure of oleate anions in the second adsorption layer of the double layer on magnetite particles, the measured values become more negative than that in the alkaline pH region in the absence of surfactant. CA is adsorbed only in monolayer binding to the magnetite surface via 1 or 2 carboxylate groups even in its high excess (0.65 mmol CA/g), and so the overcharging of particles is limited; hence the negative charge density on CA-covered magnetite is smaller than that of the oleate double layer-coated particles as explained above. Therefore, all of the measured electrophoretic mobilities of citrate-coated magnetite are smaller negative values than that of the oleate-stabilized sample (fig. 6 left). The pH dependence of mobility shows that the charge state of citrate-coated magnetite is almost independent of pH, except for a significant change below pH ~5 due to the suppressed dissociation of carboxylic groups on CA ($pK_1 = 2.86$, $pK_2 = 4.47$, and $pK_3 = 5.82$). The pH-dependent charge state of PAA-coated MNPs is definitely similar to that of citrated magnetite particles.

The pH-dependent particle aggregation was measured in parallel by dynamic light scattering (DLS), which can be used even in coagulating systems, and the hydrodynamic size data provide information on the system colloidal stability. Some data are shown in figure 6 (right). The formation of large aggregates was observed near the IEPs of each system having imperfect stabilization such as naked and poorly coated magnetite particles at the smaller loadings, where positive charges of magnetite surface were partially neutralized, and so the colliding particles adhered due to the lack of electrostatic repulsion. Far from the pH of the IEP, the efficiency of particle adhesion in random collisions was much worse and only smaller aggregates could form in both acidic and alkaline regions during the given time of measurements. When perfectly coated anionic particles were dispersed at either oleate (2 mmol/g), citrate (0.65 mmol/g), or polyacrylate (1.8 mmol COOH/g) loadings, reproducible hydrodynamic size values smaller than ~150 nm were measured independently of the pH above ~6. The outermost hydrophilic shell on the double layer-coated magnetite particles contains carboxylate groups. Therefore, pH-dependent negative charges exist due to the dissociation of bound groups, $S\text{-}COOH \Leftrightarrow S\text{-}COO^- + H^+$ (pK ~4) on the surface of the oleate double layer-coated magnetite particles. Fully dissociated carboxylate groups above pH ~6 provide electrostatic repulsion between colliding particles besides the steric hindrance of the double layer. However, the dissociation degree starts to decrease with decreasing pH below ~6, and so not only does the electrostatic repulsion decline but also particles become less hydrophilic in the acidic region. In the same way, a significant decline in the stability of citrate- and polyacrylate-coated magnetite samples occurs only below pH ~5, where dissociation of COOH groups begins to suppress. Therefore, a significant decrease in colloidal stability, i.e. formation of aggregates, can be predicted in acidic medium for the carboxylated MFs.

Colloidal stability requirements for biocompatible MFs comprise not only the given range of pH relevant to biological media but also the relatively high concentration of salts, not to mention interaction with proteins at all, which is beyond our expertise. The salt tolerance of dilute MFs stabilized by different agents was investigated in dilute systems at pH ~6 to test particle aggregation in time and whether the combined electrostatic and steric stabilization provides suitable colloidal stability under physiological conditions. We should note that the salt tolerance of naked magnetite particles is very low (~0.001 mol/l NaCl) under these conditions. Coagulation kinetics measurements were performed to quantitatively determine the efficiency of different agents in the stabilization of magnetite particles dis-

Table 1. The CCC values determined in dilute magnetite sols stabilized by increasing loadings of different stabilizers at pH ~6 (concentration of physiological salt solution 150 mmol/l)

Magnetite	Amount added mmol/g	Electrophoretic mobility 10^{-8} m²/Vs	CCC (NaCl) mmol/l
Naked	0	1.31	~1
CA coated	0.13	-1.54	~5
	0.3	-2.86	~40
	0.65	-2.96	~70
PAA coated	0.56	-1.8	~120
	0.83	-3.53	~200
	1.8	-3.68	~450
Oleate coated	1	-2.04	~5
	2	-4.52	~200

persed in salt solutions of various concentrations. The size evolution of aggregates in time was followed by DLS, and the coagulation rate was calculated from the slope of kinetic curves as previously explained [17]. The stability ratio (W) was calculated and the lowest salt concentration inducing fast coagulation, i.e. the critical coagulation concentration (CCC), was determined. The different stabilizers were applied in high excess and the CCC values for NaCl determined at their different loadings are summarized in table 1. The data show an increasing resistance against salt with increased loading of citrate; however, it was not possible to provide a suitable salt tolerance expected under physiological conditions since the measured CCC values (~5–70 mmol/l remained far below 150 mmol/l even when applying it in huge excess. It should be noted that such high CA loadings enhance the dissolution of magnetite crystals to a great extent. The thick polyacrylate layer and the oleate double layer can effectively hinder the aggregation of magnetite particles due to the combined steric and electrostatic stabilization [8, 20]. The latter is supported by the greater electrophoretic values of oleate-loaded and PAA-coated samples compared to that of citrate-containing systems presented in table 1. The resistance of magnetite sols stabilized by oleate double layers and a thick PAA layer against electrolytes is enhanced above the critical salt tolerance (>0.150 mol/l) expected under physiological conditions.

Besides several unique problems of particular applications, the colloidal stability of MFs under physiological conditions is of great practical importance. Although the dispersion of separated MNPs in various organic and mainly low-polarity solvents is a well-proven technique, MFs with highly polar carriers (especially water) are always characterized by the appearance of noncontrolled aggregation. Depending on biomedical tasks, particle aggregation in the magnetic field must be excluded (in applications where the risk of blood clots in blood vessels exists) or can be used (e.g. in magnetic cell separation and hybrid organic/inorganic nano- or even mesoclusters of a controlled size and shape with a high magnetic moment in an applied magnetic field are of particular interest). Thus, the combined characterization of magnetic particles and their aggregates (clusters) is an important feature of modern diagnostics of MFs. The dilution stability of water-based MFs for biomedical applications is especially important. In figure 7 the results of a DLS investigation are presented; the particle cluster size shows a significant increase in highly diluted samples.

Small-angle neutron scattering (SANS) is one of the most efficient methods of nanostructural investigation of MFs [19, 23]. During SANS experiments a widening of the neutron beam passed through the sample is studied in terms of the differential scattering cross section, i.e. the effective number of neutrons scattered in a given spatial angle referred to the unit incident neutron flux. The latter is a function of a vector, with the module $q = (4\pi/\lambda)\sin(\theta/2)$, where λ is the incident neutron wavelength and θ is the scattering angle. This dependence is quite sensitive to structural features of the studied system at the scale of 1–100 nm. Indeed, the sizes of magnetic particles in

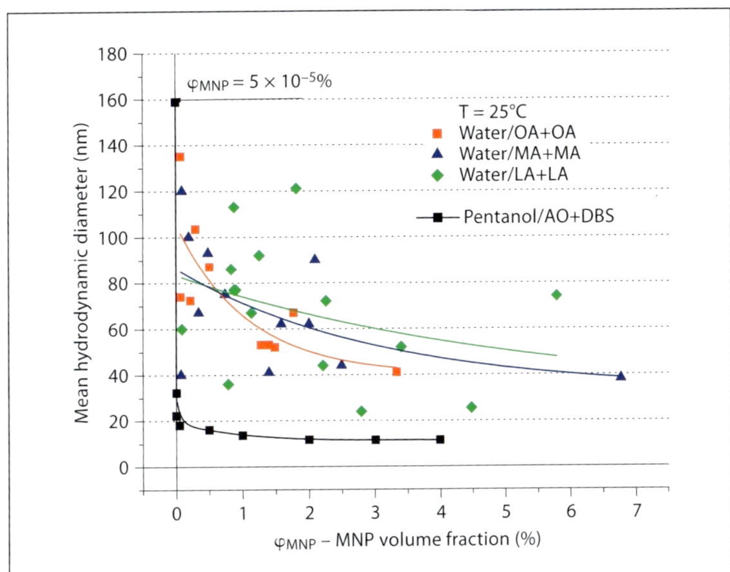

Fig. 7. MNP volume fraction dependence of the cluster D_H for water-based magnetic nanofluids stabilized by carboxylic acid double layers. Comparison with a high-colloidal stability organic polar MF (MF/pentanol) – the smooth lines are the exponential decay fits. T = Temperature.

MFs, as well as the characteristic correlation length between particles, are mostly in this dimensional range. SANS investigations including specific techniques (contrast variation and scattering of polarized neutrons) giving information about particle structure (size, surfactant shell thickness, composition of the core and shell, solvent rate penetration in the surfactant layer, and micelles), magnetic structure (magnetic size and composition), particle interaction (interparticle potential, magnetic moment correlation, and phase separation), and cluster formation (aggregation and chain formation). Major advantages in SANS in comparison with other structural methods are connected with possibilities of contrast variation in the studied systems by hydrogen/deuterium substitution, as well as with magnetic scattering of neutrons. The other important feature of this method is a high penetration depth of neutrons into the sample, which makes it possible to investigate nonmodified bulk MFs in a wide range of magnetic volume fractions of 1–20%.

An example of the SANS analysis is given in figure 8, where water-based MFs with magnetite stabilized by coating with a double layer of myristic (MA+MA) and lauric (LA+LA) acids are studied [23]. The initial concentrated fluids (ϕ_m ~10%) were diluted with different mixtures of light/heavy water so that the D_2O volume fraction in the final samples was varied in the range of 0–90% at 1 particle concentration. Such a procedure allows for realization of the so-called contrast variation, where the scattering of the MFs components is matched at different rates when varying the scattering length density of the solvent keeping the structure of the system the same. Comparatively large (characteristic size 30–40 nm) but stable (in the absence of an external magnetic field) aggregates are revealed. The presence of polydisperse aggregates does not allow us to apply the classical Guinier analysis of the curves and follow precisely the change in the corresponding scattering invariants such as forward scattered intensity and radius of gyration. A recently developed method of modified basic functions in the SANS contrast variation for polydisperse systems [23, 24] makes it possible to effectively compare the scattering from the averaged

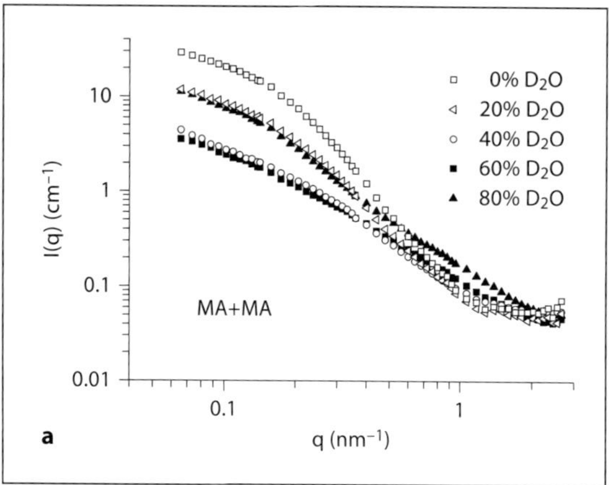

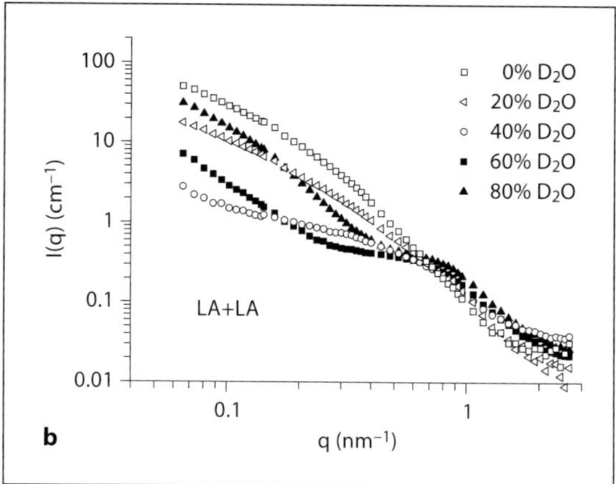

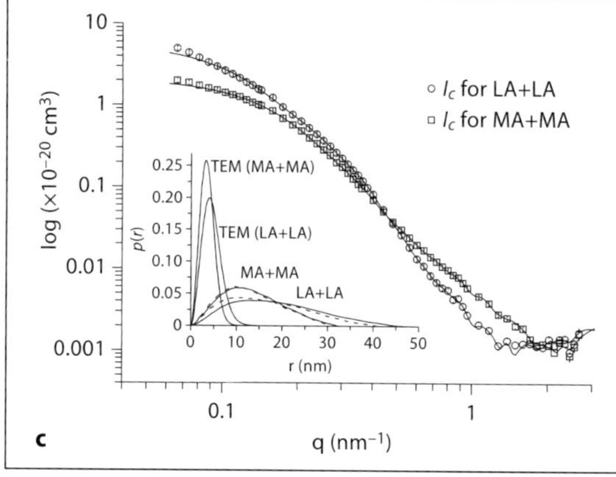

Fig. 8. SANS analysis of aggregate structure in water-based MFs with sterical stabilization of magnetite by double coating with myristic (MA+MA) and lauric (LA+LA) acids. Contrast variation with a different content of D_2O in carrier (**a**, **b**) is applied to reveal averaged scattering form factors of whole aggregates (magnetite plus surfactant, I_c). **c** The indirect Fourier transform, *p*(*r*), of I_c is compared with the *p*(*r*) for scattering from MFs in pure H_2O (scattering from magnetite only; dashed lines) and with the *p*(*r*) calculated from data of TEM taking into account separate particles. It follows that, in contrast to the LA+LA sample, in the MA+MA sample a significant part of aggregates are noncoated magnetite.

particle shape, which includes surfactant and the scattering from pure magnetite. As illustrated in figure 8c, the indirect Fourier transform is performed additionally to show the important difference between the 2 kinds of scattering. The surfactant contribution in the case of LA+LA is significantly larger and shows the presence of the expected stabilizing shell with a thickness of about 3.5 nm, which is in agreement with the double length of the LA molecule (1.4 nm). The maximal size of the aggregates from the 2 kinds of scattering coincides with that for the magnetite component, which shows that magnetite in the aggregates is not fully coated with the surfactant in the last case. This indicates that free surfactants with a larger tail length agglomerate more easily during the preparation of MFs, which prevents their adsorption on the magnetite surface and is consistent with the fact that for longer surfactants in the series of saturated monocarboxylic acids (such as palmitic and stearic acids) dispersing nonaggregated magnetite particles in water is practically impossible.

Magnetic Nanocomposites for Biomedical Applications

In a hybrid MNP-polymer system the specific composition of the polymeric material allows for functionalization, different morphology design, and control of the MNP properties. Smart mate-

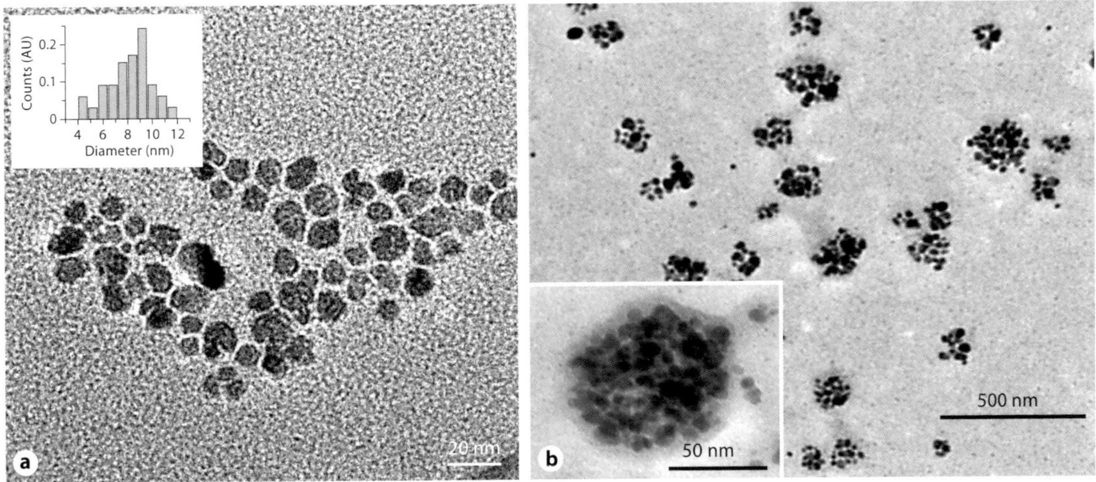

Fig. 9. TEM image of hybrid magnetic nanostructures. **a** MPy-co-DP prepared with the magnetic nanofluid MF/W/LA+LA and functionalized pyrrole copolymer. **b** PNIPA magnetic microgel.

rials based on MNPs and stimuli-responsive biocompatible polymers and block copolymers present an increasing interest for applications in biotechnology and nanomedicine. These smart composites can be designed in a multitude of architectures (dispersed core-shell nanoparticles, gels, and self-assembled structures).

Core-Shell Nanostructures and Microgels

Magnetic nanostructures designed either as core-shell nanoparticles or as microgels may be obtained by simple procedures using water-based magnetic nanofluids and different polymers as starting materials.

Magnetic core-shell type nanostructures were obtained by coating the magnetite nanoparticles of the MF samples with functionalized pyrrole copolymer shells. The pyrrole monomer with attached glycyl-leucine (Py-Gly-Leu) and magnetic nanofluid MF/W/LA+LA were used as starting materials to obtain the functionalized core-shell magnetic nanostructures (MPy-co-DP) [25].

Also a water-based magnetic nanofluid, MF/W/OA+OA was used for the synthesis of smart magnetic microgels containing magnetite nanoparticles encapsulated into a thermoresponsive polymer, i.e. poly(N-isopropylacrylamide) (PNIPA). The polymerization of the monomer N-isopropylacrylamide was carried out in aqueous solution in the presence of a cross-linking agent N,N′-methylenbisacrylamide, APS, and magnetic nanofluid. Stable suspensions in water of magnetic microgels were obtained.

The formation of hybrid nanostructures based on magnetic nanofluid and either pyrrole copolymer functionalized with dipeptide, MPy-co-DP, or thermoresponsive polymer M-PNIPA was evidenced by Fourier transform infrared spectroscopy [25].

The TEM image of the hybrid magnetic nanostructures obtained by the copolymerization of pyrrole and functionalized pyrrole Py-Gly-Leu in the presence of magnetic nanofluid is shown in figure 9a. The diameter histogram shown in the inset of figure 9a indicates a mean diameter of around 9 nm. This size is only slightly different from that of the magnetite nanoparticles obtained from the MF/W/LA+LA (about 7 nm) [23]. A different morphology was observed for the magnetic microgels based on magnetic nanofluid

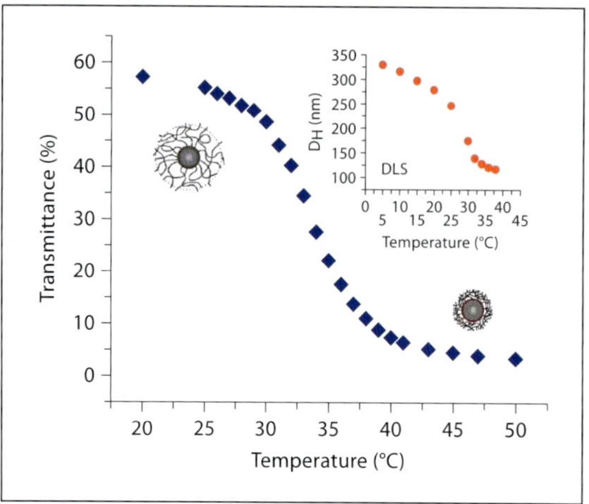

Fig. 10. Variation of transmittance at 800 nm of PNIPA magnetic microgel as a function of temperature. **Inset** Variation of the D_H of PNIPA magnetic microgels as a function of temperature.

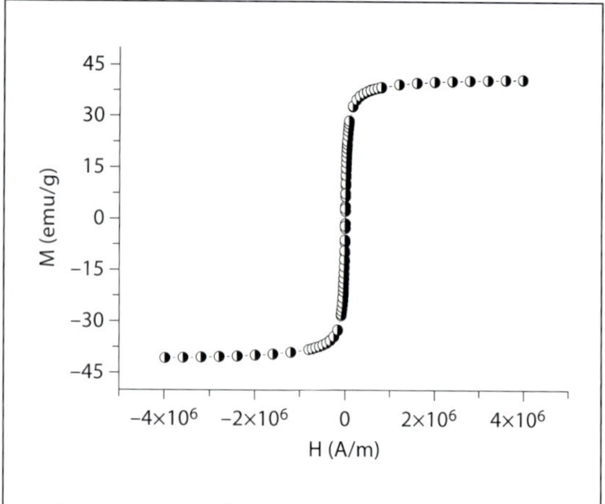

Fig. 11. Magnetization curve of magnetic microgel M-PNIPA at room temperature; dried sample.

and a cross-linked polymer, PNIPA (fig. 9b). Clusters of MNPs stabilized with a double layer of oleic acid are encapsulated into polymeric PNIPA spheres having diameters in the range of 50–100 nm.

Among stimuli-responsive smart polymeric materials which can respond in volume changes to small external stimuli, PNIPA is the most widely studied thermosensitive polymer. It has a transition from the swollen state to the collapsed state in water at a temperature of around 32°C, which is well known as the lower critical solution temperature (LCST). The encapsulation of magnetite nanoparticles into PNIPA microspheres results in a dual-responsive smart microgel sensitive to temperature and to magnetic fields.

The transition of PNIPA magnetic microgels between the swollen and collapsed states by temperature variation was evidenced by DLS and UV-visible spectroscopy (fig. 10). The decrease in transmittance is observed at temperatures above the LCST appearing at around 30°C due to the increase in Rayleigh scattering in the collapsed state of the microgel.

DLS measurements at different temperatures in the range of 10–40°C show a decrease in the D_H of the PNIPA microgel spheres at temperatures higher than the LCST appearing at around 30°C due to the polymer transition to a collapsed state (inset in fig. 10).

The magnetization curve of the PNIPA magnetic microgel dried sample at room temperature (fig. 11) shows superparamagnetic behavior, as expected for magnetite nanoparticles of small sizes (less than 10 nm), stabilized by a double layer of surfactant. Moreover, this fact indicates that the magnetite nanoparticles are still well separated in the polymer matrix. The value of the M_S, i.e. 41.4 emu/g for the magnetic microgels prepared by our procedure, is relatively high as compared with the values reported for MNPs encapsulated into responsive polymers obtained by other procedures, e.g. 24.8 emu/g [6]. High magnetization values mean that our synthesis procedure favors close packing of surfacted magnetite nanoparticles and allows the encapsulation of a relatively high concentration of magnetite nanoparticles into the PNIPA spheres.

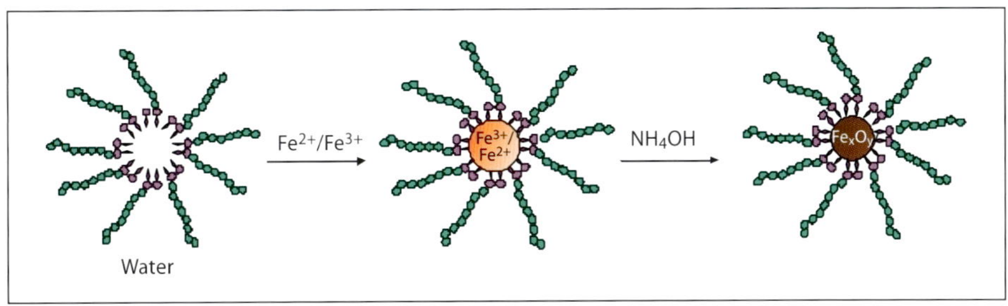

Fig. 12. Synthetic pathway followed for the preparation of SPION-containing micelles.

The pyrrole copolymer functionalized core-shell magnetic nanostructure (MPy-co-DP) also shows superparamagnetic behavior and an M_s of 65 emu/g Fe_3O_4, which is only slightly lower than that of the magnetite stabilized with LA+LA (M_s = 75.5 emu/g Fe_3O_4).

The superparamagnetic properties, the relatively high magnetization values, and the magneto- and thermoresponsive properties of polymer-coated/encapsulated MNPs make these composite materials suitable for further applications like magnetic bioseparation, drug delivery, and molecular recognition of biomolecules.

Superparamagnetic Polymer-Based Micelles

Well-defined amphiphilic block copolymers (ABCs) possessing hydrophilic and functional hydrophobic block segments may act as effective stabilizers for SPIONs in aqueous solutions. ABCs capable of generating micellar nanomorphologies in aqueous media may provide improved stabilization and control over nanoparticle growth via the block forming the micellar corona, whereas the functional block segment constructing the core is responsible for the coordination and solubilization of the inorganic compound.

In most investigations dealing with the synthesis of novel core-shell nanohybrids consisting of functional ABCs and SPIONs, carboxylic acid-containing block copolymers have been mainly employed. A new class of well-defined diblock copolymers as functional coordination copolymers were synthesized based on 2-(acetoacetoxy) ethyl methacrylate (AEMA) [26]. The acetoacetoxy moieties of AEMA-containing polymers can act as strong bidentate ligands for different metal ions yielding sterically stabilized colloidal hybrid nanomaterials. Well-defined ABCs belonging to this new family, based on AEMA and poly(ethylene glycol)methyl ether methacrylate (PEGMA; water-soluble, biocompatible, and thermoresponsive) have been recently synthesized employing a controlled radical polymerization method and evaluated regarding their ability to act as stabilizers for iron oxide nanoparticles in aqueous solutions [27].

A 3-step synthetic methodology (fig. 12) was employed for the preparation of magnetoresponsive polymer-based micelles involving the chemical coprecipitation of iron chloride (II) and (III) salts in the presence of $PEGMA_x$-*b*-$AEMA_y$ micelles under weak basic conditions. By varying the block length ratios of the PEGMA and AEMA segments in the $PEGMA_x$-*b*-$AEMA_y$ diblock copolymers, micelles of tunable diameters ranging from ~10 to 70 nm were obtained [27]. This ability, i.e. to synthetically control the micellar size, is extremely important since size manipulation is a crucial parameter affecting various biomedical properties such as circulation times of nanoparticles in the blood stream.

As previously mentioned, the use of nanosized micelles as templates for the formation of MNPs

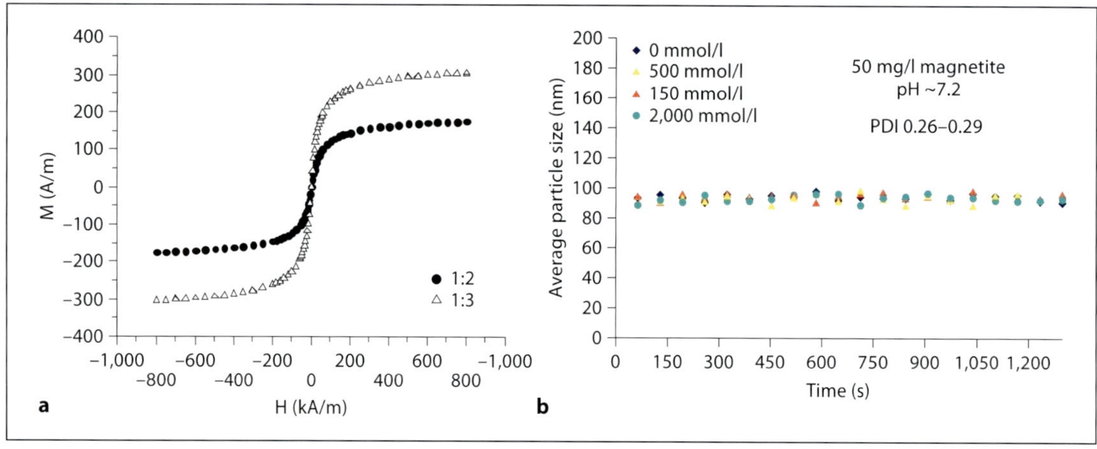

Fig. 13. a Vibrational sample magnetometry curves of the PEGMA$_{70}$-*b*-AEMA$_{16}$/Fe$_x$O$_y$ hybrids characterized by different magnetic loadings. **b** Coagulation kinetics measurements performed by dynamic light scattering on the PEGMA$_x$-*b*-AEMA$_y$/Fe$_x$O$_y$ micellar systems at a concentration of 50 mg/l magnetite in the presence of various NaCl concentrations, demonstrating the salt tolerance of these systems. PDI = Polydispersity index.

promotes generation of iron oxide nanoparticles inside the micellar core characterized by very small diameters. Such ultrasmall iron oxide nanoparticles exhibit superparamagnetic behavior. The latter is preferred in MRI applications since upon removal of the magnetic field no net magnetization is retained. As shown in figure 13a, the PEGMA$_x$-*b*-AEMA$_y$/Fe$_x$O$_y$ hybrid micelles demonstrated superparamagnetic behavior at room temperature.

Regarding their stability in aqueous media, these systems were highly stable not only in pure water but also in saline solutions even at high NaCl concentrations of up to 2 mol/l (fig. 13b).

As far as biocompatibility is concerned, cell growth assays on a prostate cancer cell line led to statistically insignificant differences on cell viability when cells were treated with the PEGMA$_x$-*b*-AEMA$_y$/Fe$_x$O$_y$ micelles, pure water, and Resovist®, a clinically approved dextran-based MRI contrast agent, of comparable size. Additionally, their in vitro uptake by macrophage cells was found to be lower compared to that of Resovist, suggesting a prolonged blood circulation time and enhanced in vivo efficacy [27].

Conclusively, these novel systems, exhibiting magnetoresponsive (superparamagnetic) properties, high stability, and in vitro biocompatibility, combined with the presence of a thermoresponsive (PEG-containing) block segment on their exterior may potentially find applicability as new MRI contrast agents or as magnetothermally triggered drug delivery systems.

Conclusions

Chemical coprecipitation and gas-phase laser pyrolysis proved to be highly efficient procedures in preparing relatively large quantities of biocompatible MNPs, in particular surfacted/functionalized magnetite and surface-passivated iron nanoparticles in the size range of 3–15 nm, as well as of water-based magnetic nanofluids. Comparative investigations of the adsorbtion of various surfactants (CA, C12 to C18 carboxylic acids, and polyacrilic acid) and stabilization mechanisms of dispersed MNPs in aqueous medium revealed that the resistance of magnetite sols stabilized by oleate double layers and a thick

PAA layer against electrolytes is enhanced above the critical salt tolerance (>0.150 mol/l) expected under physiological conditions. Structural analysis by SANS and DLS of the particle clusters in water-based nanofluids evidenced different surface coating efficiencies depending on the surfactant chain length. Using longer surfactants in the series of saturated monocarboxylic acids (such as palmitic and stearic acids), dispersion of nonaggregated magnetite particles in water is practically impossible, in contrast with unsaturated oleic acid, one of the most efficient stabilizers under physiological conditions. Magnetic core-shell type nanostructures were obtained by coating the surfacted magnetite nanoparticles of water-based MF with functionalized pyrrole copolymer shells, resulting in functionalized core-shell magnetic nanostructures (MPy-co-DP). Also water based nanofluids were used to synthesize PNIPA magnetic microgels with high magnetic loading, which can be further functionalized with PAA. The transition of PNIPA magnetic microgels between the swollen and collapsed states by temperature variation was evidenced by DLS and UV-visible spectroscopy, with the LCST appearing at around 30°C. Well-defined ABCs based on AEMA and PEGMA (water-soluble, biocompatible, and thermoresponsive) proved to be efficient stabilizers of iron oxide nanoparticles in aqueous solutions and conducted to magnetoresponsive polymer-based micelles. The novel magnetic hybrid nanostructured materials described in this paper have superparamagnetic behavior and exhibit magneto- and thermoresponsive properties, high stability, and in vitro biocompatibility; consequently, they may find applicability as new MRI contrast agents, as magnetic carriers in bioseparation equipments, or as magnetothermally triggered drug delivery systems.

Acknowledgements

This work was partly supported by the Romanian Academy through the program 'Magnetically Controllable Fluids: Structure, Properties, Complex Flows, Modeling and Applications'. The financial support received from the EU FP7 project MagPro2Life (CP-IP 229335-2) (L. Vékás, R. Turcu, I. Morjan), the Cyprus and Romania Research Promotion Foundations (bilateral collaboration program, KY-POY/0407/09) and the project NANOMA-FP7-ICT-2007-2 (T. Krasia-Christoforou), the project HRJRG-016 of the Helmholtz-RFBR program Joint Research Groups (M.V. Avdeev) and from the NKTH-OTKA (Hungary; A7-69109/2007) grant (E. Tombacz) is gratefully acknowledged.

Disclosure Statement

The authors have nothing to disclose.

References

1 Torchilin VP: Multifunctional nanocarriers. Adv Drug Deliv Rev 2006;58: 1532–1555.
2 Veiseh O, Gunn JW, Zhang M: Design and fabrication of magnetic nanoparticles for targeted drug delivery and imaging. Adv Drug Deliv Rev 2010;62: 284–304.
3 Boyer C, Whittaker MR, Bulmus V, Liu J, Davis TP: The design and utility of polymer-stabilized iron-oxide nanoparticles for nanomedicine applications. NPG Asia Mater 2010;2:23–30.
4 Mistlberger G, Klimant I: Luminescent magnetic particles: structures, syntheses, multimodal imaging, and analytical applications. Bioanal Rev DOI: 10.1007/s12566-010-0017-7.
5 Bucak S, Jones DA, Laibinis PE, Hatton TA: Protein separations using colloidal magnetic nanoparticles. Biotechnol Prog 2003;19:477–484.
6 Shang H, Chang WS, Kan S, Majetich SA, Lee G: Synthesis and characterization of paramagnetic microparticles through emulsion-templated free radical polymerization. Langmuir 2006;22: 2516–2522.
7 Pankhurst QA, Connolly J, Jones SK, Dobson J: Applications of magnetic nanoparticles in biomedicine. J Phys D Appl Phys 2003;36:R167–R181.
8 Pankhurst QA, Than NKT, Jonhson SK, Dobson J: Progress in applications of magnetic nanoparticles in biomedicine. J Phys D Appl Phys 2009;42: 224001–224015.
9 Hajdu A, Tombacz E, Illes E, Bica D, Vekas L: Magnetic nanoparticles stabilized under physiological conditions for biomedical applications. Progr Colloid Polym Sci 2008;135:20–37.

10 Bica D, Vekas L, Avdeev MV, Marinica O, Socoliuc V, Balasoiu M, Garamus VM: Sterically stabilized water based magnetic fluids: synthesis, structure and properties. Journal Magn Magn Mater 2007;311:17–21.
11 Massart R: Preparation of aqueous magnetic liquids in alkaline and acidic media. IEEE TMagn 1981;17:1247–1248.
12 Massart R: Dispersion of the particles in the base liquid; in Berkovski B, Bashtovoy V (eds): Magnetic Fluids and Applications Handbook. New York, Begell House, 1996, pp 24–32.
13 Charles SW: Preparation of magnetic fluids; in Odenbach S (ed): Ferrofluids: Magnetically Controllable Fluids and Their Applications – Lecture Notes in Physics. Heidelberg, Springer, 2002, vol 594, pp 3–18.
14 Bica D: Preparation of magnetic fluids for various applications. Rom Rep Phys 1995;47:265–272.
15 Popovici E, Dumitrache F, Morjan I, Alexandrescu R, Ciupina V, Prodan G, Vekas L, Bica D, Marinica O, Vasile E: Iron/iron oxides core-shell nanoparticles by laser pyrolysis: structural characterization and enhanced particle dispersion. Appl Surf Sci 2007;54: 1048–1052.
16 Morjan I, Alexandrescu R, Dumitrache F, Birjega R, Fleaca C, Soare I, Luculescu CR, Filoti G, Kuncser V, Vekas L, Popa NC, Prodan G: Iron oxide-based nanoparticles with different mean sizes obtained by the laser pyrolysis: structural and magnetic properties. J Nanosci Nanotechnol 2010;10:1223–1234.
17 Illés E, Tombácz E: The effect of humic acid adsorption on pH-dependent surface charging and aggregation of magnetite nanoparticles. J Colloid Interface Sci 2006;295:115–123.
18 Tombácz E, Illés E, Majzik A, Hajdu A, Rideg N, Szekeres M: Ageing in the inorganic nanoworld: example of magnetite nanoparticles in aqueous medium. Croat Chem Acta 2007;80:503–515.
19 Vékás L, Avdeev MV, Bica D: Magnetic nanofluids: synthesis and structure; in Shi D (ed): NanoScience in Biomedicine. New York, Springer, 2009, pp 645–704.
20 Wooding A, Kilner M, Lambrick DB: Studies of the double surfactant layer stabilization of water-based magnetic fluids. J Colloid Interface Sci 1991;144: 236–242.
21 Tombácz E, Bica D, Hajdu A, Illés E, Majzik A, Vékás L: Surfactant double layer stabilized magnetic nanofluids for biomedical application. J Phys Condens Matter 2008;20:204103.
22 Goodarzi A, Sahoo Y, Swihart MT, Prasad PN: Aqueous Ferrofluid of citric acid coated magnetite particles. Mater Res Soc Symp Proc 2004;789:6.6.1–6.6.6.
23 Avdeev MV, Mucha B, Lamszus K, Vekas L, Garamus VM, Feoktystov AV, Marinica O, Turcu R, Willumeit R: Structure and in vitro biological testing of water-based ferrofluids stabilized by monocarboxylic acids. Langmuir 2010; 26:8503–8509.
24 Avdeev MV: Contrast variation in small-angle scattering experiments on polydisperse and superparamagnetic systems: basic function approach. J Appl Crystallogr 2007;40:56–70.
25 Nan A, Turcu R, Bratu I, Leostean C, Chauvet O, Gautron E, Liebscher J: Novel magnetic core-shell Fe_3O_4 polypyrrole nanoparticles functionalized by peptides or albumin. ARKIVOC 2010, pp 185–198.
26 Schlaad , Krasia T, Patrickios CS: Controlled synthesis of coordination block copolymers with beta-dicarbonyl ligating segments. Macromolecules 2001; 34:7585–7588.
27 Papaphilippou P, Loizou L, Popa NC, Han A, Vekas L, Odysseos A, Krasia-Christoforou T: Superparamagnetic hybrid micelles, based on iron oxide nanoparticles and well-defined diblock copolymers possessing beta-ketoester functionalities. Biomacromolecules 2009;10:2662–2671.

L. Vekas
Laboratory of Magnetic Fluids
CFATR, Romanian Academy-Timisoara Branch
24 Mihai Viteazu street, RO–300223 Timisoara (Romania)
Tel. +40 256 403 700, E-Mail vekas@acad-tim.tm.edu.ro

Alexiou C (ed): Nanomedicine – Basic and Clinical Applications in Diagnostics and Therapy.
Else Kröner-Fresenius Symp. Basel, Karger, 2011, vol 2, pp 53–70

Nanotechnology-Based Spatiotemporal Controlled Drug Delivery Strategies

Sanchita Biswas · Challa S.S.R. Kumar

Center for Advanced Microstructures and Devices, Louisiana State University, Baton Rouge, La., USA

Abstract

The last two decades have witnessed an exponential growth of investigations into the application of nanomaterials for drug delivery. A majority of these investigations have focused on either passive or controlled or targeted drug delivery. However, a combination of these attributes and, most importantly, spatiotemporal drug delivery systems are needed in order to achieve the required therapeutic efficiency. In this review article, we provide insight into nanotechnology-based spatiotemporal controlled drug delivery (SCD) strategies. Taking few examples from the recent literature, we make a case for the systematic development of SCD vehicles by considering complex disease characteristics, the pharmacokinetics of the drug delivery system as well as the physicochemical characterizations of drug, and finally toxicity issues.

Drugs are administered to the human body through various means: the enteral route (oral, buccal, sublingual, and rectal), the parenteral route (intramuscular, intravenous, subcutaneous), and other routes (topical and inhalation). Typical routes for drug delivery are pictorially represented in figure 1. Even though the field of drug delivery has progressed dramatically in the last couple of decades, a number of problems continue to exist in all types of drug delivery [1–3]. Some of the serious lacunae in the current drug delivery systems are as follows: (1) A reduced loading and therapeutic effect due to partial degradation of the drug before it reaches its destination is a major concern. For example, a drug administered orally may undergo changes in the stomach due to the action of gastric enzymes. Drugs administered through enteral routes undergo a first-pass effect, which is the term used for the hepatic metabolism of a pharmacological agent when it is absorbed from the gut and delivered to the liver via the portal circulation. These processes lead to premature drug loss due to drug metabolism and rapid clearance from the body. (2) A lack of targeting ability is yet another major disadvantage of many drugs. When a drug is administered parenterally, it is distributed to all parts of the human body through the bloodstream, rather than to the specific target organ which needs the pharmacological treatment. Biochemical and physiological barriers of certain organs also limit drug delivery to the specified organ, like the human brain. (3) The cytotoxicity of many drugs for cancer is like-

Fig. 1. Typical routes of drug delivery represented pictorially.

ly to destroy healthy tissue along with carcinomatous tissue. The cytotoxic effect of these chemotherapeutic drugs appears to be greatest in organs like bone marrow, gonads, hair follicles, and the digestive tract that contain rapidly proliferating cells. The adverse effects of chemotherapy include fatigue, nausea, vomiting, alopecia (loss of hair), gastrointestinal disturbances, impaired fertility, impaired ovarian function, and bone marrow suppression resulting in anemia, leucopenia, and thrombocytopenia. (4) Drug interaction with other pharmacological agents may also alter the efficacy of a particular drug.

Nanoparticles with many unique properties, such as a high surface area-to-volume ratio, magnetic, optical properties compared to the bulk material, have been generating huge interest among scientists for decades. The advent of nanotechnology shows great promise to overcome some of the above mentioned disadvantages and limiting factors in currently available drug delivery methods. The application of nanotechnology tools is likely to pave the way for the discovery and use of efficient and improved nanoparticle-based drug delivery systems. With the help of such sophisticated drug delivery systems, one could anticipate a deployment of medication in the desired form to a specific target in the human body without causing any adverse effects to the healthy tissues [4]. A number of nanomaterials that show encouraging effects both as diagnostic agents and as drug delivery systems have already been reported [4]. Moreover, recent trends in stimuli-responsive drug carriers for several life-threatening diseases appear to be very promising [5(a, b), 6].

Some of the specific advantages of nanotechnology-based drug delivery systems are given below. They provide positive attributes for any drug, in comparison to 'free' drugs, such as improved solubility, in vivo stability, and a better biodistribution. They can also alter the unfavorable pharmacokinetics of some 'free' drugs. They facilitate the uptake and transport of therapeutically active

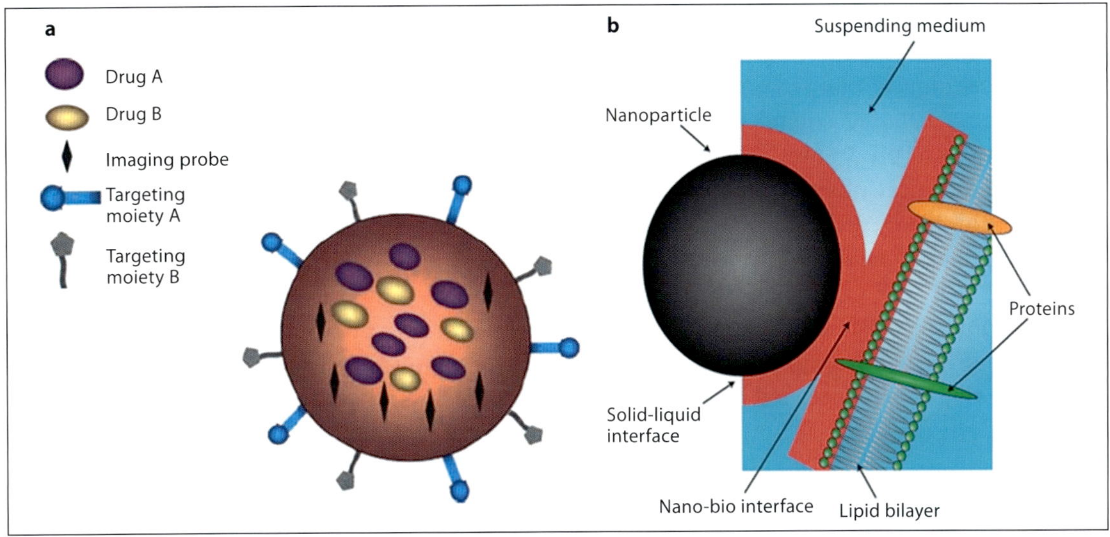

Fig. 2. a A typical multifunctional nanodrug delivery system [7]. **b** Representation of the interface between a nanoparticle and a lipid bilayer [13].

molecules with improved pharmacokinetics. They also facilitate the creation of dose differentials between the site of disease and the rest of the body, thus maximizing the therapeutic effect while minimizing the nonspecific uptake and side effects. Therefore, huge loading of drugs within a nanocarrier can render 'drug reservoirs' for controlled and sustained release in order to maintain the required drug level within a therapeutic window. Nanodrug delivery systems, stimulated by passive and/or active targeting strategies, can enhance the intracellular concentration of drugs in cancer cells while avoiding toxicity in normal cells. A typical nanodrug delivery system is pictorially represented in figure 2. In this figure one can see a multifunctional nanocarrier encapsulating therapeutic and diagnostic pay loads. The figure also shows the typical interaction between such multifunctional nanocarriers and lipid bilayers surrounding the cells. Furthermore, when they bind to specific receptors and enter a cell, they are usually enveloped by endosomes via receptor-mediated endocytosis, thereby bypassing the recognition of P-glycoprotein, one of the main chemicals involved in drug resistance mechanisms. Besides, they can be loaded/functionalized with both therapeutic agents and recognition units as well as with an imaging agent within a single system so that they will eventually be capable of detecting malignant cells (active targeting moiety), visualizing their location in the body (real-time in vivo imaging), killing the cancer cells without side effects by saving normal cells (active targeting and controlled drug released system), and monitoring the treatment effect in real time [7–9]. More specifically, there is a possibility for nanodrug delivery systems to deliver drugs in a controlled manner which can be manipulated both spatially and temporally. Drug delivery systems that have both spatial and temporal control are called spatiotemporal controlled drug delivery (SCD) systems.

There are quite a few recent reviews and publications which cover a number of types of nanotechnology-based controlled drug delivery systems [5, 10–12]. Many of these reviews focus on stimuli-responsive (temporal) or just spatially controlled drug release strategies using several kinds of techniques, biomarkers, stimuli, and materials of a different composition, size, and shape. Some of the procedures have already been implemented clinically for the treatment of many diseases, and to date almost more than two dozen nano-based medicines are available on the market [13]. For example, Doxil/Caelyx, a PEGylated liposome-based doxorubicin, was approved for advanced ovarian cancer, metastatic breast cancer, and AIDS-related Kaposi's sarcoma [14]. In spite of the initial clinical success, the nanocarriers capable of spatial targeting, time-specific sustained drug release to the binding site, and simultaneous diagnosis/monitoring of the process, within an integrated system for controlled drug delivery, are very few [6, 15]. As nanotechnology-based tools are further implemented to enhance drug delivery systems, recent reports show promising results for SCD. With the increasing number of investigations into SCD, there is a need for reviewing this field and to the best of our knowledge there is no comprehensive review to date.

In this review, our goal is to provide a summary of the recent developments in the field of SCD and to critically evaluate the opportunities and challenges in this emerging field. In the first part, we will discuss different types of drug delivery systems. In the second part, the role of nanomaterials in SCD will be discussed. In the third part, recent approaches of SCD to treat different diseases will be discussed. This part will include cases where time- and space-specific drug delivery is based mainly on stimulations by optical and magnetic responses and/or active targeting. In the fourth part, the most important applications and related challenges of SCD will be discussed. In the final part, an understanding and future outlook from the aspect of a nanomaterial-based effective drug delivery method will be discussed.

Different Types of Drug Delivery Systems

Overview of Drug Delivery Approaches

Over the past decade, several drug delivery vehicles have been designed based on different nanomaterials, such as polymers [14–16], dendrimers [17–19], liposomes [20–24], and metallic nanoparticles [21–23] of different sizes and shapes (spherical, rod, coil, bin [24], and cage [25]). While the pharmacokinetics and biodistribution [26] of a drug are predominantly governed by aspects such as the size, polarity, and number and nature of hydrophobic and hydrophilic residues, a majority of the nanomaterials under development as drug delivery systems do not take these aspects into consideration. A major approach, so far, has been to utilize nanomaterials only to encapsulate the drug molecule and thus 'hide' unfavorable interactions between the drug and the body, effectively replacing them with the properties of the nanomaterials. Some of the general approaches to prepare nanomaterials for drug delivery applications are:

- Copolymerization using monomers containing imaging agent/targeting agent/drug.
- Functional block copolymeric precursors, tagged with contrast agent/drug, by chemical conjugations.
- A noncovalent approach, incorporating the drug within a core-shell micelle type structure.
- Encapsulation of drugs within thin polymeric films [27].
- Core-shell type polymeric nanoparticles for controlled release.

In addition, the drug delivery systems developed were endowed with either passive or active targeting or both capabilities in order to improve the targeting ability and bioavailability of the drug to the desired site (fig. 3) [29]. The figure shows nanocarriers with both types of targeting reaching tumors selectively through the leaky vasculature surrounding the tumors. Upon arrival at tumor sites, nanocarriers with targeting molecules can bind to the target tumor cells or enter

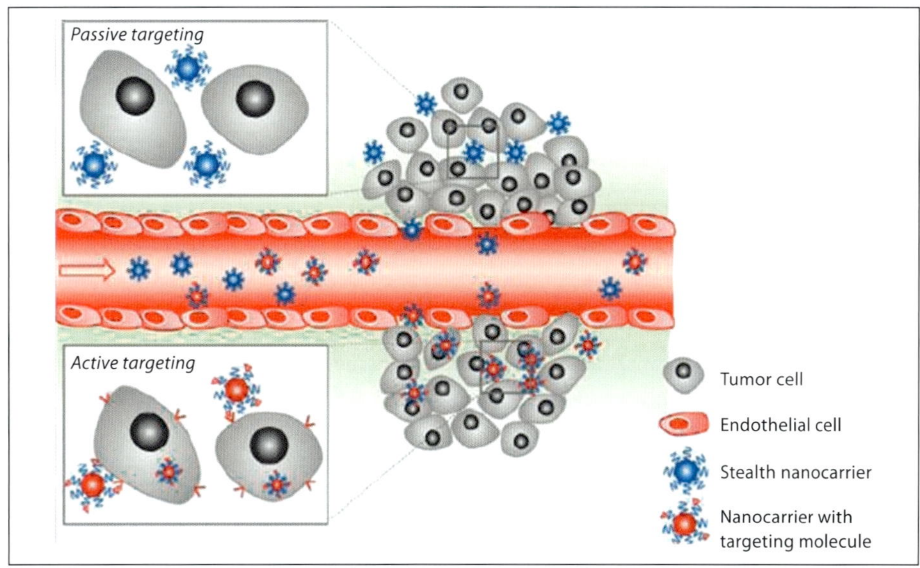

Fig. 3. A schematic representation of nanocarriers that passively or actively target tumors [28].

the cells via specific receptor (cell)-ligand (carrier) interactions, whereas stealth nanocarriers are less efficient in interacting with tumor cells [28].

Passive Targeting

Typically, the passive targeting of nanomaterial-based drug delivery systems depends on the enhanced permeability and retention (EPR) effect [30, 31]. This is particularly important in the treatment of primary tumors where defective and leaky vasculatures with pore sizes around 100–600 nm are present [29, 32]. The passive targeting effect is achieved by extravagation of the nanocarrier through the leaky vasculature at the tumor site. So, for suitable drug delivery using passive targeting the ideal nanocarrier is likely to have the following features [33, 34]: (1) it should have a size somewhere between 10 and 150 nm, avoiding renal clearance and capture by the liver but ensuring effective entrance into the vasculature; (2) it should consist of neutral or negatively charged particles to prevent elimination by the kidney [35], and (3) it should escape the reticuloendothelial system and have a longer half-life in the blood. One common approach to meeting the criteria is coating the nanoparticles with polyethylene glycol, known as PEGylation [36]. This will also ensure improved water solubility of the hydrophobic drug/diagnostic agent [37].

Some of the shortcomings in passive targeting are: (1) the pore size and uptake of the nanocarriers are highly dependent on the type and state of the tumor, and (2) the degree of angiogenesis and the elevated interstitial fluid pressure inside the tumor may inhibit efficient uptake of the nanocarrier/drug. Also, sometimes PEGylation reduces the cellular uptake of the nanocarriers as it prevents interaction with the tumors. Finally, it may lead to multidrug resistance (MDR).

Active Targeting

Active targeting requires specific ligands, with high affinities to receptors overexpressed on diseased sites, on the surface of the nanodrug delivery systems. Examples of targeting molecules that have been attracting attention include different varieties of antibodies, ligands, peptides, and nucleic acids. Some of them are tabulated in table 1.

Table 1. Commonly utilized targeting agents for nanodrug delivery systems with active targeting

Targeting agent	Application	Reference
RGD peptide (Arg-Gly-Asp)	Binds to integrin ($\alpha_v\beta_3$) receptors Is overexpreesed in metastatic angiogenesis of carcinoma cells	[38–40]
Transferrin	Transferrin receptor monoclonal antibodies (like OX26 or R17217) Is overexpressed in several tumors including blood leukemia, colon cancer, placental cancer, and lung cancer	[41–43]
Folic acid	Binds to vitamin folate Found in ovarian, lung, and uterine tumors, and human head and neck primary and metastatic tumor tissues	[44–48]
Anti HER-2-mAb, Trastuzumab	Binds to HER-2 receptor Is used to treat breast cancer, ovarian cancer, prostate cancer, and gastric cancer; measures the progression in tumorigenesis as well as metastasis	[49, 50]
VEGF	Binds to VEGF receptor Therapeutic angiogenesis for ischemic diseases (e.g. coronary artery disease and peripheral vascular disease) and solid tumor	[51, 52]
LHRH	Binds to LHRH receptors present on the majority of primary and metastatic tumor cells	[53–57]

VEGF = Vascular endothelial growth factor; LHRH = luteinizing hormone and releasing hormone.

Table 2. Examples of drug delivery systems designed based on different stimuli

Nature of the stimulus	Nanomaterials	Reference
pH	Several polymers, e.g. HEC-g-PAA, PBD-PGA, and PLE-PLL	[48, 59]
Optical	Gold nanoparticles of different sizes and shapes, quantum dots, organic dyes coated with several PEGylated polymers	[21, 25]
Magnetic	Biocompatible polymer- (e.g. dextran, chitosan) coated iron oxides, Gd nanoparticles, porous silica shell-Fe_3O_4-PS core	[5(b), 60–62]
Temperature	PNIPAAM-based polymer and its derivative, e.g. PAMPA-PNIPAAM, PLA-PNIPAAM	[58, 59]

Stimuli-Responsive Delivery

Controlled drug release by external stimuli such as pH, temperature, electric or magnetic fields, ultrasound, light radiation, and redox potential, as well as the use of different combinations of these, has attracted much attention [6, 58] (table 2). A schematic representation of various stimuli with potential for drug delivery systems is shown in figure 4a. A typical stimuli-responsive nanodrug delivery system is shown in figure 4b. Based on

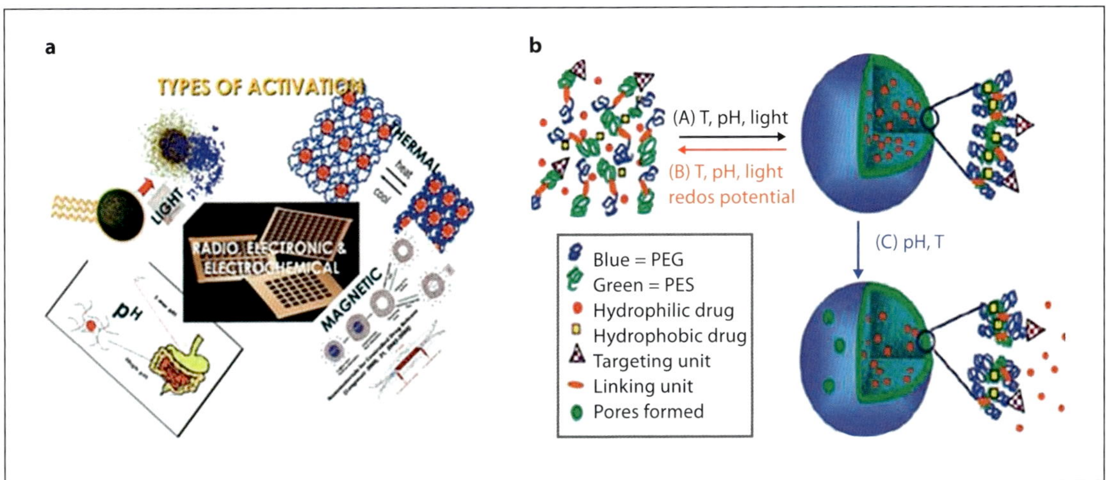

Fig. 4. a A representation of typical stimuli which can be used for externally triggered drug delivery systems. **b** A schematic representation of a stimuli-responsive polymeric system [58].

the figures, one can envisage externally triggered controlled release of the drug from within a nanocarrier. Several of these investigations have been carried out using well-known biodegradable polymers, liposomes, biocompatible block copolymers, and dendrimers. However, the delivery of drugs from functional polymers and in some cases load-bearing implants made from these materials is limited to a few standard systems and a small set of drugs [58, 63]. While such stimuli-responsive nanodrug delivery systems have the potential for temporal control, no systematic studies related to their utility in spatiotemporal controlled drug delivery have been conducted.

Spatiotemporal Controlled Drug Delivery

Spatiotemporal drug delivery is related to the controlled delivery of a drug to the desired site having both spatial extension and temporal duration, i.e in both a space-specific manner and a time-specific manner. If one considers the development of drug delivery systems as a pyramid (fig. 5), one can envision an increase in the complexity of drug delivery depending on the nature of the drug as well as the targeted disease tissue under consideration. While the least complex nanomaterial-based drug delivery system is the one that is passive, the most complex system has spatiotemporal control. SCD, however, is relevant not only in targeting specific tissues and anatomic sites but also at the cellular level. Biological macromolecules such as proteins and DNA are hydrophilic and therefore have a limited capacity to pass the cell membrane. Once those molecules enter the cell, by either passive or receptor-mediated endocytosis, they accumulate in endosomes which in most instances fuse with lysosomes where they are degraded and lose their activity. For successful delivery of a drug inside the cell, a carrier has to get the therapeutic agent across the cell membrane and in addition facilitate endosomal escape in order to prevent degradation within the lysosome.

The SCD of a number of sensitive classes of drugs such as nucleic acids, peptides, or proteins is critical as the delivery of such drugs, in tune with the physiological state of the diseased target area, can lead to increased bioactivity and efficacy as well as decreased toxicity. Nanotechnology-

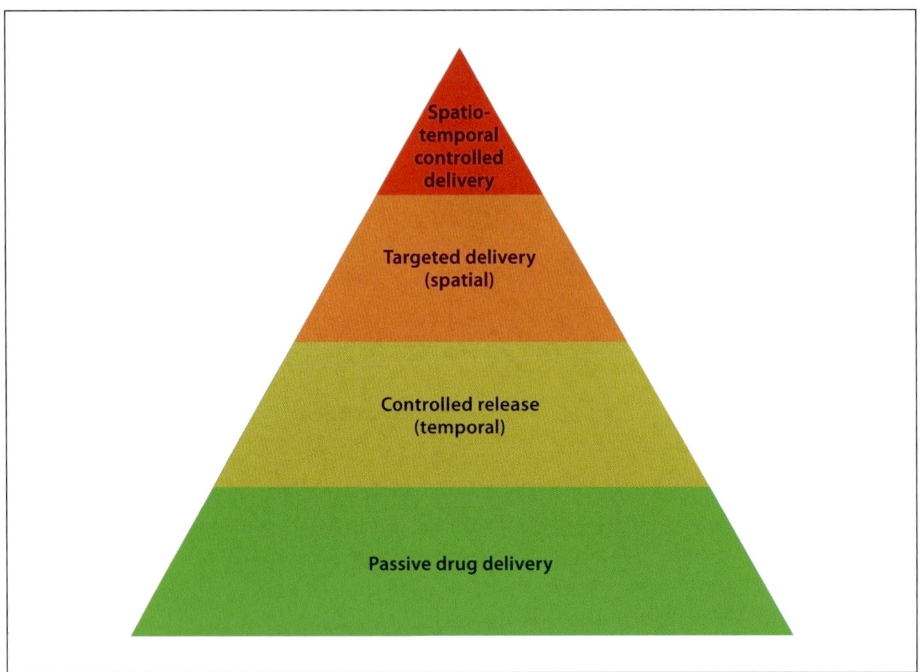

Fig. 5. A schematic representation of increased complexities in drug delivery.

based drug delivery systems offer a number of opportunities for SCD as the nanoparticle's surface can be bound to different targeting agents. Similarly, the utilization of appropriate core-shell type nanoparticles can also lead to better temporal control. In addition, a number of other attributes can be incorporated into nanoparticle-based spatiotemporal drug delivery systems. These are: (1) combined delivery of therapeutic drugs and diagnostic agents, (2) targeted delivery of drugs in a cell- or tissue-specific manner, (4) transcytosis of drugs across the tight epithelial and endothelial barriers, (5) delivery of larger macromolecule drugs to intracellular sites of action, (6) codelivery of two or more drugs or a therapeutic modality for combination therapy, (7) visualization of sites of drug delivery by combining therapeutic agents with imaging modalities, and (8) real-time determination of the in vivo efficacy of a therapeutic agent. Spatiotemporal delivery strategies have the potential to achieve similar or better results using a single-stage procedure. However, there are several factors that must be considered when designing an SCD: the toxicity of the degraded byproducts, the degradation mechanism (e.g. hydrolytic or enzymatic), the efficiency of loading, the preservation of bioactivity, and the controllability of the release rate and dose. Numerous deployment methods have been devised that make use of natural or synthetic biomaterials synthesized as porous scaffolds, hydrogels, nanoparticles, fibers, or tubes. SCD strategies include extended protein release from biomaterials and methods to bind or immobilize proteins within natural or synthetic scaffolds. A key issue is whether the delivery vehicle fabrication process denatures or otherwise deactivates sensitive proteins or drugs, thereby compromising therapeutic efficiency. We present below a few promising examples of SCD that are currently under investigation.

Some Recent Examples of Spatiotemporal Controlled Drug Delivery

pH-responsive SCD is one of the most studied drug delivery systems as the pH of different organs (the stomach has a pH of ~2 and the intestine's pH is ~5–8), even at the intracellular level, varies remarkably and they can also be targeted [59, 63]. Generally, cancerous tissues are relatively more acidic (pH ~6.8–7.2) compared to healthy ones and in blood (pH ~7.4). Even at the cellular level, the variation is enormous. For example, in endosomes it is about 5.0–6.5, while in lysosomes it is even lower (4.5–5.0). The pH-responsive copolymers have been designed in such a way that they can change from being hydrophilic to being hydrophobic in the acidic environment of the endosomal-lysosomal compartments. Such a change is believed to cause their partition into the endolysosomal membrane and lead to their disruption releasing the cargo into the cytoplasm [63]. Several polymers containing weak acid (carboxylic, phosphoric) or base- (amine) containing polymers (ionizable pKa ~3–10) are generally used for pH-triggered drug delivery. For example, Lee et al. [48] recently reported a folate-conjugated 'clickable' polymer-caged nanobin (PCN) vector to deliver a toxic, anticancer drug (doxorubicin) which has been shown to hydrolyze in the acidic environment and release its cargo (fig. 6) [52]. The highly spatioselective (for the folic acid over-expressed carcinoma cells) carrier releases the drug very efficiently as the pH is lowered during the maturation of the endosome.

The use of light as a stimulus for drug delivery has also received increased attention due to its advantages in the spatial and temporal control of drug release. Different types of organic dyes (having linear/nonlinear optical properties), quantum dots, and metal nanoparticles have generally been used as suitable materials for light-driven drug delivery vehicles. Among these, gold nanoparticles have particularly emerged as attractive candidates for the delivery of various payloads activated by light [13]. Their unique physical (photophysical and magnetic) and chemical properties – inertness, nontoxicity, and controlled size – make them an attractive tool for many applications. Some of related examples are: (1) glutathione-based drug release, (2) photodynamic therapy, (3) NO release, (4) release of biomacromolecules like proteins and genes (nonviral delivery by both the noncovalent and the covalent approach), and (5) photothermal destruction based on a hyperthermia effect at near-infrared (NIR) wavelengths [21].

Febvay et al. [64] recently reported a method for precise spatial and temporal controlled cytosolic delivery of drugs that would otherwise be cell impermeable. They prepared the drug delivery system by incorporating a light-triggered endosomal escape mechanism within a cell-targetable mesoporous silica nanoparticle carrier (fig. 7) [64]. As shown in the figure, the size-controlled mesoporous silica nanoparticles (30–200 nm) were synthesized and biofunctionalized with a targeting moiety (for spatial control) in order to target the P-glycoprotein, a transporter protein responsible for MDR in many tumors, and the cell impermeable fluorescent dye Alexa546 for the light-triggered delivery (temporal control). Laser scanning confocal microscopy investigation on the system allowed the successful observation of single vesicle disruption events along with unprecedented control over the cytosolic access of the encapsulated drug in the irradiated cells.

Another interesting piece of work by Yavuz et al. [25] reported the gold nanocages bound to smart polymer-based delivery systems. In this system, photothermal effect was exploited for the controlled release of the drug (shown in fig. 8) [25]. Gold nanocages represent a class of nanostructures with hollow interiors and porous walls. They have strong absorption (for the photothermal effect) in the NIR while maintaining a compact size. When the surface of a gold nanocage is covered with a smart polymer, the preloaded effector can be released in a controllable fashion using an NIR laser. The polymer is based

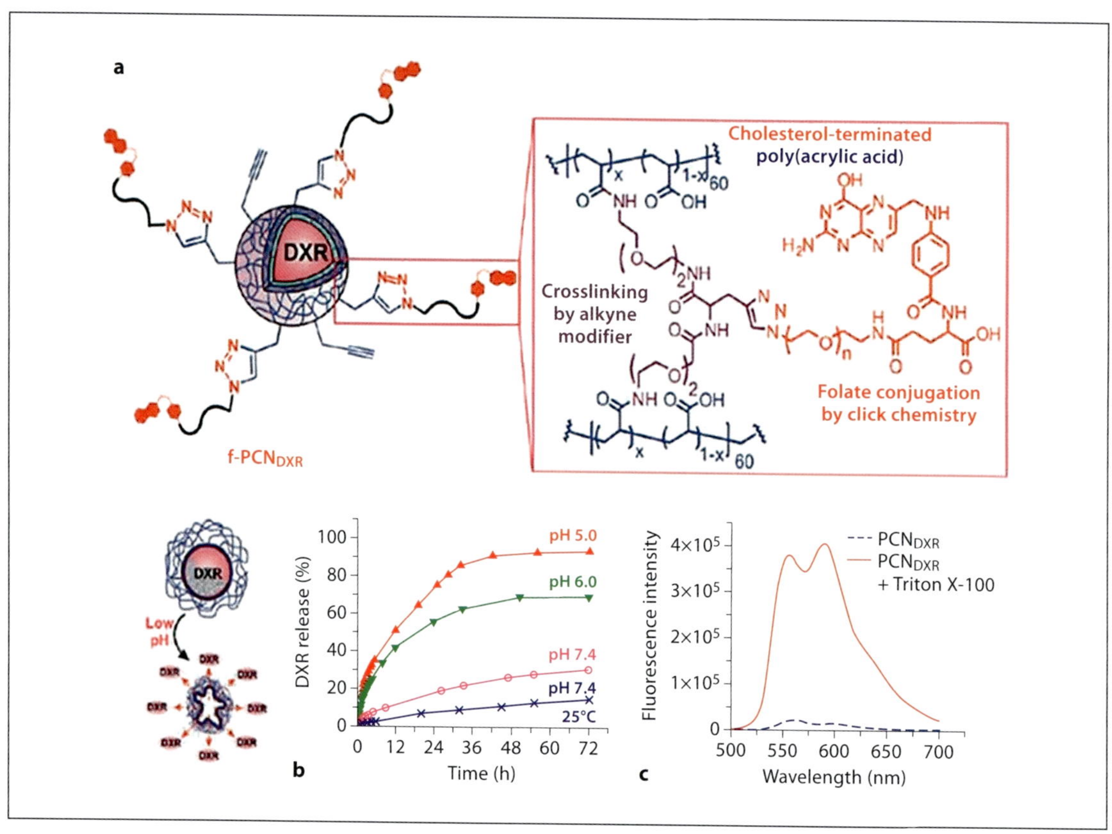

Fig. 6. a DXR-loaded, folate-conjugated PCN, (f-PCN_{DXR}). **b** Time-dependent DXR-releasing profiles of PCN (PCN_{DXR}) under pH 5.0, 6.0, and 7.4 at 37°C (closed symbols) and 25°C (open symbols). **c** Fluorescence emission spectra of PCN_{DXR} before and after treatment with Triton X-100 showing significant quenching of DXR fluorescent intensity inside the liposome [48].

on the copolymer thermoreversible poly(N-isopropylacrylamide) (pNIPAAm) and its derivative, pNIPAAm-co-pAAm (to tune its lower critical solution temperature between 32 and 50°C), which can change conformation in response to small variations in temperature. Upon exposure to a laser beam with a wavelength that matches the absorption peak of the Au nanocage, the light will be absorbed and converted into heat through photothermal effect. The heat will dissipate into the surroundings, and the rise in temperature will cause the polymer chains to collapse, exposing the pores on the nanocage and thereby releasing the preloaded effector (alizarin-PEG, Dox, lysozyme). When the laser is turned off, heating will immediately cease and the drop in temperature will bring the polymer back to its original, extended conformation, closing the pores and stopping the release. The release of the targeted drug is controlled by manipulating the power density and/or irradiation time. Also, these nanocages are easily manipulated by active targeting [65].

Liposomes have been proposed and used clinically as drug nanocarriers for decades, but their clinical applications are often limited due to slow release and poor availability of the encapsulated drug. Wu et al. [66] showed the development of

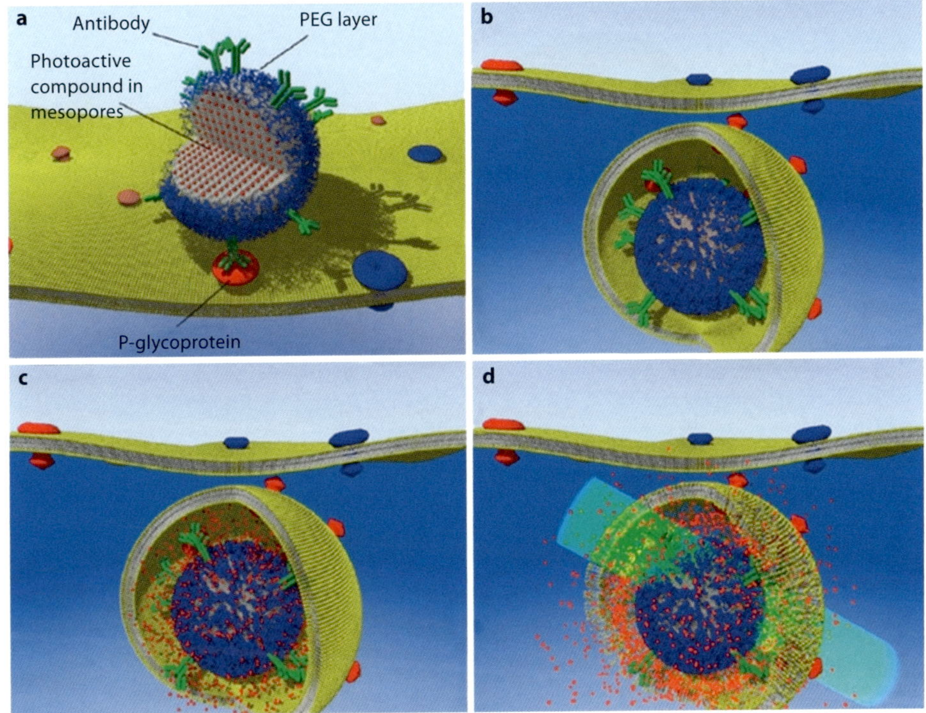

Fig. 7. A schematic of light-activated and targeted cytosolic delivery of membrane-impermeable compounds. **a** Antibody-functionalized nanoparticles are loaded with a model compound (the fluorescent dye Alexa546 in our experiments) and targeted to cells expressing P-gp-GFP (GFP bound to the P-glycoprotein transporter). After nanoparticle endocytosis (**b**), the cargo is released in the endosome (**c**). Exposure to light at the dye's excitation wavelength (546 nm) promotes ROS-mediated membrane damage (**d**) with cytosolic delivery of Alexa546 exclusively in the P-gp-expressing cells [64].

liposome based SCD where liposome release can be initiated within seconds by irradiating hollow gold nanoshells with a NIR pulsed laser to address the SCD. The NIR light penetrates into the tissue up to 10 cm, allowing these hollow gold nanoshell/liposome complexes to be administered noninvasively within a significant fraction of the human body.

Magnetically responsive drug delivery systems have also become very attractive for SCD [5(b), 67]. The first successful clinical experiment with magnetically controlled drug targeting in human patients was done by Lubbe et al. [5(a)] using polymer-coated ferrofluid bound to an antibiotic antracylin drug, epirubicin, for the treatment of advanced-stage cancer and solid tumors [68]. This technique allows delivery of a certain amount of drug to a specific targeted site and helps to overcome many blood barriers easily with the aid of a magnetic field. Many other advanced magnetic nanoparticles with higher magnetic moments, nonfouling surfaces, and multifunctionalities are now being developed for applications in the detection, diagnosis, and treatment of malignant tumors and cardiovascular and neurological diseases. Different iron oxides, gadolinium, and other metallic and bimetallic nanoparticles have been used for magneti-

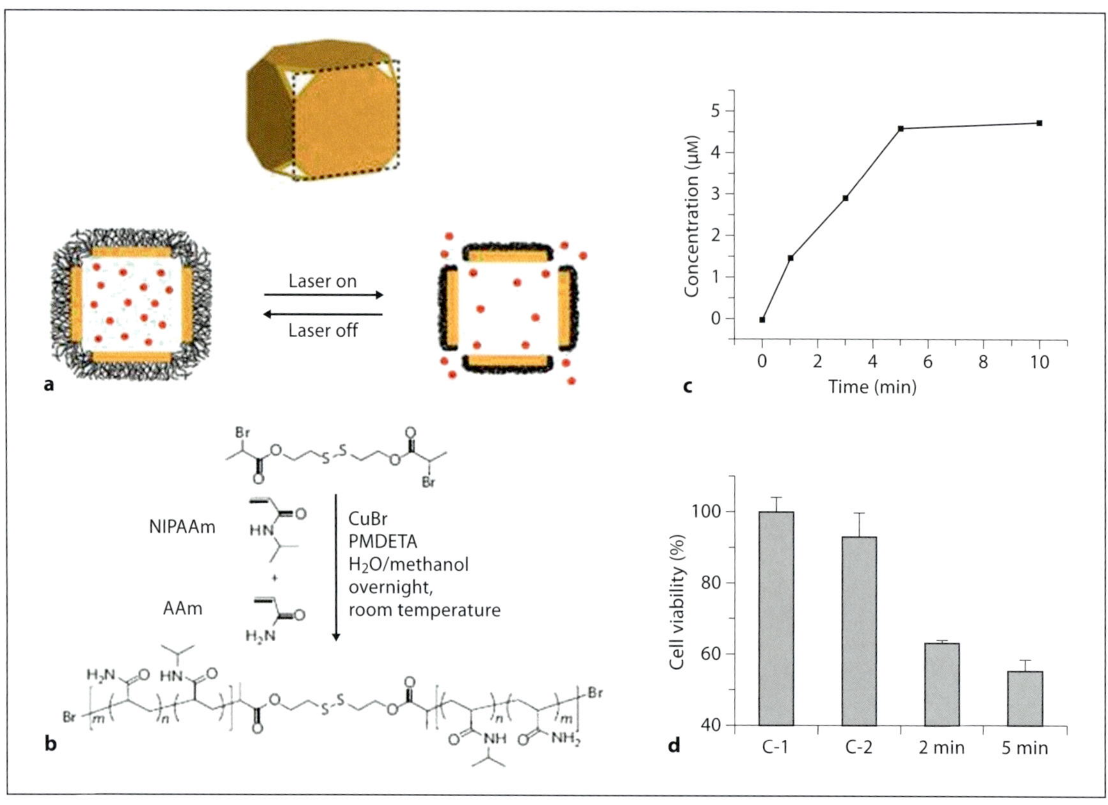

Fig. 8. A schematic illustration and characterization of the controlled-release system. **a** Schematic illustrating how the photothermal responsive system works. **b** Synthesis of the copolymer NIPAAm-Aam by atom transfer radical polymerization. **c** A plot of the concentrations of Dox released from the Au nanocages upon heating at 45°C for different periods of time. **d** Cell viability for samples after going through different treatments. C-1 = Cells irradiated with a pulsed NIR laser for 2 min in the absence of Au nanocages; C-2 = cells irradiated with the laser for 2 min in the presence of Dox-free Au nanocages; 2/5 min = cells irradiated with the laser for 2 and 5 min in the presence of Dox-loaded Au nanocages. A power density of 20 mWc^{-2} was used for all of these studies [25].

cally controlled SCD, and more details on these can be found in many recent reviews [5(b), 69].

Recently, Zhao et al. [60] reported a magnetically responsive ferrogel where SPIONs were coated with PF127, cross-linked alginate, and adipic acid dihydrazide and decorated with RGD peptide. They were utilized for the sustained release of hydrophobic drugs, and other bioactive agents (fig. 9) [60]. The results of this study suggest possibilities for successful development of an 'on-demand' and 'reversible approach' for delivering various biological agents using the newly designed magnetically stimulated macroporous ferrogels. The magnetic field-induced gel deformation resulted in water convection. The associated shear forces accelerated the release of drugs encapsulated in ferrogels or cells adherent to the pore surfaces of the macroporous ferrogels. There are also reports suggesting that the carbon nanotube-based drug delivery processes are becoming attractive tools for SCD because of their ease of synthesis and functionalization of biomarkers, low toxicity, and nonimmunogenic nature [70, 71].

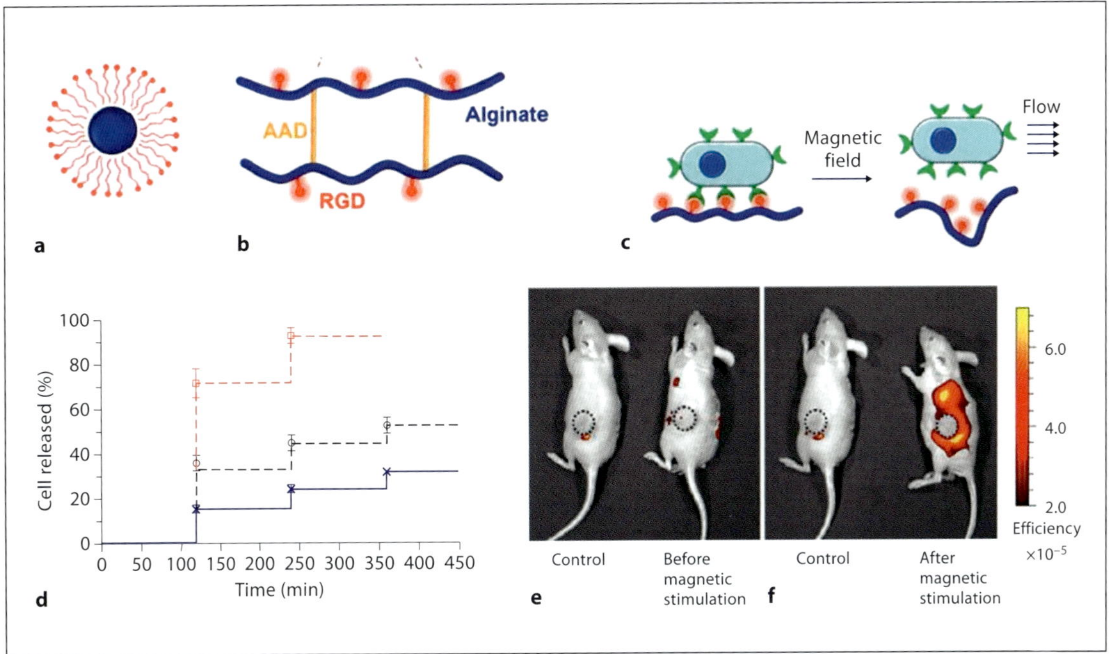

Fig. 9. Schematic plots of the nanoparticles coated with Pluronic F127 (**a**) and alginate covalently cross-linked by adipic acid dihydrazide and coupled with RGD peptides (**b**). **c** Schematic plot of gel deformation and the resulting water convection inducing cell release from macroporous gels. **d** Cumulative release profiles of fibroblasts from macroporous ferrogels with 100% (cross), 50% (circle), and 10% (square) of the baseline RGD density, following application of a cycled magnetic field. **e, f** In vivo fluorescence images of mice implanted with macroporous ferrogels containing mouse mesenchymal stem cells stained with DiOC18 before (**e**) and after (**f**) magnetic stimulation. The control case was subject to no magnetic stimulation. The positions of the gel disks are indicated by circles on the figure [60].

Finally, here is an example that truly highlights the best of SCD using nanomaterials. In this example, injured vasculature was repaired successfully by achieving spatial control using a combination of vascular antigens as targeting agents and temporal control using polymer core lipid shell nanoparticles (fig. 10) [72]. The figure shows a schematic of the synthesis of the drug-conjugated polymer complex (fig. 10a) and HPLC profiles of the drug versus drug release from the drug polymer conjugate (fig. 10b). In figure 10c, a schematic of nanoparticles with a core containing drug eluting polymer, a lipid monolayer, a PEG antibiofouling layer, and peptide ligands (hooks) to adhere to the exposed basement membrane during vascular injury is shown. The particle size and the size distribution of the core-shell nanoparticles can be seen from their TEM images (fig. 10d) and the dynamic light scattering measurements before and after peptide conjugation, respectively (fig. 10e). Information about the required surface charges on these particles was obtained using zeta potential measurements (fig. 10f). While the targeting peptide ligands provided spatially controlled drug targeting, the core-shell particles with the polymer core loaded with slow-eluting conjugates of paclitaxel for controlled ester hydrolysis and drug release provided the temporal control for the drug release over a period of 12 days. This is an elegant example of a nanomaterial-based spatiotemporally controlled delivery vehicle which

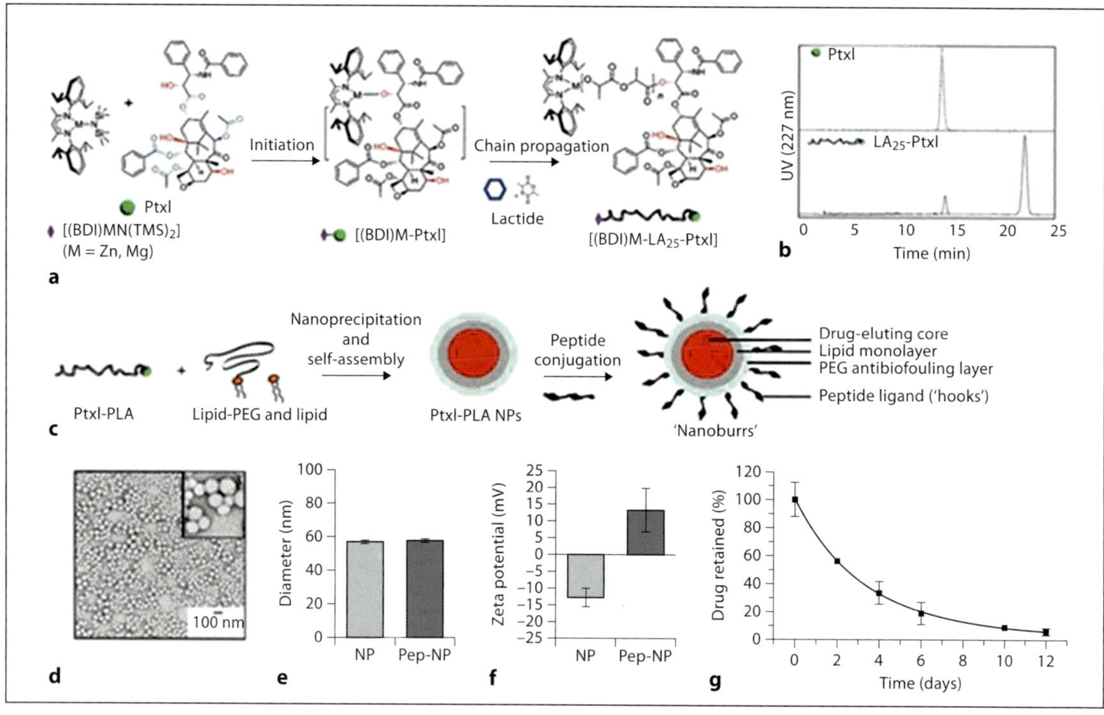

Fig. 10. Core-shell nanoparticles for spatiocontrolled delivery for the treatment of injured vasculature [reproduced from 72]. Ptxl = Paclitaxel.

demonstrates localization of the drugs at the sites of injured vasculature and exhibits controlled drug release.

Current Challenges and Opportunities in Spatiotemporal Controlled Drug Delivery

A short review of the recent literature, as presented above, on nanotechnology-based drug delivery systems indicates that nanomaterials do offer potential advantages as drug carrier systems with spatiotemporal control. SCD has certainly generated a lot of interest among a number of researchers spanning multidisciplinary fields. Some of the investigations are in preclinical as well as clinical trials. Among these, the treatment of cancers, cardiovascular diseases, and immunological diseases (of both diagnostic and therapeutic importance) utilizing SCD is attracting the most attention. A majority of the commercially available nanotechnology-based drug delivery systems mostly rely on nontargeted nanocarriers, depending on the vasculatures of the solid tumors and the EPR effect. However, targeted delivery is more important in vascular endothelial targeting for oncology, cardiovascular, and immunological tissue where the EPR effect is not under consideration. Recently, CALAA-01 (a transferrin-targeted RNAi nanotherapeutic) to reduce the expression of the M2 subunit of ribonucleotide reductase (R2) was approved for a phase I clinical test for solid tumor therapy (Calando Pharmaceuticals; http://www.insertt.com). SCD is also likely to offer better treatment options for the treatment of many other diseases besides cancer. For example, therapeutic neovascularization/angiogenesis is an attractive approach for the treatment of ischemic muscles resulting from the obstruction of blood flow in the cells because of lack of nutrients [56, 60]. Kim et al. [50] demonstrated that VEGF-modified fluorescent silica nanoparticles can be utilized for ischemic muscle treatment taking advantage of imaging and therapeutic angiogenesis [60]. Similarly, SCD is also likely to play an important role in clinical procedures reported for regenerative medicine for restoration of damaged tissues [57] and treatment of large segmental bone defects. The purification of bone marrow cells from contaminated carcinoma cells using immunomagnetic beads has become a well-established method clinically [27]. Also, the potential of nanoparticles as a platform for vaccine development by targeting lymph node-residing dendritic cells is another important research area currently under investigation.

While there is clear potential for SCD, there still exists a need to address important issues such as poor oral bioavailability, instability of the drug during circulation, inadequate tissue distribution, and toxicity. Especially in the context of personalized medicine, a long road is ahead for nanotechnology-based drug delivery systems in general and spatiotemporal controlled systems in particular [73, 74]. Some of the challenges and opportunities that the emerging field of nanotechnology-based SCD presents are listed below.

- The principal limitation of SCD is the ability to deliver drugs to the cytoplasm. SCD systems need to be designed taking into consideration issues such as biocompatibility, physicochemical properties, immunogenicity, and toxicity within and without.
- Once an SCD system is designed, there are number of concerns related to their reproducible synthesis and large scale production as their multifunctionality adds to their complexity. This is particularly important for the commercialization of all nanotechnology-based drug delivery systems.
- Molecular imaging is a powerful tool to study the mechanism of biological processes in vitro and in vivo which is equally applicable for monitoring drug delivery. Molecular imaging using multifunctional drug delivery systems will reveal insights into the mechanisms of drug delivery and the efficacy of the drug. Currently available imaging modalities for this purpose are optical imaging, γ-ray scintigraphy, single-photon emission computed tomography, positron emission tomography, and magnetic resonance imaging [55]. Cellular uptake, subcellular trafficking, pharmacokinetics, biodistribution, and tumor targeting efficiency have also been visualized using some of these techniques. Therefore, the next stage in the design of SCD systems is to incorporate features for real-time diagnosis or monitoring during the treatment. This will be the first step that needs to be taken in order to move in the direction of personalized medicine.
- Optimizing the ligand density of a targeting agent around the SCD is important in order to facilitate the balance between tissue penetration and cellular uptake, resulting in optimal therapeutic efficacy.
- SCD also needs to take into consideration the heterogeneity of expression on many disease targets. Therefore, a single targeting agent is not sufficient for the required spatial control. In some cases, a range of ligands that target across a range of diseases in a consistent and reproducible manner is required [72].
- In designing SCD systems, yet another crucial factor is the nature of the interaction between the SCD and the drug itself. This needs to be investigated carefully from the point of view of drug stability as well as the overall toxicity of the system.
- Drug loading continues to be a major problem in SCD. The question is how drug encapsulation can be maximized without triggering the 'burst effect'.

- The temporal control demonstrated to date in SCD is only related to increasing the time required for release of the drug. However, more complex release profiles will be required for the optimal treatment of many diseases. For example, the SCD system should have the option of drug release that is either sequential or pulsatile or, if necessary, of a complete stop in the case of drug overdose.
- SCD systems are usually multifunctional nanomaterials and therefore will have complex pharmacokinetics. A systematic investigation into their pharmacokinetics as a whole as well as as individual components is necessary.
- The multifunctionality of SCD systems also results in complicating their toxicological evaluation. Once again, there is a need for a step-by-step systematic evaluation.
- Finally, while it is too early to draw any conclusions on the potential impact of SCD, based on limited literature on MDR and immunogenicity, one has to keep these issues in mind during the design process itself.

Conclusions and Future Perspectives

The importance of spatiocontrol in drug delivery systems cannot be overstated, though only recently have experimental investigations begun in this direction. Thanks to nanotechnology tools, the search for SCD systems has begun to take into consideration the required complex targeting and complicated drug release profiles. SCD is required not only to enhance the viability of existing effective drugs but also to deliver newer drugs for the treatment of complex diseases. Even though nanomaterials have been explored for their utility in drug delivery for more than two decades, a majority of these investigations have been carried out without considering the total picture involving complex pharmacokinetics and toxicity. Optimization of spatiotemporal delivery strategies is a challenge that will clearly require tremendous interdisciplinary cooperation among cell biologists, biomaterial scientists, tissue engineers, and clinicians.

Acknowledgements

C.S.S.R.K. is thankful to the Else Kröner-Fresenius Foundation and to Prof. Dr. med. Christoph Alexiou for the invitation to the Second Else Kröner-Fresenius Symposium. The inspiration for this book chapter came from the invited talk. The funding from the NIH (1RO1CA142-01A1) is gratefully acknowledged.

Disclosure Statement

The authors have nothing to disclose.

References

1 Allen TM, Cullis PR: Drug delivery systems: entering the mainstream. Science 2004;303:1818–1822.
2 Colombo P, Sonvico F, Colombo G, Bettini R: Novel platforms for oral drug delivery. Pharm Res 2009;26:601–611.
3 Nikander K: Challenges and opportunities in respiratory drug delivery devices. Expert Opin Drug Deliv 2010;7: 1235–1238.
4 Rothenfluh DA, Hubbell JA: Integration column: biofunctional polymeric nanoparticles for spatio-temporal control of drug delivery and biomedical applications. Integr Biol (Camb) 2009; 1:446–451.
5 (a) Lubbe AS, Alexiou C, Bergemann C: Clinical applications of magnetic drug targeting. J Surg Res 2001;95:200–206. (b) Kumar CS, Mohammad F: Magnetic nanomaterials for hyperthermia-based therapy and controlled drug delivery. Adv Drug Del Rev 2011;63:789–808.
6 Motornov M, Roiter Y, Tokarev I, Minko S: Stimuli-responsive nanoparticles, nanogels and capsules for integrated multifunctional intelligent systems. Prog Polym Sci 2010;35:174–211.

7 Cho K, Wang X, Nie S, Chen ZG, Shin DM: Therapeutic nanoparticles for drug delivery in cancer. Clin Cancer Res 2008;14:1310–1316.
8 Reif R, Wang M, Joshi S, A'Amar O, Bigio IJ: Optical method for real-time monitoring of drug concentrations facilitates the development of novel methods for drug delivery to brain tissue. J Biomed Opt 2007;12:034036.
9 Saito R, Krauze MT, Bringas JR, Noble C, McKnight TR, Jackson P, Wendland MF, Mamot C, Drummond DC, Kirpotin DB: Gadolinium-loaded liposomes allow for real-time magnetic resonance imaging of convection-enhanced delivery in the primate brain. Exp Neurol 2005;196:381–389.
10 Kumar CSSR: Nanomaterials for Medical Diagnosis and Therapy. Weinheim, Wiley-VCH, 2007.
11 Kumar CSSR: Nanomaterials for Cancer Therapy. Weinheim, Wiley-VCH, 2007.
12 Kumar CS, Mohammad F: Magnetic nanomaterials for hyperthermia-based therapy and controlled drug delivery. Adv Drug Deliv Rev 2011, E-pub ahead of print.
13 Nel AE, Mädler L, Velegol D, Xia T, Hoek EM, Somasundaran P, Klaessig F, Castranova V, Thompson M: Understanding biophysicochemical interactions at the nano-bio interface. Nat Mater 2009;8:543–557.
14 Kim S, Kim JH, Jeon O, Kwon IC, Park K: Engineered polymers for advanced drug delivery. Eur J Pharm Biopharm 2009;71:420–430.
15 Soppimath KS, Aminabhavi TM, Kulkarni AR, Rudzinski WE: Biodegradable polymeric nanoparticles as drug delivery devices. J Control Release 2001;70:1–20.
16 van Dongen SF, de Hoog HP, Peters R, Nallani M, Nolte RJ, van Hest JC: Biohybrid polymer capsules. Chem Rev 2009;109:6212–6274.
17 Astruc D, Boisselier E, Ornelas C: Dendrimers designed for functions: from physical, photophysical, and supramolecular properties to applications in sensing, catalysis, molecular electronics, photonics, and nanomedicine. Chem Rev 2010;110:1857–1959.
18 Lee CC, MacKay JA, Fréchet JM, Szoka FC: Designing dendrimers for biological applications. Nat Biotechnol 2005; 23:1517–1526.
19 Medina SH, El-Sayed ME: Dendrimers as carriers for delivery of chemotherapeutic agents. Chem Rev 2009;109: 3141–3157.
20 Laverman P, Boerman OC, Oyen WJ, Dams ET, Storm G, Corstens FH: Liposomes for scintigraphic detection of infection and inflammation. Adv Drug Deliv Rev 1999;37:225–235.
21 Ghosh P, Han G, De M, Kim CK, Rotello VM: Gold nanoparticles in delivery applications. Adv Drug Deliv Rev 2008; 60:1307–1315.
22 Vivero-Escoto JL, Slowing II, Wu CW, Lin VS: Photoinduced intracellular controlled release drug delivery in human cells by gold-capped mesoporous silica nanosphere. J Am Chem Soc 2009;131:3462–3463.
23 Neuberger T, Schöpf B, Hofmann H, Hofmann M, Von Rechenberg B: Superparamagnetic nanoparticles for biomedical applications: possibilities and limitations of a new drug delivery system. J Magn Magn Mater 2005;293: 483–496.
24 Lee SM, Song Y, Hong BJ, MacRenaris KW, Mastarone DJ, O'Halloran TV, Meade TJ, Nguyen SB: Modular polymer-caged nanobins as a theranostic platform with enhanced magnetic resonance relaxivity and pH-responsive drug release. Angew Chem Int Ed Engl 2010;122:10156–10160.
25 Yavuz MS, Cheng Y, Chen J, Cobley CM, Zhang Q, Rycenga M, Xie J, Kim C, Song KH, Schwartz AG: Gold nanocages covered by smart polymers for controlled release with near-infrared light. Nat Mater 2009;8:935–939.
26 Li SD, Huang L: Pharmacokinetics and biodistribution of nanoparticles. Mol Pharm 2008;5:496–504.
27 Oh JK, Drumright R, Siegwart DJ, Matyjaszewski K: The development of microgels/nanogels for drug delivery applications. Prog Polym Sci 2008;33: 448–477.
28 Gullotti E, Yeo Y: Extracellularly activated nanocarriers: a new paradigm of tumor targeted drug delivery. Mol Pharm 2009;6:1041–1051.
29 Dreher MR, Liu W, Michelich CR, Dewhirst MW, Yuan F, Chilkoti A: Tumor vascular permeability, accumulation, and penetration of macromolecular drug carriers. J Natl Cancer Inst 2006;98:335–344.
30 Maeda H: The enhanced permeability and retention (EPR) effect in tumor vasculature: the key role of tumor-selective macromolecular drug targeting. Adv Enzyme Regul 2001;41:189–207.
31 Torchilin V: Tumor delivery of macromolecular drugs based on the EPR effect. Adv Drug Deliv Rev 2011;63:131–135.
32 Yuan F, Dellian M, Fukumura D, Leunig M, Berk DA, Torchilin VP, Jain RK: Vascular permeability in a human tumor xenograft: molecular size dependence and cutoff size. Cancer Res 1995; 55:3752–3756.
33 Alexis F, Pridgen E, Molnar LK, Farokhzad OC: Factors affecting the clearance and biodistribution of polymeric nanoparticles. Molecular Pharm 2008;5:505–515.
34 Mitragotri S, Lahann J: Physical approaches to biomaterial design. Nat Mater 2009;8:15–23.
35 Guasch A, Deen WM, Myers BD: Charge selectivity of the glomerular filtration barrier in healthy and nephrotic humans. J Clin Invest 1993;92: 2274–2282.
36 Vlerken L, Vyas T, Amiji M: Poly-(ethylene glycol)-modified nanocarriers for tumor-targeted and intracellular delivery. Pharm Res 2007;24:1405–1414.
37 Biswas S, Wang X, Morales AR, Ahn HY, Belfield KD: Integrin-targeting block copolymer probes for two-photon fluorescence bioimaging. Biomacromolecules 2011;12:441–449.
38 Murphy EA, Majeti BK, Barnes LA, Makale M, Weis SM, Lutu-Fuga K, Wrasidlo W, Cheresh DA: Nanoparticle-mediated drug delivery to tumor vasculature suppresses metastasis. Proc Natl Acad Sci USA 2008;105:9343.
39 Liu Z, Chen K, Davis C, Sherlock S, Cao Q, Chen X, Dai H: Drug delivery with carbon nanotubes for in vivo cancer treatment. Cancer Res 2008;68:6652.
40 Pike DB, Ghandehari H: HPMA copolymer-cyclic RGD conjugates for tumor targeting. Adv Drug Deliv Rev 2010;62: 167–183.
41 Gan CW, Feng SS: Transferrin-conjugated nanoparticles of poly(lactide)-D-alpha-tocopheryl polyethylene glycol succinate diblock copolymer for targeted drug delivery across the blood-brain barrier. Biomaterials 2010;31:7748–7757.
42 Qian ZM, Li H, Sun H, Ho K: Targeted drug delivery via the transferrin receptor-mediated endocytosis pathway. Pharmacol Rev 2002;54:561–587.

43 Ulbrich K, Hekmatara T, Herbert E, Kreuter J: Transferrin- and transferrin-receptor-antibody-modified nanoparticles enable drug delivery across the blood-brain barrier (BBB). Eur J Pharm Biopharm 2009;71:251–256.
44 Leamon CP, Reddy JA: Folate-targeted chemotherapy. Adv Drug Deliv Rev 2004;56:1127–1141.
45 Pan J, Feng SS: Targeting and imaging cancer cells by folate-decorated, quantum dots (QDs)-loaded nanoparticles of biodegradable polymers. Biomaterials 2009;30:1176–1183.
46 Saba NF, Wang X, Müller S, Tighiouart M, Cho K, Nie S, Chen ZG, Shin DM: Examining expression of folate receptor in squamous cell carcinoma of the head and neck as a target for a novel nanotherapeutic drug. Head Neck 2009;31:475–481.
47 Low PS, Kularatne SA: Folate-targeted therapeutic and imaging agents for cancer. Curr Opin Chem Biol 2009;13: 256–262.
48 Lee SM, Chen H, O'Halloran TV, Nguyen SB: 'Clickable' polymer-caged nanobins as a modular drug delivery platform. J Am Chem Soc 2009;131: 9311–9320.
49 Tai W, Mahato R, Cheng K: The role of HER2 in cancer therapy and targeted drug delivery. J Control Release 2010; 146:264–275.
50 Kim J, Cao L, Shvartsman D, Silva EA, Mooney DJ: Targeted delivery of nanoparticles to ischemic muscle for imaging and therapeutic angiogenesis. Nano Lett 2011;11:694–700.
51 Chen RR, Silva EA, Yuen WW, Mooney DJ: Spatio-temporal VEGF and PDGF delivery patterns blood vessel formation and maturation. Pharm Res 2007; 24:258–264.
52 Jain RK: Molecular regulation of vessel maturation. Nat Med 2003;9:685–693.
53 Branca RT, Cleveland ZI, Fubara B, Kumar CS, Maronpot RR, Leuschner C, Warren W, Driehuys B: Molecular MRI for sensitive and specific detection of lung metastases. Proc Natl Acad Sci USA 2010;107:3693–3697.
54 Meng J, Fan J, Galiana G, et al: LHRH-functionalized superparamagnetic iron oxide nanoparticles for breast cancer targeting and contrast enhancement in MRI. Mater Sci Eng C Mater Biol Appl 2009;29:1467–1479.
55 Leuschner C, Kumar CS, Hansel W, Soboyejo W, Zhou J, Hormes J: LHRH-conjugated magnetic iron oxide nanoparticles for detection of breast cancer metastases. Breast Cancer Res Treat 2006;99:163–176.
56 Zhou J, Leuschner C, Kumar C, Hormes JF, Soboyejo WO: Sub-cellular accumulation of magnetic nanoparticles in breast tumors and metastases. Biomaterials 2006;27:2001–2008.
57 Leuschner C: LHRH conjugated magnetic nanoparticles for diagnosis and treatment of cancers; in Kumar CSSR (ed): Nanomaterials for Cancer Diagnosis. Weinheim, Wiley-VCH, 2006.
58 Ruel-Gariépy E, Leroux JC: In situ-forming hydrogels – review of temperature-sensitive systems. Eur J Pharm Biopharm 2004;58:409–426.
59 Meng F, Zhong Z, Feijen J: Stimuli-responsive polymersomes for programmed drug delivery. Biomacromolecules 2009;10:197–209.
60 Zhao X, Kim J, Cezar CA, Huebsch N, Lee K, Bouhadir K, Mooney DJ: Active scaffolds for on-demand drug and cell delivery. Proc Natl Acad Sci USA 2011; 108:67–72.
61 Kong SD, Zhang W, Lee JH, Brammer K, Lal R, Karin M, Jin S: Magnetically vectored nanocapsules for tumor penetration and remotely switchable on-demand drug release. Nano Lett 2010, E-pub ahead of print.
62 Lu Z, Prouty MD, Guo Z, Golub VO, Kumar C, Lvov YM: Magnetic switch of permeability for polyelectrolyte microcapsules embedded with Co@ Au nanoparticles. Langmuir 2005;21: 2042–2050.
63 Schmaljohann D: Thermo- and pH-responsive polymers in drug delivery. Adv Drug Deliv Rev 2006;58:1655–1670.
64 Febvay S, Marini DM, Belcher AM, Clapham DE: Targeted cytosolic delivery of cell-impermeable compounds by nanoparticle-mediated, light-triggered endosome disruption. Nano Lett 2010; 10:2211–2219.
65 Chen J, Saeki F, Wiley BJ, Cang H, Cobb MJ, Li ZY, Au L, Zhang H, Kimmey MB, Li X: Gold nanocages: bioconjugation and their potential use as optical imaging contrast agents. Nano Lett 2005;5:473–477.
66 Wu G, Mikhailovsky A, Khant HA, Fu C, Chiu W, Zasadzinski JA: Remotely triggered liposome release by near-infrared light absorption via hollow gold nanoshells. J Am Chem Soc 2008;130: 8175–8177.
67 Senyei A, Widder K, Czerlinski G: Magnetic guidance of drug-carrying microspheres. J Appl Phys 2009;49:3578–3583.
68 Lübbe AS, Bergemann C, Riess H, Schriever F, Reichardt P, Possinger K, Matthias M, Dörken B, Herrmann F, Gürtler R: Clinical experiences with magnetic drug targeting: a phase I study with 4′-epidoxorubicin in 14 patients with advanced solid tumors. Cancer Res 1996;56:4686.
69 Sun C, Lee JS, Zhang M: Magnetic nanoparticles in MR imaging and drug delivery. Adv Drug Deliv Rev 2008;60: 1252–1265.
70 Bianco A, Kostarelos K, Prato M: Applications of carbon nanotubes in drug delivery. Curr Opin Chem Biol 2005;9: 674–679.
71 Bhirde AA, Patel V, Gavard J, Zhang G, Sousa AA, Masedunskas A, Leapman RD, Weigert R, Gutkind JS, Rusling JF: Targeted killing of cancer cells in vivo and in vitro with EGF-directed carbon nanotube-based drug delivery. ACS Nano 2009;3:307–316.
72 Chan JM, Zhang L, Tong R, Ghosh D, Gao W, Liao G, Yuet KP, Gray D, Rhee JW, Cheng J: Spatiotemporal controlled delivery of nanoparticles to injured vasculature. Proc Natl Acad Sci USA 2010;107:2213–2218.
73 Farokhzad OC, Langer R: Impact of nanotechnology on drug delivery. ACS Nano 2009;3:16–20.
74 Kumar C: Nanotechnology tools in pharmaceutical R&D. Mater Today 2010;12:24–30.

Challa S.S.R. Kumar
Center for Advanced Microstructures and Devices
Louisiana State University, 6980 Jefferson Hwy
Baton Rouge, LA 70806 (USA)
Tel. +1 225 578 9320, E-Mail ckumar1@lsu.edu

Alexiou C (ed): Nanomedicine – Basic and Clinical Applications in Diagnostics and Therapy.
Else Kröner-Fresenius Symp. Basel, Karger, 2011, vol 2, pp 71–79

Preclinical Efficacy and Toxicity Testing of Engineered Nanomaterials

Jennifer H. Grossman · Scott E. McNeil

Nanotechnology Characterization Laboratory, Advanced Technology Program, SAIC-Frederick, Inc., NCI-Frederick, Frederick, Md., USA

Abstract

As engineered nanomaterials (ENMs) – man-made products between 1 and approximately 100 nm in size – are increasingly used in consumer, industrial, and medical products, it becomes more important than ever to understand ENM interactions with biology and the environment. Recent results show that seemingly small changes in ENM chemistry can cause dramatic differences in potency and safety, making thorough characterization critical for the manufacture and use of ENMs. Characterization is especially important for ENMs used in medical products. In the pharmaceutical industry, ENMs are being used to reformulate drugs to 'engineer out' disqualifying characteristics such as poor solubility and/or liver and spleen accumulation. In contrast to the development of new molecular entities, using ENMs to reformulate and salvage previously discontinued drugs greatly reduces the time, risk, and investment required for new drug development.

Importance of Characterization

Preclinical characterization of engineered nanomaterials (ENMs) should include physicochemical characterization, sterility and pyrogenicity assessment, biodistribution (absorption, distribution, metabolism, and excretion; ADME) studies, and toxicity testing, which includes both in vitro characterization and in vivo animal studies [1] (fig. 1).

Physicochemical Characterization

Thorough physicochemical characterization of an ENM drug formulation involves characterization/quantitation of all components of the formulation (e.g. drug loading and measurement of 'free' or un-ENM-bound drug, targeting moieties, and any impurities) and evaluation of properties such as stability, drug release, and activity under a variety of conditions [2]. The primary difference between the physicochemical characterization of ENM-based pharmaceuticals and that of traditional nonnanotech drugs is due to the complexity of the ENM: ENM-based pharmaceuticals often contain a complex mixture of molecular species that must work in concert to achieve functionality (fig. 2). Because of this complexity, even a 'pure' ENM formulation may exhibit significant heterogeneity and polydispersity. This means that ENM purity must be defined by an acceptable range that affords the necessary safety and efficacy profile for the formulation rather

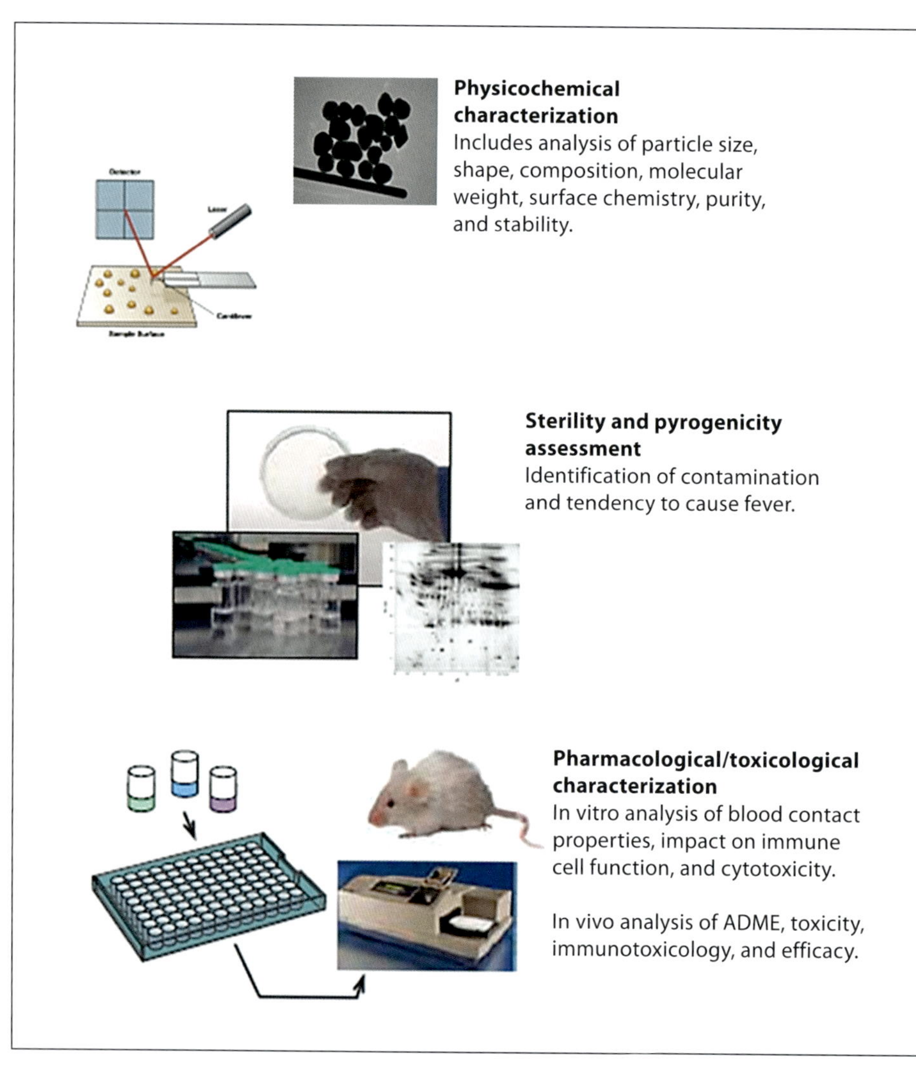

Fig. 1. Tiered approach to ENM characterization. A rational preclinical characterization strategy for nanoparticles includes physicochemical characterization, sterility and pyrogenicity assessment, and pharmacological/toxicological characterization, which includes both in vitro tests and in vivo animal studies. This particular characterization scheme reflects the assay cascade developed by the Nanotechnology Characterization Laboratory (NCL), part of the National Cancer Institute's Alliance for Nanotechnology in Cancer (http://ncl.cancer.gov/working_assay-cascade.asp).

than an absolute purity standard. For example, an ENM-based formulation with a targeting ligand may be able to maintain efficacy with 5–20 targeting molecules per nanoparticle rather than a specified absolute number of ligands per particle. As long as the ligand distribution is well characterized in a sufficiently large random sampling of the material, and is reproducible from batch to batch of synthesis, this range may afford the necessary limits for reliable lot release.

Preclinical physicochemical characterization of an ENM includes measurement of size and

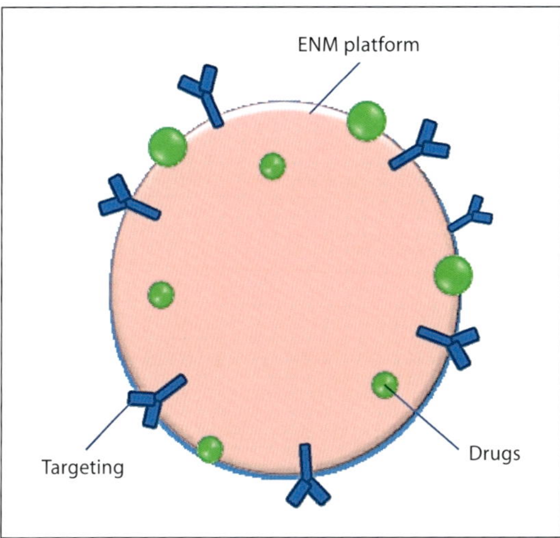

Fig. 2. The primary difference between the physicochemical characterization of ENM-based pharmaceuticals and that of traditional nonnanotech drugs is due to the complexity of the ENM. ENM-based pharmaceuticals often contain a complex mixture of molecular species that must work in concert to achieve functionality. A nanoparticle's surface can be functionalized with hydrophilic polymers (e.g. PEG) to improve solubility or help the particle evade uptake by the immune system, targeting molecules (e.g. antibodies), drugs, and imaging contrast agents for diagnostics. The interior core can be solid (e.g. quantum dots) or liquid (e.g. liposomes) or contain an encapsulated drug. Because of this complexity, even a 'pure' ENM formulation may exhibit significant heterogeneity and polydispersity.

shape, surface chemistry, and aggregation/agglomeration state [3]. Particle size characterization is of particular importance to nanomedicine since small changes in particle size can dramatically affect ENM biodistribution [4, 5]. ENM size characterization is often complicated by the polydispersity of samples, but characterizing nanoparticle size using multiple orthogonal methods, such as microscopy and light scattering [6], can provide a detailed understanding of the size of an ENM formulation.

Another important and challenging area of nanoparticle characterization is measurement under conditions that mimic in vivo conditions. Many properties of nanoparticles are environment and condition dependent. For example, the size distribution and zeta potential at physiological pH and ionic strength may differ from their values in water. Oftentimes, it is impossible to characterize the samples in vivo. For electron-dense nanoparticles, EM images of ex vivo tissues from animal studies can provide a wealth of information on nanoparticle disposition into cells, tissues, and organs. For ENMs that incorporate elements not usually found in tissues (e.g. titanium, platinum, gold, etc.), energy-dispersive X-ray spectroscopy or inductively coupled plasma mass spectrometry can be used to identify the ENM in tissue.

Sterility

Nanoparticles must be tested for sterility and pyrogenicity (tendency to induce fever) as contamination will skew the results of biological studies. The FDA currently recommends 2 types of pyrogen tests: a limulus amebocyte lysate (LAL)-based assay for endotoxin, sensitive to even picogram quantities of endotoxin, and the so-called rabbit pyrogen test (an in vivo test which has the capability to identify various pyrogenic substances). Many ENMs interfere with the LAL test. For example, catalytic ENMs, such as some dendrimers, may cause false positives in the LAL test by activating the LAL proteolytic cascade. Other ENMs may adsorb endotoxin on their surfaces, obscuring it from detection leading to false-negative results. A review of these issues and a decision tree for dealing with potential ENM endotoxin contamination is presented in a recent publication by Dobrovolskaia et al. [7].

In vitro

Even all of the physicochemical characterizations discussed above may not always detect subtle batch-to-batch variations that may affect ENM safety and potency. Because of this, functional bioassays should be used whenever possible to account for batch-to-batch variations that may affect product performance. This is the same reasoning that necessitated the recommendation of the use of potency bioassays in the current FDA guidance for antibody-based pharmaceuticals [8]. Such bioassays can be used to evaluate the function of an ENM (e.g. receptor binding, antitumor efficacy, etc.) and can make use of in vitro models. An in vitro bioassay should be validated using ENMs known to be efficacious and safe (e.g. standard reference materials) and established as predictive of in vivo results; only then can it be used to evaluate batch-to-batch consistency.

In vitro experiments are not surrogates for in vivo studies but can provide an initial cost-effective screen of the toxicity and efficacy of an ENM and inform the design of animal studies. In vitro studies also enable the elucidation of biochemical mechanisms under controlled conditions not achievable in in vivo studies. The fewer variables and amplified reactions of in vitro assays generally make their results more straightforward to interpret than those of in vivo experiments. Unfortunately, ENMs frequently interfere with conventional in vitro pharmacologic assays. For instance, many nanoparticles aggregate or adsorb proteins in in vitro assays. Still other nanoparticles, such as gold colloids, scatter light and invalidate colorimetric assays that rely on absorbance measurements. Other nanoparticles, such as dendrimers, can have catalytic properties that often interfere with standardized enzymatic tests, such as the LAL test for endotoxin contamination.

In vitro assays to evaluate blood contact properties have been shown to be particularly useful for determining the biocompatibility of an ENM with blood [9]. However, care must be taken to avoid assay interference. For example, hemolysis (damage to red blood cells) is a toxicity that can lead to life-threatening conditions such as hemolytic anemia, jaundice, and renal failure. There are several examples of ENM interference with standard assays to evaluate hemolysis. For example, nanoparticles may damage red blood cells but then absorb hemoglobin, aggregate, and be removed from the supernatant during centrifugation, yielding a false-negative result. Alternatively, the optical properties of the nanoparticles may cause a false-positive result (e.g. gold colloids, some nanoemulsions, doxorubicin-loaded liposomes, and water-soluble fullerene derivatives absorb light at the assay detection wavelength). Metal-containing particles (e.g. quantum dots and colloids) may oxidize cyanmethemoglobin, shifting its absorbance peak maximum away from the detection wavelength. Finally, some ENMs induce rapid blood coagulation, entirely protecting erythrocytes from exposure to the ENMs [10].

In vivo

Though testing in in vitro physiological models can give an initial estimate of formulation toxicity, realistically, it is not possible for the laboratory bench to match the complex set of conditions found in vivo. It is therefore necessary to characterize the efficacy, biodistribution, and toxicity of a drug formulation in animal models. The FDA offers guidelines for in vivo studies [11–13], and specific considerations for ENM samples are detailed in the literature [1].

One challenge for ADME animal studies of ENMs is tracking the individual components of a multicomponent entity. If the therapeutic (drug) moiety of a particle disassociates from the ENM platform upon administration or during circulation in the blood, is degraded inside the particle, or fails to disassociate in targeted tissue, the formulation may be ineffective [14]. In vivo studies which image/trace only the drug or only the ENM platform may therefore be inadequate for understanding the therapeutic efficacy and toxicity of a multicomponent nanoparticle. Radiolabelled studies which use multiple isotopes or multiple imaging methods with nonoverlapping signals to track the various components of a nanoparticle are more reliable.

Each Engineered Nanomaterial Is Unique

All of this characterization is necessary because subtle changes in ENM chemistry can cause dramatic differences in safety and potency. For example, a small change in ENM size can dramatically change the in vivo distribution [4]. A slight change in the molar ratio of polyethylene glycol (PEG) on the surface of a nanoparticle may also influence the particle's recognition by the reticuloendothelial system (RES) [5]. Similarly, a small increase in zeta potential (related to surface charge) can significantly impact a nanoparticle's blood contact properties. Given the nearly infinite potential for variation in each of these parameters and the combinatorial possibilities, each nanoparticle formulation is virtually unique and it is difficult to extrapolate results based simply on the class of material. Since small changes in formulation can drastically influence outcomes, it is important to understand which nanoparticle attributes contribute to biocompatibility. Ultimately, this will allow scientists to engineer around toxicities and to develop predictive tools (such as modeling and simulation).

Numerous studies from the literature have demonstrated that small changes in chemistry can dramatically influence in vivo outcomes. For example, in one study, 3 sizes (80, 170, and 240 nm) of poly methoxypolyethylene glycol cyanoacrylate-co-n-hexadecyl cyanoacrylate-TNF (PEG-PHDCA-TNF) ENMs were compared [15]. The 80-nm ENM exhibited a significantly longer circulation time, decreased liver uptake, and greater accumulation in tumors as compared to larger particle sizes (170 and 240 nm). This was due to decreased serum protein adsorption and phagocytic uptake of the smaller particle.

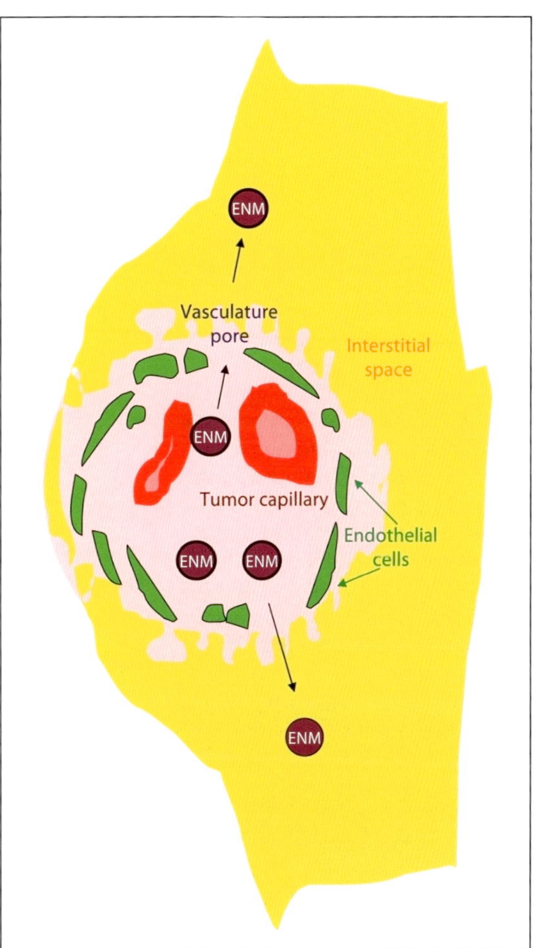

Fig. 3. EPR effect. In the EPR effect, ENMs enter a tumor through gaps in the tumor vasculature endothelium. These gaps are much larger (100 nm to 2 μm) in tumor endothelium than in healthy tissue (2–6 nm) and provide a route for ENM extravasation. Tumors also lack functional lymphatic drainage, which means that ENMs that have permeated the tumor vasculature are likely to be retained, whereas small molecules can diffuse freely back into the blood stream.

Size is not the only property which determines the in vivo fate of an ENM. Surface properties can increase the stability of an ENM and prolong its circulation in the blood, which can increase passive accumulation in tumors via the enhanced permeability and retention (EPR) effect (fig. 3) [16, 17]. Surface effects can also dramatically influence opsonization (protein adsorption facilitating uptake) and uptake by cells of the immune system. Coating nanoparticles with synthetic polymers (e.g. PEG) can significantly reduce the binding of plasma proteins, interaction with op-

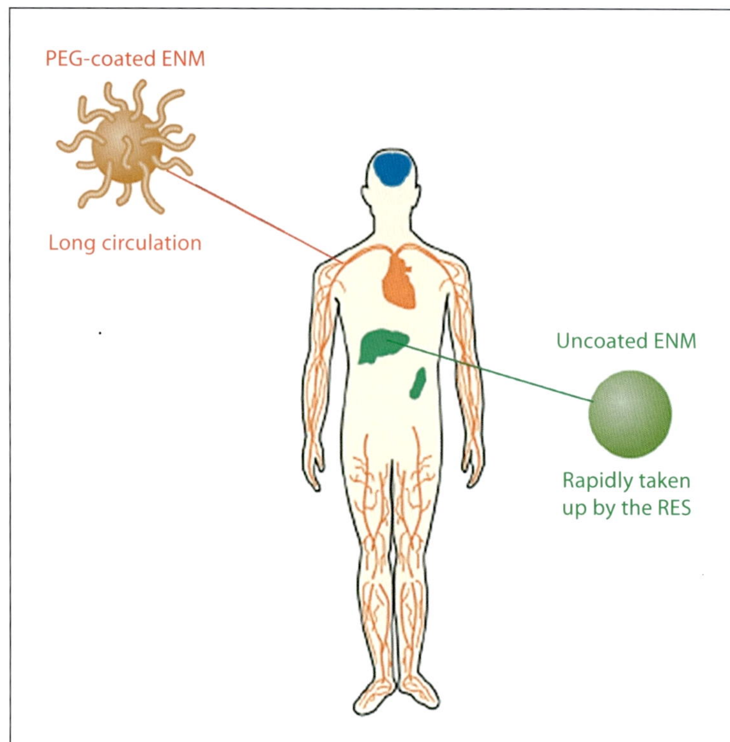

Fig. 4. Biodistribution is influenced by ENM size and surface characteristics. Uncoated ENMs are opsonized and taken up by the immune system (especially by cells of the RES). PEG-coated ENMs avoid opsonization and uptake by the RES system and are long circulating relative to uncoated particles [4]. PEG-coated ENMs under about 300 nm in diameter extravasate through tumor vasculature and can selectively accumulate in tumor tissue via the EPR effect.

sonins, and clearance by the RES system [18] (fig. 4). For example, PEGylated liposomal doxorubicin (an ENM-based cancer drug marketed as Doxil in the USA and as Caelyx in the EU) has an enhanced blood circulation time compared to uncoated liposomes such as Daunosome and Myocet [19].

Small changes in the drug release rate can also make big differences in in vivo efficacy. For encapsulating ENMs, different ratios of polymer or lipid can significantly change the degradation rate and control the rate of drug release from the ENM [20]. For ENMs covalently conjugated to drugs, the properties of the covalent linker modulate the rate of drug release. If a linker is too stable, there will be little or no drug release at the tumor. Other linkers are too unstable and the drug dissociates before the ENM can reach the site of disease. For example, this was the case in a phase I trial with an ENM polymer-conjugated paclitaxel prodrug. The ENM-based formulation was inefficacious due to rapid hydrolysis of the linker [21].

A Case Study

Our laboratory conducted an animal study to determine the safety of a polymer-coated gold nanoparticle intended as a cancer therapy. We injected laboratory rats with the ENMs and found that the animals developed lung lesions. The lesions were unexpected; the ENM manufacturer's previous studies had not resulted in lung lesions. To further examine what was happening, we repeated the same experiment with a freshly synthesized batch of ENM sample, and the rats had no lesions. A fairly rigorous battery of testing found the 2 batches of ENMs to be essentially indistinguishable – they were produced in the same synthetic process, had equivalent size and surface charge, and appeared similar under electron microscopy, atomic force microscopy, and dynamic light scattering. Finally, we looked at the particles' polymer coatings. A sample of the fresh batch had a slightly higher molar ratio of polymer on its surface than the older batch. We found that the polymer on the ENM in the older batch had been displaced by ions over time while it was stored in solution. There was less than a 50% difference in the amounts of PEG on the particles in the different batches, but this difference caused one batch to be toxic while the other was benign. The manufacturer was alerted to the difference between the fresh sample and the older sample. The small difference in the concentration of coating caused a large difference in the in vivo results – ultimately making the difference between an ENM that was safe and one that was not.

Salvaging Discontinued Drugs

It costs over USD 1 billion on average to develop a new molecular entity through clinical trials and onto the market in the USA [22]. Part of this expense comes from the many drugs that are discontinued during the process. Such 'discontinued drugs' represent large sunk costs for pharmaceutical companies since these drugs will not provide a return on the investment made in their development. Drugs are discontinued during pharmaceutical development due to undesirable properties, such as poor solubility, liver accumulation, etc. A common rule of thumb is that 4 out of 5 drugs that enter clinical trials will fail due to toxicity or poor pharmacokinetics.

Nanotechnology offers drug companies an opportunity to reformulate these discontinued drugs and potentially recoup costs. Desirable properties can be enhanced with ENM-based formulations, while adverse properties can literally be engineered out. For example, there is now considerable evidence that reformulation of cancer drugs using ENMs can increase the amount of active pharmaceutical ingredient delivered to tumors by 2 orders of magnitude through the EPR effect [16, 17]. ENM platforms can be used to increase solubility and bioavailability, enhance targeting, and provide for the controlled release of a variety of therapeutics, potentially making discontinued drugs viable again.

ENM-based reformulations of chemotherapeutics have been particularly successful, delivering strong medicines to tumors while reducing adverse effects on healthy tissue. Conventional chemotherapy usually resorts to agents that disrupt cell replication. Since many healthy tissues are proliferating, chemotherapeutics may have many adverse side effects, such as nausea, neuropathy, hair loss, and fatigue. These side effects are often so severe they limit the amount of drug that can be administered. The side effects can be reduced if the cytotoxic drug (the chemotherapeutic) is attached to an ENM, which reduces systemic exposure and concentrates the drug in tumor tissue. Such ENM-based therapeutics can have increased efficacy and decreased toxicity compared to traditional chemotherapeutic cancer drugs.

ENM-based cancer drugs are improvements over traditional chemotherapeutics for 2 main reasons: First, the larger size of nanoparticles relative to small-molecule drugs lends them an advantage for systemic/intravenous administration since nanoparticles can avoid first-pass clearance from the bloodstream. This causes certain nanoparticles to be 'long circulating' relative to most small molecules. Tumors are heavily vascularized, and the longer a drug circulates in the bloodstream, the greater its chances are of accumulating in the tumor. Secondly, nanoscale particles less than about 300 nm in diameter extravasate through tumor vasculature and selectively accumulate in tumor tissue via the EPR effect (fig. 3). This is called 'passive targeting' and is a result of the defective 'leaky' vasculature and reduced lymphatic drainage systems that are characteristic of tumors. Nanoparticle passive targeting can be enhanced by 'active targeting', where specific ligands (peptides, antibodies, aptamers, or small molecules) are attached to the nanoparticle surface to bind receptors/antigens overexpressed on cancer cells. Targeting increases accumulation in tumors and lessens exposure to healthy tissue, which can improve a toxic drug's therapeutic index and reduce adverse side effects.

There are already several FDA-approved nanoparticle-based cancer therapies in clinical trials. In most therapeutic ENM-based nanomedicines, a toxic drug is either encapsulated within the core of a nanoparticle or conjugated to the surface. ENM-based products in the clinical stages of development include Nanospectra Biosciences, Inc.'s AuroShell and Aurimune from CytImmune Sciences, Inc. AuroShell is a gold-coated nanoshell that uses passive targeting to reach tumor sites and absorbs near-infrared laser light to heat and thermally destroy tumors and the surrounding blood vessels without significant damage to healthy tissue. Aurimune is nanosized gold bound to tumor necrosis factor (TNF). TNF is an exceptionally potent chemotherapeutic that was tested in clinical trials in the 1990s but had to be discontinued due to severe adverse side effects. In the recent phase I clinical trial of Aurimune, 3 times what had previously been a lethal dose of TNF (LD_{50}) was administered to patients with almost no ill effect, illustrating how ENM-based formulations can greatly reduce the toxicity of chemotherapeutics [23].

Two nanotech reformulations of cancer drugs are currently FDA approved: Abraxane and Doxil. Abraxane is an albumin-bound reformulation of paclitaxel and has been shown to be both more efficacious and less toxic than its legacy counterpart, Taxol. Doxil is a nanosized reformulation of doxorubicin that has decreased cardiotoxicity compared to the free drug. Although both of the active pharmaceutical ingredients are off patent, Doxil annual sales were USD 650 million in 2009 and Abraxane sales exceeded USD 320 million.

Market reports suggest that the market for ENM-enabled drugs will be USD 26 billion by 2012 and potentially USD 220 billion by 2015.

Conclusions

ENMs are being used successfully in medicine, specifically for drug delivery and for reformulation of discontinued drugs. As ENM use in pharmaceuticals becomes widespread, it becomes more important than ever to thoroughly characterize ENM interactions with biology. Here, we've reviewed the importance of ENM characterization, how small changes in ENM chemistry can cause dramatic differences in in vivo potency and safety, and the myriad uses for ENMs for pharmaceutical development.

Acknowledgements

This project has been funded in whole or in part with federal funds from the National Cancer Institute, National Institutes of Health, under contract HHSN261200800001E. The content of this publication does not necessarily reflect the views or policies of the Department of Health and Human Services, nor does mention of trade names, commercial products, or organizations imply endorsement by the U.S. Government.

Disclosure Statement

The authors have nothing to disclose.

References

1 Patri A, Dobrovolskaia M, Stern ST, McNeil S: Preclinical characterization of engineered nanoparticles intended for cancer therapeutics; in Amiji MM (ed): Nanotechnology for Cancer Therapy. New York, Taylor and Francis, 2007, pp 105–138.

2 Brown SC, et al: Nanoparticle characterization for cancer nanotechnology and other biological applications; in Grobmeyer SR, Moudgil BM (eds): Methods in Molecular Biology. Clifton, Springer, 2010, pp 39–65.

3 Powers KW, et al: Research strategies for safety evaluation of nanomaterials. 4. Characterization of nanoscale particles for toxicological evaluation. Toxicol Sci 2006;90:296–303.

4 Aggarwal P, Hall JB, McLeland CB, Dobrovolskaia MA, McNeil SE: Nanoparticle interaction with plasma proteins as it relates to particle biodistribution, biocompatibility and therapeutic efficacy. Adv Drug Deliv Rev 2009;61:428–437.

5 Dobrovolskaia MA, Aggarwal P, Hall JB, McNeil SE: Preclinical studies to understand nanoparticle interaction with the immune system and its potential effects on nanoparticle biodistribution. Mol Pharm 2008;5:487–495.

6 Hall JB, Dobrovolskaia MA, Patri AK, McNeil SE: Characterization of nanoparticles for therapeutics. Nanomedicine (Lond) 2007;2:789–803.

7 Dobrovolskaia MA, et al: Ambiguities in applying traditional limulus amebocyte lysate tests to quantify endotoxin in nanoparticle formulations. Nanomedicine (Lond) 2010;5:555–562.

8 US Food and Drug Administration: Points to Consider in the Manufacture and Testing of Monoclonal Antibody Products for Human Use. 1997. http://www.fda.gov/downloads/Biologics-BloodVaccines/GuidanceCompliance-RegulatoryInformation/OtherRecommendationsforManufacturers/UCM153182.pdf.

9 Dobrovolskaia MA, McNeil SE: Immunological properties of engineered nanomaterials. Nat Nanotechnol 2007;2:469–478.

10 Dobrovolskaia MA, et al: Method for analysis of nanoparticle hemolytic properties in vitro. Nano Lett 2008;8:2180–2187.

11 US Food and Drug Administration: Developing Medical Imaging Drug and Biological Products – Part 1: Conducting Safety Assessments. 2004. http://www.fda.gov/downloads/Drugs/GuidanceComplianceRegulatoryInformation/Guidances/ucm071600.pdf.

12 US Food and Drug Administration: Estimating the Maximum Safe Starting Dose in Initial Clinical Trials for Therapeutics in Adult Healthy Volunteers. 2005. http://www.fda.gov/downloads/Drugs/GuidanceComplianceRegulatoryInformation/Guidances/ucm078932.pdf.

13 US Food and Drug Administration: Content and Format of Investigational New Drug Applications (INDs) for Phase 1 Studies of Drugs, Including Well-Characterized, Therapeutic, Biotechnology-Derived Products. 1995. http://www.fda.gov/downloads/Drugs/GuidanceComplianceRegulatoryInformation/Guidances/ucm071597.pdf.

14 Zolnik BS, et al: Rapid distribution of liposomal short-chain ceramide in vitro and in vivo. Drug Metab Dispos 2008;36:1709–1715.

15 Fang C, et al: In vivo tumor targeting of tumor necrosis factor-alpha-loaded stealth nanoparticles: effect of MePEG molecular weight and particle size. Eur J Pharm Sci 2006;27:27–36.

16 Maeda H: Tumor-selective delivery of macromolecular drugs via the EPR effect: background and future prospects. Bioconjug Chem 2010;21:797–802.

17 Greish K: Enhanced permeability and retention (EPR) effect for anticancer nanomedicine drug targeting; in Grobmeyer SR, Moudgil BM (eds): Methods in Molecular Biology. Clifton, Springer, 2010, pp 25–37.

18 Papahadjopoulos D, et al: Sterically stabilized liposomes: improvements in pharmacokinetics and antitumor therapeutic efficacy. Proc Natl Acad Sci USA 1991;88:11460–11464.

19 Matsumura Y, et al: Phase I clinical trial and pharmacokinetic evaluation of NK911, a micelle-encapsulated doxorubicin. Br J Cancer 2004;91:1775–1781.

20 Woodle MC, et al: Versatility in lipid compositions showing prolonged circulation with sterically stabilized liposomes. Biochim Biophys Acta 1992;1105:193–200.

21 Meerum Terwogt JM, et al: Phase I clinical and pharmacokinetic study of PNU166945, a novel water-soluble polymer-conjugated prodrug of paclitaxel. Anticancer drugs 2001;12:315–323.

22 DiMasi JA, Hansen RW, Grabowski HG: The price of innovation: new estimates of drug development costs. J Health Econ 2003;22:151–185.

23 Libutti SK, et al: Phase I and pharmacokinetic studies of CYT-6091, a novel PEGylated colloidal gold-rhTNF nanomedicine. Clin Cancer Res 2010;16:6139–6149.

Scott E. McNeil
SAIC-Frederick, Inc., NCI at Frederick
Nanotechnology Characterization Laboratory
National Cancer Institute at Frederick Attn
PO Box B Building 469, 1050 Boyles Street
Frederick, MD 21702-1201 (USA)
Tel. +1 301 846 6939, E-Mail ncl@mail.nih.gov

Alexiou C (ed): Nanomedicine – Basic and Clinical Applications in Diagnostics and Therapy.
Else Kröner-Fresenius Symp. Basel, Karger, 2011, vol 2, pp 80–87

Interactions of Carbon Nanotubes with the Immune System: Focus on Mechanisms of Internalization and Biodegradation

Bengt Fadeel[a, b] · Anna A. Shvedova[c, d] · Valerian E. Kagan[e]

[a]Division of Molecular Toxicology, Institute of Environmental Medicine, Karolinska Institutet, and [b]Childhood Cancer Research Unit, Department of Women's and Children's Health, Karolinska University Hospital, Karolinska Institutet, Stockholm, Sweden; [c]Pathology and Physiology Research Branch, Health Effects Laboratory Division, National Institute for Occupational Safety and Health, [d]Department of Physiology and Pharmacology, West Virginia University, Morgantown, W. Va., and [e]Department of Environmental and Occupational Health, University of Pittsburgh, Pittsburgh, Pa., USA

Abstract

Carbon nanotubes (CNT) are cylinders of one or several coaxial graphite layer(s) with a diameter on the nanometer scale. CNT can be readily functionalized, and exciting studies on the use of CNT as excipients for drug delivery and imaging of disease processes have been reported. On the other hand, CNT were also shown to induce oxidative stress, inflammation, and fibrosis, and animal studies have suggested similarities between the pathogenic properties of certain multiwalled CNT and those of asbestos fibers. Recent studies have disclosed that CNT are susceptible to enzymatic biodegradation, and this observation could hold the key to the safe application of these nanomaterials in biomedicine. Here, we provide a brief overview of pertinent toxicological and biomedical investigations of CNT including recent work on the interaction of CNT with immune-competent cells, focusing on cellular recognition of nanotubes and their enzymatic degradation.

Nanotechnology presents many opportunities and benefits for new materials with significantly improved properties as well as revolutionary applications in the fields of energy, environment, and medicine. Carbon nanotubes (CNT) are among the most studied nanomaterials to date and are currently of interest for a variety of uses in technological as well as biomedical applications, including as drug delivery devices and contrast agents in medical imaging. At the same time, there is considerable concern about the potential adverse effects of these materials, and several recent studies have pointed to asbestos-like properties of CNT. It is important to note that 'CNT' come in many flavors, and careful physicochemical characterization of each sample is required. Indeed, some CNT may very well be harmful to human health, and exposure should be avoided, whereas other CNT preparations could be harnessed for biomedical use.

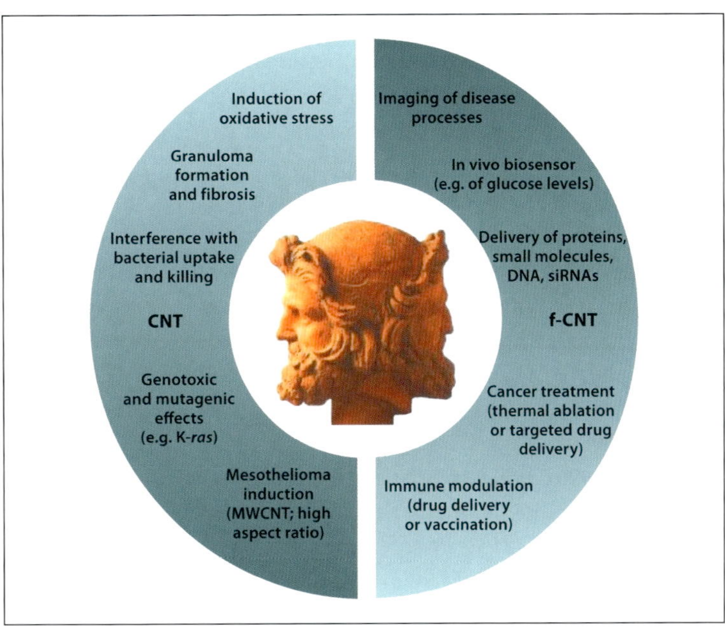

Fig. 1. CNT – engineered nanomaterials with Janus-like properties. CNT have been reported to exert toxic effects in cell culture as well as in vivo. This schematic diagram summarizes some of the toxicities of CNT reported to date. On the other hand, CNT are readily functionalized and this may result in enhanced dispersibility and improved biocompatibility, making these materials versatile and attractive for various biomedical applications. Selected examples of pharmaceutical and biomedical applications of functionalized CNT (f-CNT) are depicted. Adapted from Shvedova et al. [3] with permission from Elsevier.

Paradoxically, the novel characteristics of nanomaterials that are essential for successful and innovative applications might also lead to negative health impacts [1, 2] (fig. 1). Therefore, toxicological investigations are needed to determine the relationship between the physicochemical properties of nanoparticles and their cellular interactions, reactivity, and biological/systemic consequences. In particular, it may be important to understand how nanoparticles including CNT interact with or interfere with immune-competent cells and whether the immune system is capable of handling nanomaterials or nanomaterials simply 'fly under the radar' and thus escape immune surveillance. This could have considerable implications for the biodistribution of nanomaterials in the body. Furthermore, studies on the biopersistence or biodegradability of nanomaterials, and in particular of CNT, are also of importance. Nanomaterials intended for biomedical use should be either excretable or degradable (or biologically inert, but few materials are inert in the long term) as they may otherwise accumulate in tissues and exert undesirable effects.

Toxicological Studies

Considerable attention has been focused on in vitro (cell culture) assessment of CNT toxicity. Several groups have reported that exposure to single-walled CNT (SWCNT) or multiwalled CNT (MWCNT) causes cytotoxicity in various cell lines [for a review, see 3]. Importantly, we and others have reported low cytotoxicity for purified CNT (low iron content) versus raw, nonpurified CNT which were found to induce oxidative stress [4, 5], thus leading to the conclusion that metal traces associated with the nanotube samples are responsible for the toxicological effects. Recently, several investigators have cautioned that CNT can interfere with commonly used dye-based viability assays [6, 7]. These findings strongly suggest verifying cytotoxicity data using independent test systems. Moreover, evidence has been provided that an increase in the agglomeration state of CNT appears to be related to the cytotoxicity of these materials [8]. Therefore, it may be prudent to conduct in vitro cytotoxicity testing of CNT using well-dispersed suspensions. In sum, if one considers interference with dye-based viability assays, agglomeration issues related to the method of dispersion, and oxidative stress due to metal impurities, the available data favor the conclusion that well-dispersed, purified CNT exhibit relatively low cytotoxicity in vitro. When it comes to in vivo studies, however, one has to also consider the context of exposure: needless to say, for any material that is intended for biomedical application (i.e. intentional exposure), safety issues are of utmost importance, and the materials to be administered should be well defined in terms of composition, dispersion, and so on. On the other hand, in the case of unintentional exposure in the occupational setting, the issue of whether CNT samples are purified or nonpurified, or well dispersed versus entangled, is somewhat academic: the material is what it is, and exposure may or may not occur. Hence, it may be argued that studies conducted on as-produced CNT are as important as studies on purified or functionalized samples. This distinction between safety assessment in the biomedical setting versus the assessment of the potential risk for workers or consumers accidentally exposed to CNT (or, for that matter, any other class of engineered nanomaterial) is often overlooked.

Pulmonary toxicity of SWCNT has been reported in several studies [9–12]. The inflammatory response to SWCNT delivered to the lung via pharyngeal aspiration is characterized by a brisk acute phase response followed by an early onset of lung fibrosis [12]. Some argue that intratracheal or pharyngeal aspiration of SWCNT in a single bolus dose is an artificial exposure. Moreover, aspiration/intratracheal studies reported thus far have utilized relatively high dose exposures, which may not be relevant to the chronic low dose exposure occurring in an occupational setting. Inhalation of SWCNT more closely mimics occupational and environmental scenarios. The recent application of a specially designed aerosolization technology resulted in the ability to obtain stable SWCNT aerosols and has enabled studies of the inhalation of nonpurified SWCNT (iron content of 17.7% by weight) at 5 mg/m^3, 5 h/day, for 4 days versus pharyngeal aspiration of varying doses (5–20 μg/mouse) of the same SWCNT. The chain of pathological events in both exposure routes was realized through interactions of an early inflammatory response and oxidative stress culminating in the development of multifocal granulomatous pneumonia and interstitial fibrosis [13]. SWCNT inhalation was more effective than aspiration in causing inflammatory response, oxidative stress, collagen deposition, and fibrosis as well as mutagenic effects in the lungs of C57BL/6 mice [13]. The observed toxic outcomes of exposure of C57BL/6 mice to aerosolized respirable SWCNT (5 mg/m^3, 5 h/day, 4 days) resulted in a calculated lung burden of 5 μg/mouse. This lung burden would be achieved by workers exposed for less than 1 year at the peak airborne concentrations measured in an occupational setting [14].

The pathogenic fiber paradigm predicts that fibers that are long and thin (high aspect ratio) and biopersistent are more pathogenic than short fibers [15]. Interestingly, in a recent study so-called nanoscale dispersions of SWCNT were produced using a biocompatible block copolymer, Pluronic F 108NF (Pluronic) [16]. The well-dispersed CNT did not induce granulomas or fibrosis in the airways of mice, and the authors concluded that the toxicity of SWCNT is attributable to aggregation of the nanomaterial rather than the high aspect ratio of the individual nanotubes.

Animal studies of MWCNT have not produced consistent results. The pulmonary toxicity of MWCNT in rats exposed by intratracheal instillation to intact or ground (short) MWCNT showed that both MWCNT samples caused acute pulmonary inflammation at 3 and 15 days postexposure and pulmonary fibrosis at 60 days postexposure [17]. Others have reported that inhalation exposure to MWCNT does not induce significant lung toxicity but does cause immunosuppression [18]. These discrepancies may be attributed to differences in the CNT samples tested and rodent species used and differences in observation period in the two different studies. Even more controversial is the question of whether CNT may possess asbestos-like pathogenic properties [15]. Poland et al. [19] reported that abdominal injection of long but not short MWCNT caused an inflammatory reaction of the abdominal lining in mice which was similar to that seen after exposure of mice to asbestos. Takagi et al. [20] reported that MWCNT trigger mesothelioma formation when administered intraperitoneally to p53 heterozygous mice that have been reported to be sensitive to asbestos. However, high (nonrealistic) doses were used. Others have reported that MWCNT do not elicit any carcinogenic response in a 2-year bioassay in the peritoneal cavity of the rat [21]. The length of the CNT samples tested was suggested to be an important factor in determining the pathogenic potential of these materials.

Recent, important studies have provided evidence that MWCNT reach the subpleura in mice after a single inhalation exposure of 30 mg/m^3 for 6 h. Nanotubes were embedded in the subpleural wall and within subpleural macrophages. Subpleural fibrosis increased after 2 and 6 weeks following inhalation. None of these effects was seen in mice that inhaled carbon black nanoparticles or a lower dose of CNT (1 mg/m^3) [22]. Notwithstanding, an issue which remains is whether the acute responses to MWCNT would persist and progress to mesothelioma (there is a lag time of up to 30 years for mesothelioma formation in humans exposed to pathogenic asbestos fibers).

Cellular Uptake of Carbon Nanotubes

Cellular uptake is an important factor in determining the toxicity of engineered nanomaterials. Long CNT may induce 'frustrated phagocytosis' in macrophages, while short CNT appear to be handled much more effectively by the immune system [19]. Internalization of CNT by cells is also a critical determinant in different biomedical applications such as the delivery of therapeutically active molecules, including proteins, peptides, and genes. Functionalization of SWCNT and MWCNT using the 1,3-dipolar cycloaddition reaction, thus rendering the CNT water soluble, allows for the internalization of CNT by a wide range of cell types [23]. Porter et al. [24] reported that human monocyte-derived macrophages are able to internalize SWCNT; however, the process apparently took between 2 and 4 days, a long time for a macrophage to ingest a meal. Nevertheless, careful imaging revealed that the majority of SWCNT were located within phagosomes and lysosomes, suggestive of active, phagocytic uptake of the nanotubes. Some SWCNT were also found to translocate across the membrane into the cytoplasm [24]. Indeed, it has been proposed that CNT could penetrate the lipid bilayer by virtue of their 'needle-like' structure. The notion of spontaneous piercing through the membrane induced only by thermal motion was tested in a recent study in which the energy cost associated with the insertion of a CNT into a model phospholipid bilayer was calculated using the single-chain mean field theory [25]. The authors arrived at the conclusion that the energy cost of the bilayer rupture is high compared to that of the energy of thermal motion, which may support other energy-dependent translocation mechanisms, such as endocytosis. In a recent study, Neves et al. [26] reported that oxidized double-walled CNTs wrapped in RNA are taken up by prostate and cervical cancer cell lines in vitro but are then released after 24 h without discernable cellular stress. The mechanism of exocytosis of CNT remains to be understood.

Cells that are programmed to die by apoptosis typically expose the anionic phospholipid, phosphatidylserine (PS) on their surface and this serves as a specific recognition signal for neighboring phagocytes. Utilizing the same principle, we have recently shown that coating of SWCNT with PS makes the nanotubes recognizable in vitro by primary monocyte-derived human macrophages and dendritic cells [27]. Cellular uptake was suppressed by the PS-binding protein Annexin V, demonstrating that uptake was specific. In line with this, in vivo aspiration of PS-coated SWCNT stimulated their uptake by lung alveolar macrophages in mice, while SWCNT coated with the phospholipid phosphatidylcholine, which is not specifically recognized by phagocytes, were not as effectively taken up. Furthermore, Dutta et al. [28] reported that CNT avidly adsorb albumin when suspended in serum-containing medium. Upon binding to the surface of CNT, albumin is structurally altered and is recognized by scavenger receptors, which could facilitate the cellular uptake of CNT. Taken together, it seems apparent that the presence or absence of specialized signals determines the recognition and subsequent interactions of CNT with cells. Overall, pristine CNT are poorly recognized, whereas chemically modified CNT or CNT with a corona of adsorbed macromolecules (e.g. proteins and lipids) are more readily recognized and engulfed by cells.

Biodegradation of Carbon Nanotubes

Nondegradable nanomaterials can accumulate in cells and tissues where they can exert detrimental effects. Indeed, it has been shown that intravenously injected pristine (nonfunctionalized) SWCNT are highly enriched in the liver, lung, and spleen in mice and remain in the body over an extended period of time [29]. Controlling the biodistribution of CNT through surface functionalization will be instrumental in transforming these materials into candidates for drug delivery and in vivo imaging. Enzymatic degradation of SWCNT was recently demonstrated by incubating SWCNT in a cell-free system with horseradish peroxidase (HRP) and low amounts of H_2O_2 [30]. Others have reported that SWCNT with carboxylated surfaces undergo a 90-day degradation in a phagolysosomal simulant [31]. Unmodified, ozone-treated, and aryl-sulfonated nanotubes did not degrade under the same conditions. Oxidative HRP-dependent degradation was also reported recently for MWCNT after 2 months, although not to completeness [32]. In contrast, we have demonstrated complete biodegradation of CNT by human myeloperoxidase (hMPO) [33]. Biodegradation was shown in a cell-free system but we also provided evidence for hMPO-driven biodegradation in neutrophils isolated from normal healthy donors (fig. 2). Degradation occurred within 12 h. Macrophages were less proficient at biodegrading SWCNT, in line with the fact that these cells express much lower amounts of hMPO when compared to neutrophils. Importantly, SWCNT fully biodegraded by hMPO in vitro did not elicit typical inflammatory and oxidative stress responses characteristic of CNT after pharyngeal aspiration in mice [33]. These results open new opportunities for controlled regulation of CNT in an occupational or biomedical setting through natural (enzymatic) degradation by immune-competent cells (neutrophils).

Biomedical Applications of Carbon Nanotubes

Several recent developments, particularly in the synthesis of chemically modified CNT resulting in their enhanced 'solubilization' (dispersibility) and improved biocompatibility, make them more versatile and show potential pathways for their utilization in medicine [34]. Due to their strong optical absorbance, CNT have become candidates for biophysical approaches to destroy cancer cells using, for instance, hyperthermia. In principle, functionalization of CNT with tumor-specific epitopes followed by uptake only by cancerous cells affords an opportunity to locally heat the CNT using infrared laser radiation, resulting in thermal destruction of cancer cells without harm to surrounding cells. Using this approach, Kam et al. [35] showed that SWCNT functionalized with folic acid were specifically internalized by cancer cells expressing folic acid receptors on their surface. Hyperthermic treatment resulted in the killing of cancer cells in vitro without harm to by-

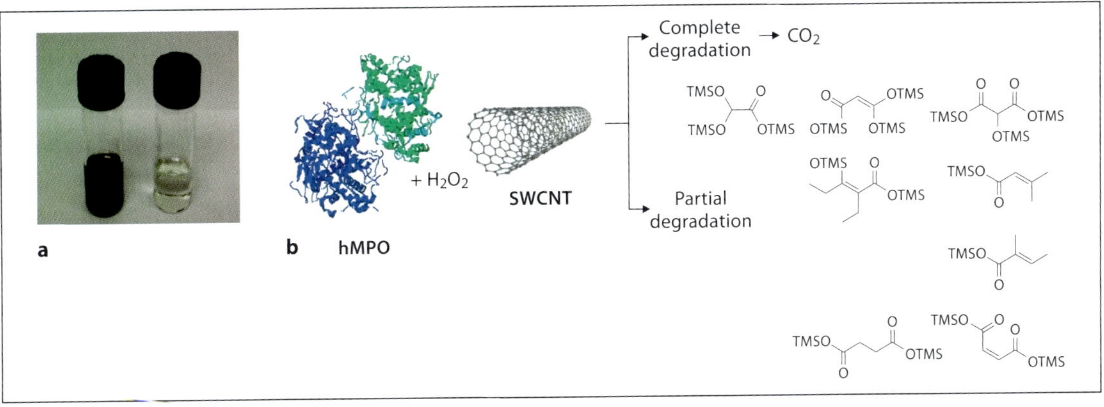

Fig. 2. Biodegradation of CNT. Recent studies have provided evidence of enzymatic degradation of SWCNT following incubation with HRP or hMPO. Detailed analysis of the biodegradation products showed that biodegradation produces CO_2. **a** Photograph showing carboxylated SWCNT (left) which were biodegraded upon 10 days of incubation with HRP and H_2O_2 (right). **b** Schematic of the biodegradation of SWCNT by neutrophil-derived hMPO. The chemical formulas depict short-chained tri-, di-, and monocarboxylated alkanes and alkenes. Reprinted from Kagan et al. [40] with permission from the American College of Occupational and Environmental Medicine.

stander cells. Similarly, Chakravarty et al. [36] used a combination of antibody-directed targeting and thermal ablation to demonstrate that SWCNT can be used for selective killing of human B lymphoma cells in vitro. Podesta et al. [37] reported that amino-functionalized MWCNT could be used to deliver therapeutic siRNA sequences in mice bearing a human lung carcinoma xenograft.

CNT have a tunable near-infrared emission that responds to changes in the local dielectric function but remains stable to permanent photobleaching. Barone et al. [38] reported the synthesis of solution phase, near-infrared sensors, with glucose sensing as a model system, using SWCNT that modulate their emission in response to the adsorption of specific biomolecules. By inserting a capillary tube containing such functionalized CNT into human epidermal tissue ex vivo, they could show that the CNT senses glucose concentrations in the range commonly found in the blood of diabetic patients. CNT could thus potentially serve as biosensors as they emit fluorescence in the near-infrared, a region of the spectrum not occupied by other organic molecules and where tissue is transparent.

Photoacoustic imaging has the potential to image organs in vivo with simultaneous high contrast and high spatial resolution, and De la Zerda et al. [39] have provided recent evidence that SWCNT conjugated with arginine-glycine-aspartic acid-containing peptides can be used as a contrast agent for photoacoustic imaging in tumor-bearing mice. Importantly, a greater spatial resolution was achieved by photoacoustic imaging using CNT as compared with fluorescence imaging using tumor-targeted quantum dots. These elegant preclinical studies suggest new approaches to noninvasive imaging and monitoring of disease processes, taking advantage of the specific physicochemical properties of CNT.

Concluding Remarks

CNT have been ascribed various toxic effects in vitro and in vivo, and some studies have even indicated similarities between certain CNT (multiwalled, high aspect ratio) and asbestos fibers. On the other hand, recent studies have also demonstrated that CNT could potentially be harnessed for biomedical purposes, as vectors for drug delivery, and for imaging. The important task that lies ahead will thus be to determine which physicochemical characteristics of CNT drive the toxic responses and to utilize this knowledge in the design of CNT that are biocompatible and safe. Furthermore, some of the unwanted responses of CNT may, in fact, lead to desirable outcomes depending on the specific context. For instance, the induction of CNT-dependent cellular toxicity could serve as a novel and effective approach to combat cancer cells, provided that the excretion or biodegradation of CNT can be controlled. Overall, a detailed understanding of the pharmacological and toxicological properties of CNT and a critical and balanced evaluation of risk/benefit ratios is required when considering the medical applications of these novel materials.

Acknowledgements

The authors were supported, in part, by the Swedish Research Council, the 7th Framework Program of the European Commission (EC-FP-7-NANOMMUNE-214281), the National Institute for Occupational Safety and Health (NIOSH), and the National Institutes of Health (NIH).

Disclosure Statement

The authors have nothing to disclose.

Disclaimer

The conclusions reported herein are those of the authors and do not necessarily represent the views of the National Institute for Occupational Safety and Health.

References

1 Shvedova AA, Kagan VE, Fadeel B: Close encounters of the small kind: adverse effects of man-made materials interfacing with the nano-cosmos of biological systems. Annu Rev Pharmacol Toxicol 2010;50:63–88.

2 Feliu N, Fadeel B: Nanotoxicology: no small matter. Nanoscale 2010;2:2514–2520.

3 Shvedova AA, Kisin ER, Porter D, Schulte P, Kagan VE, Fadeel B, Castranova V: Mechanisms of pulmonary toxicity and medical applications of carbon nanotubes: two faces of Janus? Pharmacol Ther 2009;121:192–204.

4 Kagan VE, Tyurina YY, Tyurin VA, Konduru NV, Potapovich AI, Osipov AN, Kisin ER, Schwegler-Berry D, Mercer R, Castranova V, Shvedova AA: Direct and indirect effects of single walled carbon nanotubes on RAW 264.7 macrophages: role of iron. Toxicol Lett 2006;165:88–100.

5 Pulskump K, Drabate S, Krug HF: Carbon nanotubes show no sign of acute toxicity but induce intracellular reactive oxygen species in dependence on contaminants. Toxicol Lett 2007;168:58–74.

6 Monteiro-Riviere NA, Inman AO: Challenges for assessing carbon nanomaterial toxicity to skin. Carbon NY 2006; 44:1070–1078.

7 Wörle-Knirsch JM, Pulskamp K, Krug HF: Oops they did it again! Carbon nanotubes hoax scientists in viability assays. Nano Lett 2006;6:1261–1268.

8 Wick P, Manser P, Limbach LK, Dettlaff -Weglikowska U, Krumeich F, Roth S, Stark WJ, Bruinink A: The degree and kind of agglomeration affect carbon nanotube toxicity. Toxicol Lett 2007;168:121–131.

9 Warheit DB, Laurence BR, Reed KL, Roach DH, Reynolds GA, Webb TR: Comparative pulmonary toxicity assessment of single wall carbon nanotubes in rats. Toxicol Sci 2003;26:2281–2200.

10 Lam CW, James JT, McCluskey R, Hunter RL: Pulmonary toxicity of single-wall carbon nanotubes in mice 7 and 90 days after intratracheal instillation. Toxicol Sci 2004;77:126–134.

11 Li Z, Hulderman T, Salmen R, Chapman R, Leonard SS, Young SH, Shvedova A, Luster MI, Simeonova PP: Cardiovascular effects of pulmonary exposure to single-wall carbon nanotubes. Environ Health Perspect 2007; 115:377–382.

12 Shvedova AA, Kisin ER, Mercer R, Murray AR, Johnson VJ, Potapovich AI, Tyurina YY, Gorelik O, Arepalli S, Schwegler-Berry D, Hubbs AF, Antonini J, Evans DE, Ku BK, Ramsey D, Maynard A, Kagan VE, Castranova V, Baron P: Unusual inflammatory and fibrogenic pulmonary responses to single-walled carbon nanotubes in mice. Am J Physiol 2005;289:L698–L708.

13 Shvedova AA, Kisin E, Murray AR, Johnson VJ, Gorelik O, Arepalli S, Hubbs AF, Mercer RR, Keohavong P, Sussman N, Jin J, Yin J, Stone S, Chen BT, Deye G, Maynard A, Castranova V, Baron PA, Kagan VE: Inhalation versus aspiration of single walled carbon nanotubes in C57BL/6 mice: inflammation, fibrosis, oxidative stress and mutagenesis. Am J Physiol Lung Cell Mol Physiol 2008;295:L552–L565.

14 Maynard AD, Baron PA, Foley M, Shvedova AA, Kisin ER, Castranova V: Exposure to carbon nanotube material: aerosol release during the handling of unrefined single-walled carbon nanotube material. J Toxicol Environ Health A 2004;67:87–107.

15 Donaldson K, Murphy FA, Duffin R, Poland CA: Asbestos, carbon nanotubes and the pleural mesothelium: a review of the hypothesis regarding the role of long fibre retention in the parietal pleura, inflammation and mesothelioma. Part Fibre Toxicol 2010;7:5.
16 Mutlu GM, Budinger GR, Green AA, Urich D, Soberanes S, Chiarella SE, Alheid GF, McCrimmon DR, Szleifer I, Hersam MC: Biocompatible nanoscale dispersion of single-walled carbon nanotubes minimizes in vivo pulmonary toxicity. Nano Lett 2010;10:1664–1670.
17 Muller J, Huaux F, Moreau N, Misson P, Heilier JF, Delos M, Arras M, Fonseca A, Nagy JB, Lison D: Respiratory toxicity of multi-wall carbon nanotubes. Toxicol Appl Pharmacol 2005;207:221–231.
18 Mitchell LA, Lauer FT, Burchiel SW, McDonald JD: Mechanisms for how inhaled multiwalled carbon nanotubes suppress systemic immune function in mice. Nat Nanotechnol 2009;4:451–456.
19 Poland CA, Duffin R, Kinloch I, Maynard A, Wallace WA, Seaton A, Stone V, Brown S, Macnee W, Donaldson K: Carbon nanotubes introduced into the abdominal cavity of mice show asbestos-like pathogenicity in a pilot study. Nat Nanotechnol 2008;3:423–428.
20 Takagi A, Hirose A, Nishimura T, Fukumori N, Ogata A, Ohashi N, Kitajima S, Kanno J: Induction of mesothelioma in p53+/– mouse by intraperitoneal application of multiwall carbon nanotube. J Toxicol Sci 2008;33:105–116.
21 Muller J, Delos M, Panin N, Rabolli V, Huaux F, Lison D: Absence of carcinogenic response to multiwall carbon nanotubes in a 2-year bioassay in the peritoneal cavity of the rat. Toxicol Sci 2009;110:442–448.
22 Ryman-Rasmussen JP, Cesta MF, Brody AR, Shipley-Phillips JK, Everitt JI, Tewksbury EW, Moss OR, Wong BA, Dodd DE, Andersen ME, Bonner JC: Inhaled carbon nanotubes reach the subpleural tissue in mice. Nat Nanotechnol 2009;4:747–751.
23 Kostarelos K, Lacerda L, Pastorin G, Wu W, Wieckowski S, Luangsivilay J, Godefroy S, Pantarotto D, Briand JP, Muller S, Prato M, Bianco A: Cellular uptake of functionalized carbon nanotubes is independent of functional group and cell type. Nat Nanotechnol 2007;2:108–113.
24 Porter AE, Gass M, Muller K, Skepper JN, Midgley PA, Welland M: Direct imaging of single-walled carbon nanotubes in cells. Nat Nanotechnol 2007;2:713–717.
25 Pogodin S, Baulin VA. Can a carbon nanotube pierce through a phospholipid bilayer? ACS Nano 2010;4:5293–5300.
26 Neves V, Heister E, Costa S, Tilmaciu C, Borowiak-Palen E, Ciusca CE, Flahaut E, Soula B, Coley HM, McFadeen J, Silva SR: Uptake and release of double-walled carbon nanotubes by mammalian cells. Adv Funct Mater 2010;20:3272–3279.
27 Konduru NV, Tyurina YY, Feng W, Basova LV, Belikova NA, Bayir H, Clark K, Rubin M, Stolz D, Vallhov H, Scheynius A, Witasp E, Fadeel B, Kichambare PD, Star A, Kisin ER, Murray AR, Shvedova AA, Kagan VE: Phosphatidylserine targets single-walled carbon nanotubes to professional phagocytes in vitro and in vivo. PLoS One 2009;4:e4398.
28 Dutta D, Sundaram SK, Teeguarden JG, Riley BJ, Fifield LS, Jacobs JM, Addleman SR, Kaysen GA, Moudgil BM, Weber TJ: Adsorbed proteins influence the biological activity and molecular targeting of nanomaterials. Toxicol Sci 2007;100:300–315.
29 Yang S, Guo W, Lin Y, Deng X, Wang H, Sun H, Liu, YF, Wang X, Wang W, Chen M, Huang YP, Sun YP: Biodistribution of pristine single-walled carbon nanotubes in vivo. J Phys Chem C 2007;111:17761–17764.
30 Allen BL, Kichambare PD, Gou P, Vlasova II, Kapralov AA, Konduru N, Kagan VE, Star A: Biodegradation of single-walled carbon nanotubes through enzymatic catalysis. Nano Lett 2008;8:3899–3903.
31 Liu X, Hurt RH, Kane AB: Biodurability of single-walled carbon nanotubes depends on surface functionalization. Carbon NY 2010;48:1961–1969.
32 Russier J, Ménard-Moyon C, Venturelli E, Gravel E, Marcolongo G, Meneghetti M, Doris E, Bianco A: Oxidative biodegradation of single- and multi-walled carbon nanotubes. Nanoscale 2011;3:893–896.
33 Kagan VE, Konduru NV, Feng W, Allen BL, Conroy J, Volkov Y, Vlasova II, Belikova NA, Yanamala N, Kapralov A, Tyurina YY, Shi J, Kisin ER, Murray AR, Franks J, Stolz D, Gou P, Klein-Seetharaman J, Fadeel B, Star A, Shvedova AA: Carbon nanotubes degraded by neutrophil myeloperoxidase induce less pulmonary inflammation. Nat Nanotechnol 2010;5:354–359.
34 Kostarelos K, Bianco A, Prato M: Promises, facts and challenges for carbon nanotubes in imaging and therapeutics. Nat Nanotechnol 2009;4:627–633.
35 Kam NW, O'Connell M, Wisdom JA, Dai H: Carbon nanotubes as multifunctional biological transporters and near-infrared agents for selective cancer cell destruction. Proc Natl Acad Sci USA 2005;102:11600–11605.
36 Chakravarty P, Marches R, Zimmerman NS, Swafford AD, Bajaj P, Musselman IH, Pantano P, Draper RK, Vitetta ES: Thermal ablation of tumor cells with antibody-functionalized single-walled carbon nanotubes. Proc Natl Acad Sci USA 2008;105:8697–8702.
37 Podesta JE, Al-Jamal KT, Herrero MA, Tian B, Ali-Boucetta H, Hegde V, Bianco A, Prato M, Kostarelos K: Antitumor activity and prolonged survival by carbon-nanotube-mediated therapeutic siRNA silencing in a human lung xenograft model. Small 2009;5:1176–1185.
38 Barone PW, Baik S, Heller DA, Strano MS: Near-infrared optical sensors based on single-walled carbon nanotubes. Nat Mater 2005;4:86–92.
39 De la Zerda A, Zavaleta C, Keren S, Vaithilingam S, Bodapati S, Liu Z, Levi J, Smith BR, Ma TJ, Oralkan O, Cheng Z, Chen X, Dai H, Khuri-Yakub BT, Gambhir SS: Carbon nanotubes as photoacoustic molecular imaging agents in living mice. Nat Nanotechnol 2008;3:557–562.
40 Kagan VE, Shi J, Feng W, Shvedova AA, Fadeel B: Fantastic voyage and opportunities of engineered nanomaterials: what are the potential risks of occupational exposures? J Occup Environ Med 2010;52:943–946.

Prof. Bengt Fadeel
Division of Molecular Toxicology, Institute of Environmental Medicine
Nobels väg 13, Karolinska Institutet, SE–17177 Stockholm (Sweden)
Tel. +46 8 524 877 37, E-Mail bengt.fadeel@ki.se

Alexiou C (ed): Nanomedicine – Basic and Clinical Applications in Diagnostics and Therapy.
Else Kröner-Fresenius Symp. Basel, Karger, 2011, vol 2, pp 88–95

Magnetic Particle Imaging: Principles and Clinical Application

T.M. Buzug[a] · T.F. Sattel[a] · M. Erbe[a] · S. Biederer[a] · D. Finas[b] · K. Diedrich[b] · F. Vogt[c] · J. Barkhausen[c] · J. Borgert[d] · K. Lüdtke-Buzug[a] · T. Knopp[a]

[a]Institute of Medical Engineering, University of Lübeck, and Clinics for [b]Gynecology and [c]Radiology and Nuclear Medicine, UK-SH, Campus Lübeck, Lübeck, and [d]Innovative Technologies, Research Laboratories, Philips Technologie GmbH, Hamburg, Germany

Abstract

In magnetic particle imaging (MPI), superparamagnetic iron oxide nanoparticles are used as tracer materials. The nanoparticles are subjected to a sinusoidal, oscillating magnetic field, i.e. the drive field, and respond with a nonlinear change in magnetization. As a result, the acquired induction signal contains multiples of the fundamental excitation frequency which are subsequently used for determination of the particle distribution and concentration. For spatial encoding, a selection field, i.e. a magnetic gradient field, is superimposed onto the drive field such that a field-free point (FFP) is produced at a desired location within the field of view. In a simplified picture, nanoparticles located near the FFP contribute to the signal generation, whereas particles that are far from the FFP are in saturation and cannot contribute. Image reconstruction from the induction signals can be seen as the solution for the corresponding inverse problem. Herein, the current state of the art in magnetic coil design for MPI is discussed. With a new symmetrical arrangement of coils, a field-free line can be produced that promises a significantly higher sensitivity compared with the standard arrangement for an FFP. Additionally, an alternative single-sided coil assembly for applications in a sentinel lymph node biopsy scenario is presented.

In 2005, Gleich and Weizenecker [1] reported on an invention that allowed for direct quantitative measurement of the spatial distribution of superparamagnetic nanoparticles. The basic idea of the magnetic particle imaging (MPI) concept is based on the nonlinearity of the magnetization characteristics of superparamagnetic nanoparticles. If such nanoparticles are subjected to an oscillating magnetic field, they respond with a nonlinear change in magnetization that can be described by Langevins' theory of paramagnetism [2]. Basically, two different magnetic fields must be superimposed for the realization of spatial encoding and signal generation: a static gradient field, named *selection field*, and a sinusoidally oscillating field, named *drive field*. This way, direct measurement of the spatial particle distribution, as well as concentration, is enabled.

With the sinusoidal oscillating field, particle magnetization dynamics can be acquired. The temporal variation of the magnetization can be directly measured via the induced voltage in a receive coil arrangement. The change in magnetization is nonlinear due to saturation effects of the nanoparticles. In the Fourier sense, this leads to multiples of the fundamental drive field frequency, named harmonics, in the spectrum of the acquired signal. Actually, only the derivative of the change in magnetization can be measured by induction. However, the properties of the Fourier transform allow recovery of the original signal in Fourier space simply by dividing the spectrum by the frequency variable. The harmonics, which can be seen as a fingerprint of the particles, are the desired signal components. After separation of the higher harmonics from the fundamental drive field frequency, the energy of this signal can be used to recover the concentration of the nanoparticles.

For many years, magnetic contrast agents have been used in the field of magnetic resonance imaging (MRI). Resovist® (Bayer Schering Pharma), a carboxydextran-coated iron oxide nanoparticle, is one example of such a contrast agent which has been approved for human radiological MRI of the reticuloendothelial system (RES). Thanks to this, all initial proofs of concept of MPI have been carried out with Resovist. However, in contrast to MPI, MRI acquires the intrinsic magnetization of tissue, which is distorted due to the presence of magnetic nanoparticles and hence improves the contrast of the organ of interest by a significant reduction of the relaxation time.

Alternatively, measurement techniques have been tested that acquire the time constants of the magnetic field of the particles after switching off the external magnetic field. Based on this, Romanus et al. [3] proposed the use of a technique they named magnetorelaxometry to estimate the spatial distribution of nanoparticle relaxation. However, the spatial resolution of this technique is poor. In contrast to this, MPI promises submillimeter spatial resolution and high sensitivity [4]. Since the initial publication by Gleich and Weizenecker [1], MPI has taken tremendous steps towards a versatile measurement device for magnetic particle analysis and spatial nanoparticle imaging.

Beyond imaging, Weaver et al. [5] and Biederer et al. [6] reported on a magnetic particle spectrometer that can be used to measure the nonlinear magnetization fingerprint of the particles. Based on these measurements and the Langevin theory of paramagnetism, the particles' magnetic core size can be calculated by solving the corresponding inverse problem. This, in turn, allows for an adequate simulation of the imaging performance in MPI and leads to improvements in the synthesis of new magnetic nanoparticles [7].

Moreland et al. [8] reported that an alternative imaging device to that of Gleich and Weizenecker [1] can be set up with a cantilever torque magnetometer. Goodwill et al. [9] set up a scanner design similar to that of Gleich and Weizenecker [1] using permanent magnets for the selection field. They demonstrated the device as narrowband MPI, which reduces the bandwidth requirements of filters and amplifiers of the receive chain.

Mainstream design concepts of MPI consist of 1 pair of Maxwell coils that produces the selection field and at least 1 drive field coil pair for each encoding dimension [1, 10]. Here, the volume of measurement is located in the isocenter of the coil assembly. Today, the feasible dimension of the measurement volume fits small animals only [11], although simulations show that a human whole-body scanner might be possible in the future as well [3].

Recently, Sattel et al. [12] reported on an asymmetric coil design which realizes all field-generating and signal-receiving coils on one side. With this single-side design, the patient no longer needs to fit into the measurement volume. Sattel et al. [12] indicated limitations of such an approach. However, even considering the limitations in penetration depth and resolution of this first experimental system, it seems reasonable that future single-sided MPI scanners will have improved imaging characteristics and asymmetric 3-dimesional imaging will become feasible [13]. For the development of new coil topologies, a framework has been developed and presented [see 4 and 14] that allows for simulation of the complete MPI imaging chain.

Parallel to the rapid development of new scanner designs, first applications have been published as well. For instance, ideas of imaging the traces of iron-loaded stem cells have been reported [see 15–17]. In 2009, the first in vivo results were published by Weizenecker et al. [18] which revealed structures of a beating mouse heart at a high frame rate of 46 volumes per second. Also in 2009, Ruhland et al. [19] report on the idea of using MPI for sentinel lymph node detection in breast cancer.

Magnetization Dynamics of Nanoparticles

The physical principle of MPI is based on the nonlinear change in magnetization of superparamagnetic iron oxide nanoparticles (SPIONs). The magnetization curve $M(t)$ of the SPIONs can be described by the Langevin theory of paramagnetism

$$M(t) = m_s c \left(\coth\left[\frac{m_s \mu_0 H_D(t)}{k_B T} \right] - \frac{k_B T}{m_s \mu_0 H_D(t)} \right), \quad (1)$$

where μ_0 is the permeability of vacuum, k_B the Boltzmann constant, T the temperature in Kelvin, and c the particle concentration. The magnetic moment at saturation of the particles is given by $m_s = 1/6\pi d_c^3 M_s$, where M_s is the saturation magnetization of magnetite. $H_D(t)$ is given by the drive field mentioned above. When the sinusoidal drive field $H_D(t) = H_o \sin(2\pi f_0)$ with a fundamental frequency f_0 is applied, the frequency analysis of the magnetization $M(t)$ of the nanoparticles shows additional harmonics of the fundamental frequency. Detection of an induced voltage in the receive coils, which is caused by a change in magnetization, allows for determination of the magnetization (see fig. 1a).

If a second field, i.e. the above mentioned selection field, is superimposed onto the drive field, spatial encoding can be realized. The selection field is a gradient field that produces a field-free point (FFP) in the measurement field. Particles far from the FFP become saturated and do not contribute to the acquired induction voltage of the particle magnetization dynamics (see fig. 1b). However, the drive field amplitude H_o is chosen high enough to move the FFP through the field of view. The FFP movement changes the magnetization of the nanoparticles along the trajectory that can be measured in a receive coil arrangement. By solving the inverse reconstruction problem, the spatial distribution of the nanoparticle concentration can be estimated.

The drive field coil pairs in a Helmholtz configuration and selection field coils in a Maxwell arrangement are depicted in figure 1c. A visualization of the magnetic field in a grayscale lookup table in between the coil assembly is shown as well.

Magnetic Particle Imaging Coil Design

In order to create an MPI scanner, the selection field and the drive field must be combined into a single coil pair with superimposed static and alternating currents. This MPI scanner provides an FFP with 1 degree of freedom and is, therefore, named a *line scanner.* The corresponding coil assembly is illustrated in figure 2a.

Symmetric Coil Assembly for Field-Free Point Imaging

Once the FFP has been established, it is possible to increase the number of degrees of freedom of the FFP by extending the scanner by 2 additional pairs of orthogonal drive field coils. With this device it is possible to move the FFP on a user-defined trajectory and the scanner advances to a fully 3-D imaging device as shown in figure 2b.

Symmetric Coil Assembly for Field-Free Line Imaging

In a simulation study a new method of signal acquisition in MPI was introduced and validated

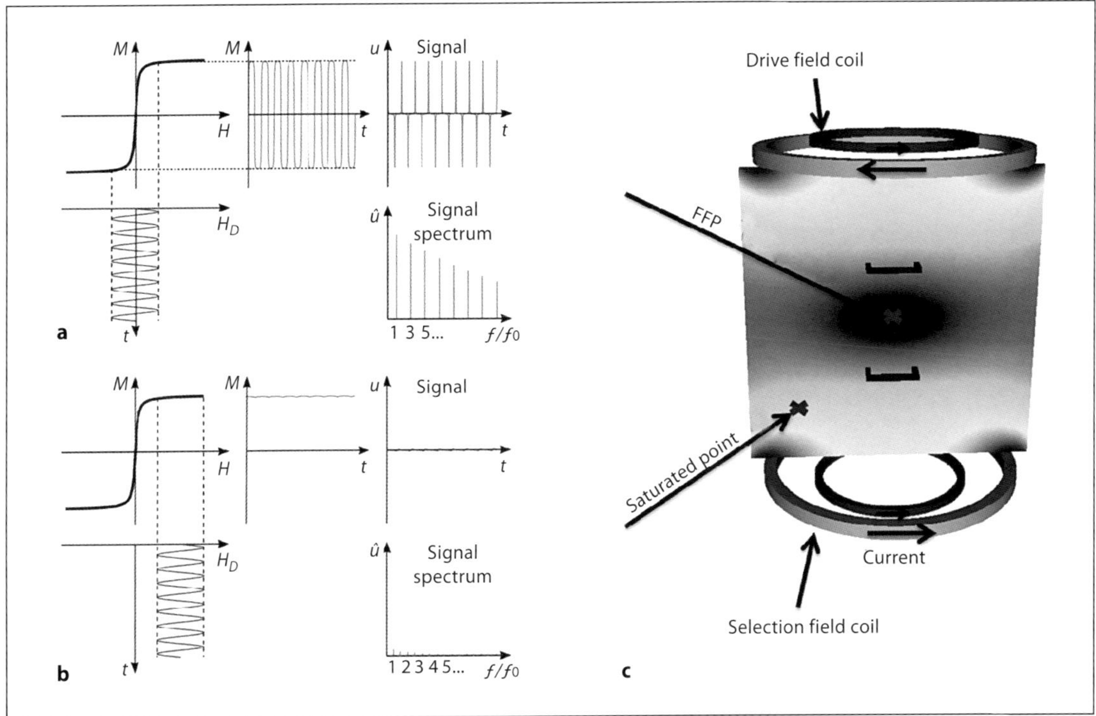

Fig. 1. a If the nanoparticle (SPION) at the FFP is subjected to a sinusoidal oscillating magnetic drive field, H_D, it reacts with a nonlinear change in magnetization, *M*. The variation in magnetization induces a voltage in the receive coil. This voltage contains multiples of the fundamental frequency of the drive signal (harmonics). **b** For nanoparticles outside the FFP the magnetization is in saturation and thus does not change. Therefore, no voltage is induced in the receive coil. **c** Drive field and selection field coil assembly that produces the encoding field.

[10]. It was shown that the sensitivity of MPI can be increased by a factor of 10 using the signal of a particle distribution located along a field-free line (FFL) instead of the conventionally used FFP discussed above. In a row of optimization steps [20, 21] the geometry of an FFL scanner device has been substantially improved regarding electrical power loss and field quality. A history of the proposed designs is illustrated in figure 3.

Asymmetric Magnetic Particle Imaging Design (Single-Sided Device)

If one of the transmit coils shown in figure 2a is scaled down such that it fits concentrically into the other transmit coil (compare to fig. 4), an FFP is established at either side of the new coil arrangement.

Again, the current in the 2 transmit coils must be opposite in direction. That way, a field is produced by superposition that annihilates at 2 points on the axis of the coil arrangement. In figure 4b, the magnetic field is simulated in a section along the coil axis around the FFP.

As mentioned above, due to symmetry 2 FFPs can be found. One is located on the front side (the patient side) and the other on the back side of the device. The first is an FFP used for imaging; for the latter it must be ensured that no magnetic material is located inside the device.

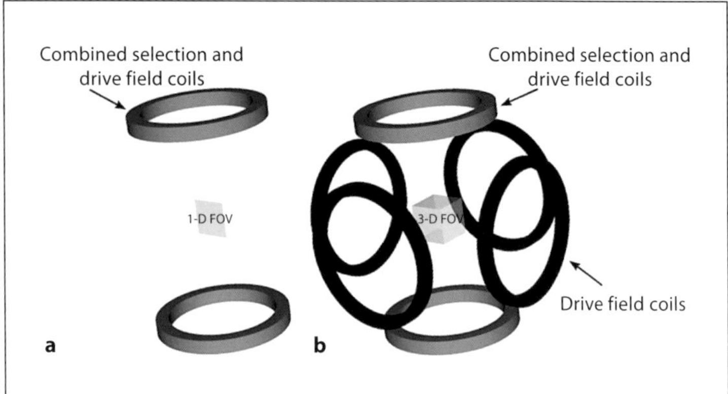

Fig. 2. Coil assemblies in MPI. **a** Line scanner with combined drive and selection field coil pairs. **b** Fully 3-D MPI imaging device with a symmetrical coil arrangement. FOV = Field of view.

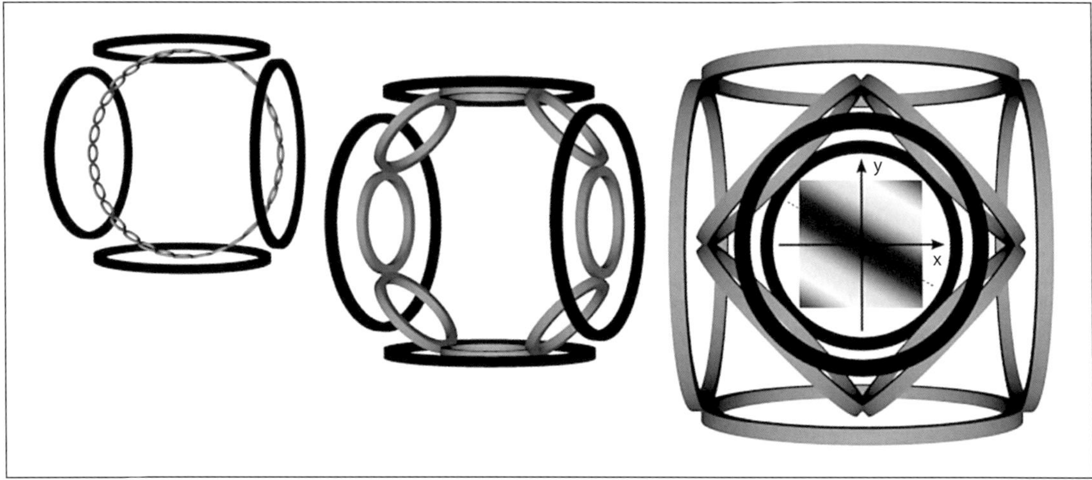

Fig. 3. Optimization steps of the FFL coil designs. The left setup consists of 16 small Maxwell coil pairs located on a circle at equidistant angles. For the scanner geometry in the middle, the number of Maxwell coil pairs has been reduced to 8, resulting in an substantial decrease in electrical power loss. On the right, the most efficient setup regarding power loss and field quality is illustrated [20, 21].

The first experimental results of a single-side MPI device have been shown for different phantoms [12]. The performance in terms of resolution and contrast differs from that of a symmetric design. Due to the strong inhomogeneity of the generated fields, the gradient and receive sensitivity is strongly position dependent within the field of view. In general, the resolution and contrast are best near the scanner front and decrease with increasing distance. Therefore, a single-sided device is best suited for high resolution imaging near the surface. However, the key benefit of this design lies in its flexibility. There is no limitation of patient size because the coil assembly can be positioned similarly to an ultrasound transducer.

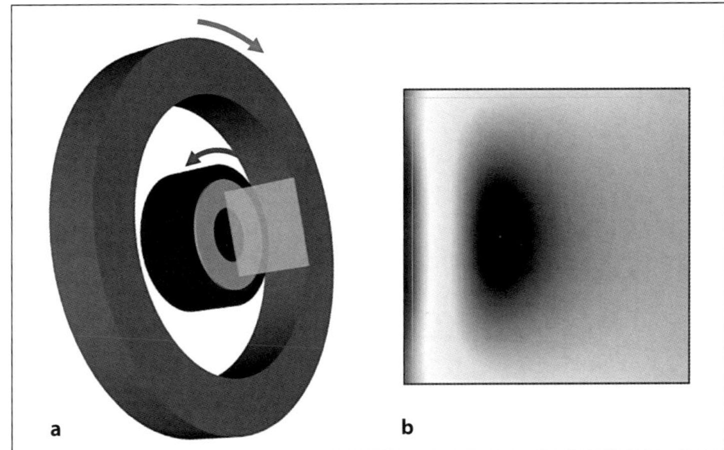

Fig. 4. Single-side MPI line scanner. **a** A flat coil arrangement generates a FFP at either side of the assembly. **b** Simulation of the expected spatial field distribution containing the FFP.

Clinical Application Scenarios

Primary solid malignant tumors often drain tumor cells through lymph vessels to local lymph nodes before they enter the chronic state and become a metastatic tumor. Thus, tumor cells can be found in lymph nodes in early-state disease. This applies, for example, to head and neck, skin, and breast tumors. Therefore, lymph node dissection is performed routinely in these tumor entities. Breast tumors are draining to the axillary lymph nodes so that axillary exploration by radical axillary lymphonodectomy is part of surgical staging [22]. Breast cancer is the most frequent cancer among females worldwide in Western industrialized countries and therefore investigation in this field is of high health significance and socioeconomic importance [23]. The individual prognosis decreases dramatically if tumor cells are present in lymph nodes. However, axillary lymphonodectomy is associated with high morbidity and a significant loss of quality of life.

To reduce morbidity due to radical surgery, the sentinel lymph node biopsy (SNLB) concept has been developed. After the introduction of routinely applied SNLB, morbidity is strongly reduced because only 1 or few lymph nodes are removed which accumulate a previous applied marker substance injected into the breast [24].

SPIONs could replace these marker substances. This would eliminate the exposure of patients and participating medical staff to ionizing radiation within in the SNLB process. Moreover, SPIONs are easy to obtain and to store. This makes the method accessible to all patients.

By using the MPI procedure, 3-D imaging and distinct localization of SPIONs can be achieved. To prove the mentioned principle of SNLB by MPI, a mouse model was applied because qualitative and quantitative enrichment of SPIONs in the axillary lymphatic tissue has been unexplored until now. Moreover, the MPI procedure has not been applied in vivo within this setting so far. To prove this concept, healthy mice were used [25] to analyze the axillary and environmental tissue for the presence and concentration of SPIONs with different techniques. After that, the results were applied to a tumor-bearing mouse model with axillary lymphatic metastases [26, 27].

To support this research, a new MPI hand probe with a unilateral solenoid arrangement designed for use in the operating theater is under construction. Intraoperative 3-D MPI imaging can allow for easy and precise detection of axillary sentinel lymph nodes [19].

In principle, the concept of SNLB by MPI can be applied to all solid tumors. Sentinel lymph node detection can be easily performed after intraoperative tracer application. Using MPI, the precise detection and localization of sentinel lymph nodes with real-time 3-D imaging will be possible.

The ultimate proof will be its use in the human model. Compared with the standard SNLB the proposed concept has the potential to be much less expensive and a less burdensome diagnostic process.

Conclusion

A wide range of applications for superparamagnetic iron oxide particles as contrast media has emerged over the last years for MRI. Owing to their small size and their hydrophilic coating, ultrasmall superparamagnetic iron oxide particles are generally able to avoid early uptake by macrophages from the mononuclear phagocyte system. Hence, these particles with an optimized T_1 relaxivity and prolonged intravascular circulation time can be used as blood pool contrast media for MR angiography.

Whereas in MRI the pharmacokinetic properties of the compounds are important, in MPI the size and the isotropy of the magnetic core are most crucial. The diameter of the core and types of specific coating for these nanoparticles influence not only the performance of this new technique but also the type of applications that would benefit from MPI. An iron core size of 30 nm and a hydrophilic coating with a diameter of less than 50 nm seem to be desirable to fully exploit MPI as an interventional tool.

In this contribution, the current state of the art in magnetic coil design for MPI has been reviewed. With a new symmetrical arrangement of coils, an FFL can be produced that promises a significantly higher sensitivity compared with the current standard arrangement for FFP imaging. With an alternative coil assembly, i.e. the single-sided concept, the application of MPI is no longer limited by patient size. This concept is most suitable for applications in SNLB.

Acknowledgements

The MPI-based sentinel lymph node scenario reported here is supported by the German Federal Ministry of Education and Research (BMBF grant No. 01EZ0912). It is also part of the university research program Imaging of Disease Processes, University of Lübeck. The authors would like to thank Philips Technologie GmbH, Innovative Technologies, Research Laboratories, Hamburg (Germany) for continuous support.

Disclosure Statement

J. Borgert is employee of Innovative Technologies, Research Laboratories, Philips Technologie GmbH. The remaining authors have nothing to disclose.

References

1 Gleich B, Weizenecker J: Tomographic imaging using the nonlinear response of magnetic particles. Nature 2005;435:1214–1217.
2 Chikazumi S, Charap SH: Physics of Magnetism. New York, Wiley, 1964.
3 Romanus E, Gross C, Koetitz R, Prass S, Lange J, Weber P, Weitschies W: Monitoring of biological binding reactions by magneto-optical relaxation measurements. Magnetohydrodynamics 2001;37:28–333.
4 Weizenecker J, Borgert J, Gleich B: A simulation study on the resolution and sensitivity of magnetic particle imaging. Phys Med Biol 2007;52:6363–6374.
5 Weaver JB, Rauwerdink AM, Charles R, Sullivan CR, Baker I: Frequency distribution of the nanoparticle magnetization in the presence of a static as well as a harmonic magnetic field. Med Phys 2008;35:1988–1994.

6 Biederer S, Knopp T, Sattel TF, Lüdtke-Buzug K, Gleich B, Weizenecker J, Borgert J, Buzug TM: Magnetization response spectroscopy of superparamagnetic nanoparticles for magnetic particle imaging. J Phys D Appl Phys 2009;42:1–7.

7 Lüdtke-Buzug K, Biederer S, Sattel T, Knopp T, Buzug TM: Preparation and Characterization of Dextran-Covered Fe_3O_4 Nanoparticles for Magnetic Particle Imaging. Proc 4th Eur Congress Med Biomed Eng. IFMBE Ser. Berlin, Springer, 2008, vol 22, pp 2343–2346.

8 Moreland J, Eckstein J, Lin Y, Liou SH, Ruggiero S: Magnetic particle imaging with a cantilever torque magnetometer. 2007. http://meetings. aps.org/link/BAPS.2007.MAR.U38.10.

9 Goodwill W, Scott C, Stang P, Conolly M: Narrowband magnetic particle imaging. IEEE Trans Med Imaging 2009; 28:1231–1237.

10 Weizenecker J, Gleich B, Borgert J: Magnetic particle imaging using a field-free line. J Phys D Appl Phys 2008; 41:105009

11 Gleich B, Weizenecker J, Borgert J: Experimental results on fast 2D-encoded magnetic particle imaging. Phys Med Biol 2008;53:N81–N84.

12 Sattel TF, Knopp T, Biederer S, Gleich B, Weizenecker J, Borgert J, Buzug TM: Single-sided device for magnetic particle imaging. J Phys D Appl Phys 2009; 42:2–5.

13 Weizenecker J, Gleich B, Rahmer J, Dahnke H, Borgert J: Three-dimensional real-time in vivo magnetic particle imaging. Phys Med Biol 2009; 54:L1–L10.

14 Knopp T, Biederer S, Sattel TF, Weizenecker J, Gleich B, Borgert J, Buzug TM: Trajectory analysis for magnetic particle imaging. Phys Med Biol 2009;56: 385–397.

15 Bulte JW, Gleich B, Weizenecker J, Bernard S, Walczak P, Markov DE, Aerts HC, Borgert J, Boeve H: Developing Cellular MPI: Initial Experience. Berkeley, ISMRM, 2008.

16 Bulte JW, Walczak P, Bernard S, Gleich B, Weizenecker J, Borgert J, Aerts H, Boeve H: Developing Cellular MPI: Initial Experience; in Buzug TM et al (eds): Magnetic Nanoparticles: Particle Science, Imaging Technology, and Clinical Applications. Singapore, World Scientific Publishers, 2010, pp 201–204.

17 Lüdtke-Buzug K, Rapoport D, Schneider D: Characterization of iron-oxide loaded adult stem cells for magnetic particle imaging in targeted cancer therapy; in 8th International Conference on the Scientific and Clinical Applications of Magnetic Carriers. Melville, AIP Conference Proceedings, 2010, vol 1311, pp 244–248.

18 Weizenecker J, Gleich B, Rahmer J, Dahnke H, Borgert J: Three-dimensional real-time in vivo magnetic particle imaging. Phys Med Biol 2009; 54:L1–L10.

19 Ruhland B, Baumann K, Knopp T, Sattel TF, Biederer S, Lüdtke-Buzug K, Buzug TM, Diedrich K, Finas D: Magnetic particle imaging with superparamagnetic nanoparticles for sentinel lymph node detection in breast cancer. Geburtshilfe Frauenheilk 2009;69:758.

20 Knopp T, Erbe M, Biederer S, Sattel TF, Buzug TM: Efficient generation of a magnetic field-free line. Med Phys 2010;37:3538–3540.

21 Knopp T, Sattel TF, Biederer S, Buzug TM: Field-free line formation in a magnetic field. J Phys A Math Theor 2010; 43:1–9.

22 Kuehn T, Bembenek A, Decker T, Munz DL, Sautter-Bihl ML, Untch M, Wallwiener D: A concept for the clinical implementation of sentinel lymph node biopsy in patients with breast carcinoma with special regard to quality assurance. Cancer 2005;103:451–461.

23 Jemal A, Siegel R, Ward E, Hao Y, Xu J, Murray T, Thun MJ: Cancer statistics, 2008. CA Cancer J Clin 2008;58:71–96.

24 Kühn T, Bembenek A, Büchels H, Decker T, Dunst J, Müllerleile U, Munz DL, Ostertag H, Sautter-Bihl ML, Schirrmeister H, Tulusan AH, Untch M, Winzer KJ, Wittekind C: Sentinel-Node-Biopsie beim Mammakarzinom: interdisziplinär abgestimmter Konsensus der Deutschen Gesellschaft für Senologie für eine qualitätsgesicherte Anwendung in der klinischen Routine. Geburtshilfe Frauenheilk 2003;63:835–840.

25 Robe A, Pic E, Lassalle HP, Bezdetnaya L, Guillemin F, Marchal F: Quantum dots in axillary lymph node mapping: biodistribution study in healthy mice. BMC Cancer 2008;8:111.

26 Ling LJ, Wang S, Liu XA, Shen EC, Ding Q, Lu C, Xu J, Cao QH, Zhu HQ, Wang F: A novel mouse model of human breast cancer stem-like cells with high $CD44^{+}CD24^{-}$/lower phenotype metastasis to human bone. Chin Med J (Engl) 2008;121:1980–1986.

27 Tsunoda N, Kokuryo T, Oda K, Senga T, Yokoyama Y, Nagino M, Nimura Y, Hamaguchi M: Nek2 as a novel molecular target for the treatment of breast carcinoma. Cancer Sci 2009;100:111–116.

Thorsten M. Buzug
Institute of Medical Engineering
University of Lübeck, Ratzeburger Allee 160
DE–23562 Lübeck (Germany)
Tel. +49 451 500 5400, E-Mail buzug@imt.uni-luebeck.de

Alexiou C (ed): Nanomedicine – Basic and Clinical Applications in Diagnostics and Therapy.
Else Kröner-Fresenius Symp. Basel, Karger, 2011, vol 2, pp 96–105

Nanoparticles in Clinical Trials

Sandeep Hedgire · Shaunagh McDermott · Mukesh Harisinghani

Department of Abdominal Imaging, Massachusetts General Hospital, Boston, Mass., USA

Abstract

Nanoparticles (NPs) with applications in tumor imaging and therapy have captivated the attention of researchers, radiologists, and clinicians in the last decade. The convergent discipline of 'thernostics' offers a plethora of applications of NPs in oncology which are much superior to the conventional modalities. Varieties of NPs have been developed either to enhance tissue contrast or to identify specific biological changes with promising results in clinical trials. Superparamagnetic iron oxide NPs are at the frontier of NP-driven imaging. This review will focus on NPs currently being used in human clinical trials and their applications in clinical cancer imaging.

Nanotechnology is the discipline dealing with development and research of objects in a scale range of 1–100 nm, and their application in medicine is known as nanomedicine. Nanomedicine is increasingly being applied in the diagnostic and therapeutic (commonly referred as thernostics) oncology realm. Imaging with biocompatible nanoparticles (NPs) has marked a new era of molecular imaging by developing cost-effective NPs with acceptable targeting efficiency, favorable pharmacokinetics, and a good safety profile [1]. Several imaging modalities differing in sensitivity and resolution can currently be applied to image these NPs. Optical imaging (fluorescence and bioluminescence detection), radionucleotide-based imaging [positron emission tomography (PET) and single-photon emission computed tomography (SPECT)] are sensitive but lack superior spatial resolution. Absorption of signals by background tissue and the inability to image the tissue at depth are limiting factors of optical imaging. By contrast, magnetic resonance imaging (MRI) and computed tomography (CT) are much less sensitive but demonstrate a resolution of less than 1 mm. MRI can image deep tissues and has the added advantage of not exposing patients to ionizing radiation. A much foretold argument of slow data acquisition is no longer the case with faster imaging sequences. For each of these imaging modalities, novel NPs have been developed either to enhance tissue contrast or to identify specific biological changes (table 1) [2].

NPs can be classified depending on their mechanism of action into the following categories:

- NPs amplifying signal changes (e.g. superparamagnetic particles used in MR imaging).

Table 1. Summary of commercially available NPs and their applications

NPs	Applications
Liposomes	Drug delivery
Dendrimers	Therapeutics
Carbon nanotubes	In vitro diagnostics, imaging
Quantum dots	In vitro diagnostics, imaging
Superparamagnetic NPs	Imaging, therapeutics
Gold NPs	In vitro diagnostics, imaging

- NPs exhibiting properties inducing phagocytosis and selective uptake by macrophages (e.g. liposomes in nuclear medicine and iron oxides in MRI).
- Visible NPs due to unique chemical properties (e.g. gold particles in electron microscopy and quantum dots in fluorescent imaging).

Ultrasmall Superparamagnetic Iron Nanoparticles

The two main formulations of ultrasmall superparamagnetic iron oxide (USPIO) NPs that have undergone clinical trials include ferumoxtran-10 and ferumoxytol (AMAG Pharmaceuticals, Inc.). Ferumoxtran-10 (Combidex™) is comprised of a monocrystalline, inverse spinel, SPIO core (2–3 or 4.3–6.0 nm) coated with polymers (low molecular weight dextran) to prolong their circulation time and to prevent aggregation [3, 4]. These are supplied as lyophilized powder which is then reconstituted with normal saline and administered over approximately 15–30 min in a weight-based intravenous dose of 2.6 mg/kg. This dose has been found to optimize signal decrease in normal lymph nodes.

Ferumoxytol is a SPIO NP designed to minimize immunologic sensitivity. The crystal size of the iron oxide core is 6.8 ± 0.4 nm in diameter as measured by X-ray diffraction [5]. The NP is formulated to be isotonic at a concentration of 30 mg Fe/ml. The physical, pharmacokinetic, and biologic properties of ferumoxytol in therapeutic trials have been reported [5, 6]. Ferumoxytol has been administered at varying doses ranging from 4 to 6 mg of Fe/kg for imaging. A 15-ml normal saline intravenous flush is given after NP administration [7].

Mechanism of Action

After intravenous injection, the lymphotrophic NPs gradually extravasate from the blood vessels and enter the interstitial space from where they are transported into the lymph nodes (fig. 1). Localization of these particles within the nodes occurs through two mechanisms. The major pathway involves direct transcapillary passage of the particles from venules into the medullary sinuses of lymph nodes. In addition, nonselective endothelial transcytosis occurs across permeable capillaries into the interstitium, after which the particles drain into the lymph nodes via the lymphatic system [8, 9]. Since this process takes 24–36 h, imaging needs to be performed 24–36 h after contrast administration (fig. 2), allowing differentiation between benign and malignant nodes irrespective of their size (fig. 3, 4) [10].

The ability of these NPs to selectively target the reticuloendothelial system led to multiple clinical trials to evaluate these NPs in cancer imaging as tumor staging, choice of therapy, and patient prognosis are highly dependent on the spread of the primary malignancy to regional and distant nodes. Regardless of the location or size of the primary tumor, finding a single metastatic node can dramatically reduce the prognosis [11, 12]. The inability of CT and MRI to determine this spread depending on the size criterion alone was concluded by multiple observers [13]. The size criterion for lymph nodes frequently overlooks metastases, particularly if only microscopic or partial metastases of the lymph node are involved, reducing the sensitivity of these modalities. Specificity may be reduced by benign inflammatory or infectious lymph node enlargement, which leads to the

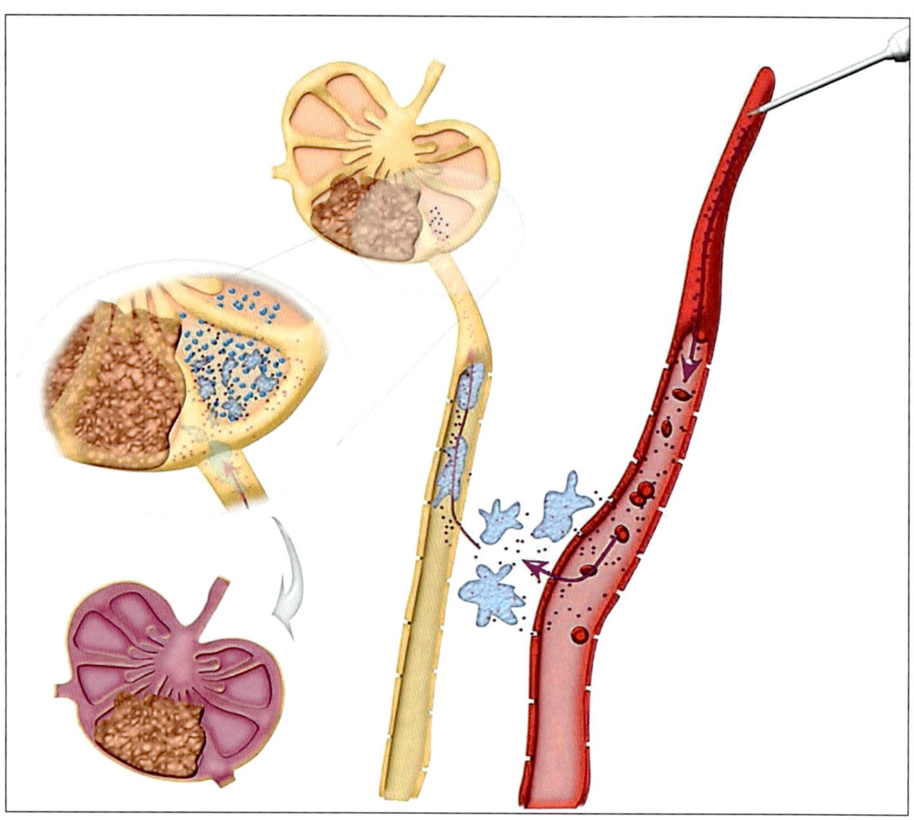

Fig. 1. The systemically injected long circulating particles gain access to the interstitium and are drained ubiquitously through lymphatic vessels. Lymph flow disturbances or disturbances of the nodal architecture by metastases lead to abnormal accumulation patterns detectable by MRI.

incorrect characterization of a benign lymph node as malignant [14]. PET and PET-CT have gained an expanding role in nodal staging, detecting the presence of recurrent malignancies and posttreatment imaging [15–17]. 18F-FDG is not a very selective tracer for tumor imaging since cell types other than tumor cells actively use glucose. Macrophages in inflammatory and infectious lesions as well as lymph nodes involved with granulomatous disease or silicosis can demonstrate an increased 18F-FDG uptake [18, 19]. False-negative outcomes may also arise in lymph nodes that are less than 1 cm in diameter, contain micrometastases [20, 21], or are involved with well-differentiated tumors and are located in close proximity to the primary tumor. Lymph node MRI (LNMRI) has been described as a highly accurate technique in the characterization of lymph nodes in various primary malignancies.

Imaging Technique

It is standard practice to perform LNMRI with 2 MRI scans performed 24–48 h apart. It has been suggested that there is no significant difference in diagnostic accuracy between paired MRI and postcontrast MRI alone for an experienced reviewer [22]. The sequences include the T1-weighted (for anatomy and spatial resolution), T2, and T2* (weighted to characterize the contrast enhancement) sequences. To select and assign a

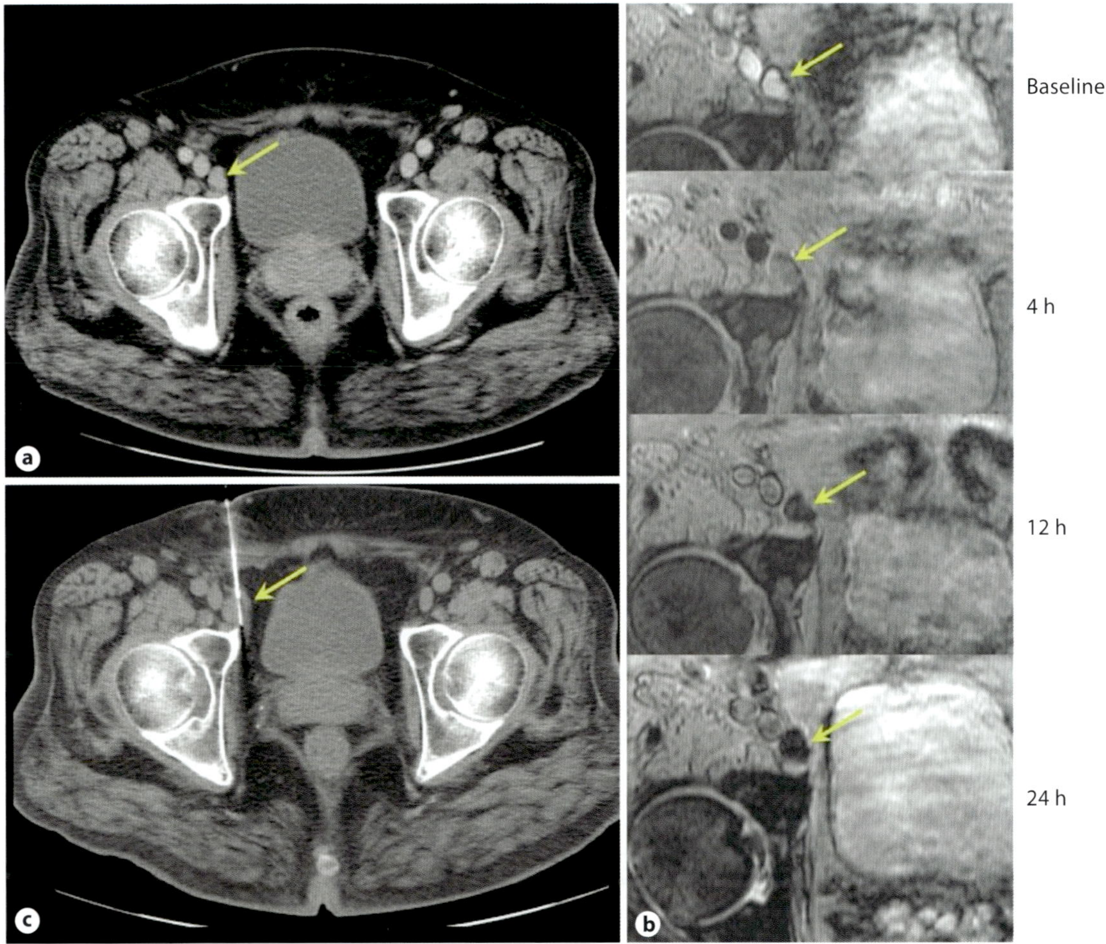

Fig. 2. Benign right external iliac node. **a** The axial contrast-enhanced CT image shows an enlarged right external iliac node. **b** Sequential images show progressive loss of signal following ferumoxytol administration. **c** CT-guided biopsy showing the biopsy needle within the enlarged lymph node. Reprinted with permission from Neoplasia Press, Inc.

particular pattern of enhancement is subjective and is a source for potential error. Lahaye et al. [23] described a method of percentage involvement whereby a node is considered metastatic if the area of high signal intensity measures more than 30% with a sensitivity of 93% and a specificity of 96%. In benign causes of bright signals such as focal nodal fibrosis, granulomatous disease, or fatty hilum, this area is usually less than 30%. Fatty hila are also characterized by 3D T1-weighted images where they are seen as bright areas [23]. Changing the echo time (TE) value is noted to have an effect on the sensitivity and specificity, with a shorter TE of 12.2 ms having a high sensitivity but lower specificity compared to a TE of 21 ms (high specificity and low sensitivity) as shown by Saksena et al. [24]. More recently another inversion recovery ON-water suppression IRON-MRI sequence has been developed to impart a positive contrast to the superparamagnetic par-

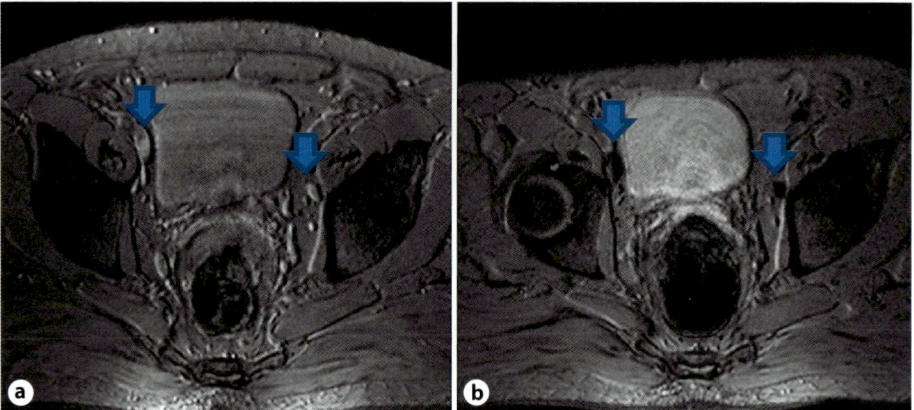

Fig. 3. Benign pelvis nodes in a patient with prostate cancer. Pre- (**a**) and post- (**b**) (blue arrows) ferumoxtran-10-enhanced images demonstrate a homogenous signal intensity drop in benign pelvic nodes.

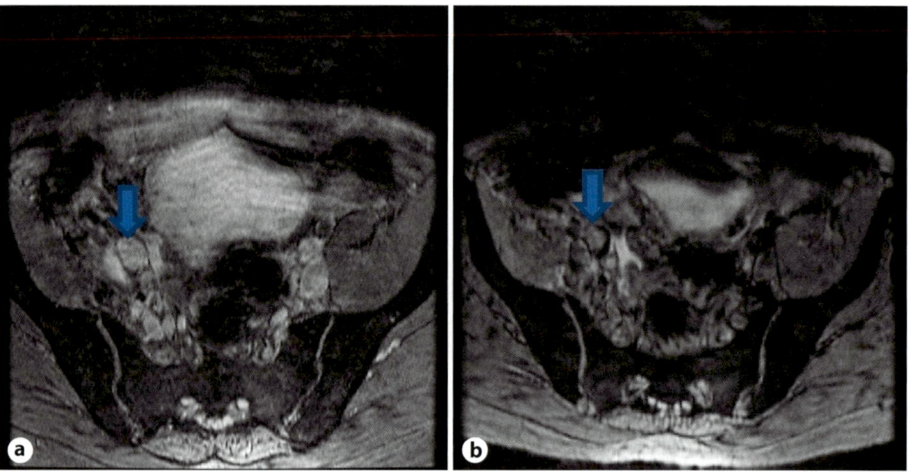

Fig. 4. Malignant lymph node in a patient with bladder cancer. Pre- (**a**) and post- (**b**) (blue arrows) ferumoxtran-10-enhanced scans show a minimal peripheral drop in signal intensity, with the bulk of the node retaining a high signal intensity indicative of malignant infiltration.

ticles [25]. Heesakkers et al. [26] reported improved image quality at 3T MRI compared to 1.5T with ferumoxtran [26].

Clinical Trials in Oncology

Since the earliest studies by Anzai et al. [27] and Bellin et al. [28], there have been various clinical trials reporting the efficacy of LNMRI. A recent meta-analysis showed that ferumoxtran-10-enhanced LNMRI had a higher diagnostic accuracy than unenhanced LNMRI, with an overall sensitivity of 88% and an overall specificity of 96% with a diagnostic odds ratio of 123.1 [29].

Body region analysis demonstrated a higher sensitivity and specificity of LNMRI in the abdomen and pelvis, which is attributable to more

Table 2. Summary of the studies done by body region and lymph node size

Body region	Studies, n	Patients, n	Lymph nodes, n	Sensitivity (95% CI)	Specificity (95% CI)
Head and neck	4	128	1,321	89% (83–93)	98% (97–99)
Chest	3	31	45	70% (55–83)	88% (76–95)
Abdomen and pelvis	9	311	1,310	92% (89–95)	96% (95–97)
Prostate only	2	130	502	91% (86–95)	97% (94–98)
Analysis by nodal size					
5–10 mm	2	98	190	96.3% (81.3–100)	98.8% (95.6–99.9)
<5 mm	1	80	185	41.2% (18.4–67.1)	98.1% (93.0–98.8)

abundant lymph flow in these regions compared to the thorax, head, and neck [30] and image degradation due to motion artifacts in the mediastinum [4]. Table 2 summarizes the studies done by region and lymph node size.

In patients with head and neck cancers, Mack et al. [31] reported a sensitivity of 96%. Use of LNMRI in head and neck cancers can help reduce the rates of radical neck lymph node dissection and thereby its complications. Similarly, postaxillary dissection lymphedema can be avoided in breast cancer patients if the metastatic nodes are identified before surgery. Michel et al. [32] reported a sensitivity of 78% and a specificity of 96% in earlier studies (2002). However, more recently LNMRI for axillary lymph node imaging in breast cancer patients showed a sensitivity of 100% and a specificity of 80% [33]. The accuracy of LNMRI in a trial for non-small cell lung cancer was initially lower (63%) than the conventional methods of size criteria (65%) [34]. Nguyen et al. [35] reported a sensitivity of 92% and a specificity of 80% in their study in 18 patients with lung cancer. Pannu et al. [36] conducted a study in 12 patients with lung cancer and demonstrated a sensitivity of 100%; however, the specificity was low at 37.5%. This low specificity could be due to a high prevalence of granulomatous infection in the mediastinum, leading to a decreased contrast uptake in inflammatory nodes. Its a notable fact that patients with lung disease have poor compliance for the breath-hold sequences, thereby raising the possibility of more artifacts, especially in GRE-weighted images which are most useful for nodal characterization.

In a study by Tatsumi et al. [37], the sensitivity was 100% and the specificity was 92.6% in patients with gastric cancer. Lymph nodes in para-aortic and retroperitoneal locations were more easily identified compared to those in the perigastric region [37]. Similar findings were noted in esophageal and rectal cancer patients.

Guimaraes et al. [38] conducted LNMRI in 9 patients with renal cell carcinoma and demonstrated that LNMRI helped not only in distinguishing primary tumors (oncocytoma vs. RCC) but also in detecting nodal metastases with a sensitivity of 100% and a specificity of 95.7% [38]. In patients with testicular cancer, LNMRI helps in differentiating stage I and stage II, which can help in selecting patients for more aggressive therapy, sparing others from unnecessary treatment related mortality. In a study by Harisinghani et al. [39] in 18 patients with testicular cancer, the sensitivity and specificity were 100% in nodes <10 mm in size. Tabatabaei et al. [40] reported similar results in 7 patients with penile cancer. It is notable that they reported a negative predictive value of 100%, meaning that it was unlikely to miss the patient who required groin dissection. For characterizing lymph nodes in prostatic cancer, the highest sensitivity reported was 100% with a specificity of 96% by Harisinghani et al. [41], with metastatic nodes outside the classical field of lymph node dissection in 11% of cases. Identifying these abnormal nodes can help in precise IMRT.

Deserno et al. [42] studied the utility of LNMRI in 58 patients with bladder cancer and found a sensitivity and specificity of 96 and 95%, respectively.

MRI of the liver using SPIOs can help to distinguish hepatocellular carcinoma and metastatic liver lesions from other liver tumors due to a lack of Kupffer cells. Ward et al. [43] studied 51 patients with colorectal metastasis to the liver and demonstrated a higher sensitivity of SPIO MRI compared to dual phase CT (79.8 vs. 75.3%). The mean accuracy (Az values) and sensitivity of SPIO-enhanced imaging (0.984; 97.2%) were found to be comparable to those of Gd-BOPTA-enhanced delayed phase imaging (0.982; 95.5%) in a study by Kim et al. [44]. Application of NPs in brain tumor imaging has been relatively underexplored. Neuwelt et al. [45] observed in their study that ferumoxtran may show areas of enhancement, even with a 0.15 T intraoperative MR, in malignant brain tumors that do not enhance with gadolinium; this is a promising fact, especially in postoperative imaging . Neuwelt et al. [45] are conducting a clinical trial to study cerebral blood volume changes in brain tumor vascularity after antiangiogenic therapy versus steroid therapy using ferumoxytol.

Other Clinical Trials

The role of USPIO MRI in vascular imaging evolved after a hypothesis by Schmitz et al. [46] that USPIOs may be used to image infiltration of macrophages in the vessel wall. Kooi et al. [47] and Trivedi et al. [48] conducted studies on patients with carotid stenosis using ferumoxtran and observed that the uptake was higher in ruptured plaques (75%) compared to stable plaques (7%), thereby speculating regarding the role of ferumoxtran as a noninvasive marker for the identification of vulnerable plaques. Tang et al. [51] showed bilateral signal loss in symptomatic patients with carotid stenosis, highlighting the systemic nature of vulnerable plaque [49, 50].

Safety Profile of Nanoparticles

Ferumoxtran is well tolerated. Patients who are hypersensitive to iron or any component in the formulation should not be administered these agents. The most common (in 4% patients) adverse effect reported in a meta-analysis was mild lumbar backache [29]. For fear of allergic reactions, these patients are typically monitored for 60 min postinfusion. The safety profile of ferumoxytol has been very good in clinical trials which allowed administration of higher doses compared to ferumoxtran.

Other Nanoparticles in Clinical Trials

Liposomes

Liposomes are phospholipid vesicles that are predominantly used as pharmaceutical agents. These can be labeled by radioisotopes to image a plethora of pathological processes which can be imaged by scintigraphy, CT, or PET.

Mechanism of Action and Imaging Techniques

After intravenous administration, these liposomes enter cells of the reticuloendothelial system as described for the superparamagnetic NPs. A half-life of 6–12 h allows the liposomes to accumulate in target cells [52]. Background clearance and achieving a higher lesion-to-blood ratio are the key factors in localizing the lesion accurately. They are cleared by urinary excretion. In a study by Harrington et al. [52] a monitored dose of PEGylated radiolabeled liposomes [65–107 MBq (1.76–2.89 mCi) diluted in 500 ml of 5% dextrose] was infused over 30–45 min and subsequently imaged using a whole-body double headed gamma camera at 0.5, 4, 24, 48, 72, 96, and 240 h.

Clinical Trials

In studies conducted by Bao et al. [53] and Harrington et al. [52], using ^{111}In-DTPA-labeled PEGylated liposomes, the highest level of uptake was seen in head and neck malignancies followed by lung cancers. The uptake in breast cancers was low [52, 53]. The utility of ^{111}In-labeled liposomes for localizing skin cancers, AIDS-related Kaposi's sarcoma, soft tissue sarcomas, glioblastomas, and metastatic brain tumors was demonstrated in multiple studies by Koukourakis et al. [54] and Harrington et al. [52]. Spinazzi et al. [55] and Leander et al. [56] studied the feasibility of liposomes as CT contrast agents and demonstrated enhancement patterns of the liver and spleen following administration of iodixanol and iomeprol liposome suspensions, respectively. Studies on characterization of focal liver lesions using iomeprol liposome are being conducted. However, the liposomal research in oncology is more focused towards anticancer pharmacology.

Safety Profile
The adverse effects of liposomes are considered minor compared to the anticancer drug to which they are tagged. Dams et al. [57] reported minor adverse effects like tightness of the chest and flushing, with reversal upon temporary suspension of the infusion. Spinazzi et al. [55] reported adverse effects like headache, asthenia, dizziness, vasodilation, back pain, tremor, and sweating with iomeprol liposome in 50% of patients who received the lowest dose of 0.5 ml/kg [55].

Gold Nanoparticles

Gold NPs are synthesized from colloidal gold and have already been proved useful for various applications in cellular biology. Gold NPs are 2–250 nm in size and can be of various shapes. Clinical trials with gold NPs are limited and comprise of their utility in imaging and as nanosensors. Kim et al. [44] proposed PEG-coated gold NPs as CT contrast agents based on observations they made in their study on animal models. Peng et al. [58] used a functionalized gold NP sensor array to differentiate between volatile organic compounds in breath samples of healthy volunteers and patients suffering from lung, breast, colorectal, and prostate cancers using gas chromatography-mass spectrometry. They identified different patterns of volatile organic compounds in different types of cancers and speculated regarding the development of a noninvasive, cost-effective diagnostic tool through single-breath analysis [58].

Future Trends

Quantum dots (semiconductor nanocrystals with a core of 2–19 nm) have been extensively studied in cellular biology. Due to their optical and electronic properties their roles in early tumor detection, lymph node mapping, blood pool imaging, and angiogenesis in animal models have been studied in detail and the results are promising; however, this success comes at the cost of toxicity due to heavy metal components such as cadmium [59]. This aspect remains to be studied in depth and is considered a major obstacle in performing human studies. Recently, Pons et al. [60] reported the development of cadmium-free quantum dots [60]. Zhou et al. [61] are conducting a clinical trial with a new type of CT scanner which uses carbon nanotubes as the X-ray source. Lanza et al. [61] at Washington University School of Medicine are conducting a clinical trial with an NP MRI contrast agent that binds to the $\alpha v\beta 3$-integrin.

The next step in NP-driven imaging in clinical trials appears to be the development of a coupling of target-specific agents with diagnostic probes, e.g. magneto-optical probes, nanosensors, and multifunctional NPs with a better safety profile.

Disclosure Statement

The authors have nothing to disclose.

References

1 Harisinghani MG: Nanoparticle enhanced imaging. Cancer Biomark 2009; 5:59.

2 Minchin RF, Martin DJ: Nanoparticles for molecular imaging – an overview. Endocrinology 2010;151:474–481.

3 Jung CW, Jacobs P: Physical and chemical properties of superparamagnetic iron oxide MR contrast agents: ferumoxides, ferumoxtran, ferumoxsil. Magn Reson Imaging 1995;13:661–674.

4 Saksena MA, Saokar A, Harisinghani MG: Lymphotropic nanoparticle enhanced MR imaging (LNMRI) technique for lymph node imaging. Eur J Radiol 2006;58:367–374.

5 Li W, Tutton S, Vu AT, et al: First-pass contrast-enhanced magnetic resonance angiography in humans using ferumoxytol, a novel ultrasmall superparamagnetic iron oxide (USPIO)-based blood pool agent. J Magn Reson Imaging 2005;21:46–52.

6 Landry R, Jacobs PM, Davis R, Shenouda M, Bolton WK: Pharmacokinetic study of ferumoxytol: a new iron replacement therapy in normal subjects and hemodialysis patients. Am J Nephrol 2005;25:400–410.

7 Reimer P, Balzer T: Ferucarbotran (Resovist): a new clinically approved RES-specific contrast agent for contrast-enhanced MRI of the liver: properties, clinical development, and applications. Eur Radiol 2003;13:1266–1276.

8 Weissleder R, Elizondo G, Wittenberg J, Rabito CA, Bengele HH, Josephson L: Ultrasmall superparamagnetic iron oxide: characterization of a new class of contrast agents for MR imaging. Radiology 1990;175:489–493.

9 Weissleder R, Elizondo G, Wittenberg J, Lee AS, Josephson L, Brady TJ: Ultrasmall superparamagnetic iron oxide: an intravenous contrast agent for assessing lymph nodes with MR imaging. Radiology 1990;175:494–498.

10 Harisinghani MG, Dixon WT, Saksena MA, et al: MR lymphangiography: imaging strategies to optimize the imaging of lymph nodes with ferumoxtran-10. Radiographics 2004;24: 867–878.

11 Anzai Y, Brunberg JA, Lufkin RB: Imaging of nodal metastases in the head and neck. J Magn Reson Imaging 1997; 7:774–783.

12 van den Brekel MW: Lymph node metastases: CT and MRI. Eur J Radiol 2000;33:230–238.

13 Tiguert R, Gheiler EL, Tefilli MV, et al: Lymph node size does not correlate with the presence of prostate cancer metastasis. Urology 1999;53:367–371.

14 Islam T, Harisinghani MG: Overview of nanoparticle use in cancer imaging. Cancer Biomark 2009;5:61–67.

15 Anderson GS, Brinkmann F, Soulen MC, Alavi A, Zhuang H: FDG positron emission tomography in the surveillance of hepatic tumors treated with radiofrequency ablation. Clin Nucl Med 2003;28:192–197.

16 Guay C, Lepine M, Verreault J, Benard F: Prognostic value of PET using 18F-FDG in Hodgkin's disease for posttreatment evaluation. J Nucl Med 2003;44: 1225–1231.

17 Karapetis CS, Strickland AH, Yip D, Steer C, Harper PG: Use of fluorodeoxyglucose positron emission tomography scans in patients with advanced germ cell tumour following chemotherapy: single-centre experience with long-term follow up. Intern Med J 2003;33:427–435.

18 Kapucu LO, Meltzer CC, Townsend DW, Keenan RJ, Luketich JD: Fluorine-18-fluorodeoxyglucose uptake in pneumonia. J Nucl Med 1998;39:1267–1269.

19 Rennen HJ, Corstens FH, Oyen WJ, Boerman OC: New concepts in infection/inflammation imaging. Q J Nucl Med 2001;45:167–173.

20 Kostakoglu L, Agress H Jr, Goldsmith SJ: Clinical role of FDG PET in evaluation of cancer patients. Radiographics 2003;23:315–340, quiz 533.

21 Ruers TJ, Langenhoff BS, Neeleman N, et al: Value of positron emission tomography with [F-18]fluorodeoxyglucose in patients with colorectal liver metastases: a prospective study. J Clin Oncol 2002;20:388–395.

22 Harisinghani MG, Saksena MA, Hahn PF, et al: Ferumoxtran-10-enhanced MR lymphangiography: does contrast-enhanced imaging alone suffice for accurate lymph node characterization? AJR Am J Roentgenol 2006;186:144–148.

23 Lahaye MJ, Engelen SM, Kessels AG, et al: USPIO-enhanced MR imaging for nodal staging in patients with primary rectal cancer: predictive criteria. Radiology 2008;246:804–811.

24 Saksena M, Harisinghani M, Hahn P, et al: Comparison of lymphotropic nanoparticle-enhanced MRI sequences in patients with various primary cancers. AJR Am J Roentgenol 2006; 187:W582–W588.

25 Stuber M, Gilson WD, Schar M, et al: Positive contrast visualization of iron oxide-labeled stem cells using inversion-recovery with ON-resonant water suppression (IRON). Magn Reson Med 2007;58:1072–1077.

26 Heesakkers RA, Futterer JJ, Hovels AM, et al: Prostate cancer evaluated with ferumoxtran-10-enhanced T2*-weighted MR imaging at 1.5 and 3.0 T: early experience. Radiology 2006;239:481–487.

27 Anzai Y, Blackwell KE, Hirschowitz SL, et al: Initial clinical experience with dextran-coated superparamagnetic iron oxide for detection of lymph node metastases in patients with head and neck cancer. Radiology 1994;192:709–715.

28 Bellin MF, Roy C, Kinkel K, et al: Lymph node metastases: safety and effectiveness of MR imaging with ultrasmall superparamagnetic iron oxide par ticles – initial clinical experience. Radiology 1998;207:799–808.

29 Will O, Purkayastha S, Chan C, et al: Diagnostic precision of nanoparticle-enhanced MRI for lymph-node metastases: a meta-analysis. Lancet Oncol 2006;7:52–60.

30 Bulte JW, Modo MM, Barentsz JO, Tekkis PP: Use of USPIOs for clinical lymph node imaging; in Ferrari M (ed): Nanoparticles in Biomedical Imaging. New York, Springer, 2008, pp 25–40.

31 Mack MG, Balzer JO, Straub R, Eichler K, Vogl TJ: Superparamagnetic iron oxide-enhanced MR imaging of head and neck lymph nodes. Radiology 2002;222:239–244.

32 Michel SC, Keller TM, Frohlich JM, et al: Preoperative breast cancer staging: MR imaging of the axilla with ultrasmall superparamagnetic iron oxide enhancement. Radiology 2002;225: 527–536.

33 Stadnik TW, Everaert H, Makkat S, Sacre R, Lamote J, Bourgain C: Breast imaging. Preoperative breast cancer staging: comparison of USPIO-enhanced MR imaging and 18F-fluorodeoxyglucose (FDC) positron emission tomography (PET) imaging for axillary lymph node staging – initial findings. Eur Radiol 2006;16:2153–2160.

34 Anzai Y, Piccoli CW, Outwater EK, et al: Evaluation of neck and body metastases to nodes with ferumoxtran 10-enhanced MR imaging: phase III safety and efficacy study. Radiology 2003;228: 777–788.

35 Nguyen BC, Stanford W, Thompson BH, et al: Multicenter clinical trial of ultrasmall superparamagnetic iron oxide in the evaluation of mediastinal lymph nodes in patients with primary lung carcinoma. J Magn Reson Imaging 1999;10:468–473.
36 Pannu HK, Wang KP, Borman TL, Bluemke DA: MR imaging of mediastinal lymph nodes: evaluation using a superparamagnetic contrast agent. J Magn Reson Imaging 2000;12:899–904.
37 Tatsumi Y, Tanigawa N, Nishimura H, et al: Preoperative diagnosis of lymph node metastases in gastric cancer by magnetic resonance imaging with ferumoxtran-10. Gastric Cancer 2006;9:120–128.
38 Guimaraes AR, Tabatabei S, Dahl D, McDougal WS, Weissleder R, Harisinghani MG: Pilot study evaluating use of lymphotrophic nanoparticle-enhanced magnetic resonance imaging for assessing lymph nodes in renal cell cancer. Urology 2008;71:708–712.
39 Harisinghani MG, Saksena M, Ross RW, et al: A pilot study of lymphotrophic nanoparticle-enhanced magnetic resonance imaging technique in early stage testicular cancer: a new method for noninvasive lymph node evaluation. Urology 2005;66:1066–1071.
40 Tabatabaei S, Harisinghani M, McDougal WS: Regional lymph node staging using lymphotropic nanoparticle enhanced magnetic resonance imaging with ferumoxtran-10 in patients with penile cancer. J Urol 2005;174:923–927, discussion 927.
41 Harisinghani MG, Barentsz J, Hahn PF, et al: Noninvasive detection of clinically occult lymph-node metastases in prostate cancer. N Engl J Med 2003;348:2491–2499.
42 Deserno WM, Harisinghani MG, Taupitz M, et al: Urinary bladder cancer: preoperative nodal staging with ferumoxtran-10-enhanced MR imaging. Radiology 2004;233:449–456.
43 Ward J, Naik KS, Guthrie JA, Wilson D, Robinson PJ: Hepatic lesion detection: comparison of MR imaging after the administration of superparamagnetic iron oxide with dual-phase CT by using alternative-free response receiver operating characteristic analysis. Radiology 1999;210:459–466.
44 Kim YK, Lee JM, Kim CS, Chung GH, Kim CY, Kim IH: Detection of liver metastases: gadobenate dimeglumine-enhanced three-dimensional dynamic phases and one-hour delayed phase MR imaging versus superparamagnetic iron oxide-enhanced MR imaging. Eur Radiol 2005;15:220–228.
45 Neuwelt EA, Varallyay P, Bago AG, Muldoon LL, Nesbit G, Nixon R: Imaging of iron oxide nanoparticles by MR and light microscopy in patients with malignant brain tumours. Neuropathol Appl Neurobiol 2004;30:456–471.
46 Schmitz SA, Taupitz M, Wagner S, Wolf KJ, Beyersdorff D, Hamm B: Magnetic resonance imaging of atherosclerotic plaques using superparamagnetic iron oxide particles. J Magn Reson Imaging 2001;14:355–361.
47 Kooi ME, Cappendijk VC, Cleutjens KB, et al: Accumulation of ultrasmall superparamagnetic particles of iron oxide in human atherosclerotic plaques can be detected by in vivo magnetic resonance imaging. Circulation 2003;10:2453–2458.
48 Trivedi RA, U-King-Im JM, Graves MJ, et al: In vivo detection of macrophages in human carotid atheroma: temporal dependence of ultrasmall superparamagnetic particles of iron oxide-enhanced MRI. Stroke 2004;35:1631–1635.
49 Buffon A, Biasucci LM, Liuzzo G, D'Onofrio G, Crea F, Maseri A: Widespread coronary inflammation in unstable angina. N Engl J Med 2002;347:5–12.
50 Mauriello A, Sangiorgi G, Fratoni S, et al: Diffuse and active inflammation occurs in both vulnerable and stable plaques of the entire coronary tree: a histopathologic study of patients dying of acute myocardial infarction. J Am Coll Cardiol 2005;45:1585–1593.
51 Tang T, Howarth SP, Miller SR, et al: Assessment of inflammatory burden contralateral to the symptomatic carotid stenosis using high-resolution ultrasmall, superparamagnetic iron oxide-enhanced MRI. Stroke 2006;37:2266–2270.
52 Harrington KJ, Mohammadtaghi S, Uster PS, et al: Effective targeting of solid tumors in patients with locally advanced cancers by radiolabeled pegylated liposomes. Clin Cancer Res 2001;7:243–254.
53 Bao A, Goins B, Klipper R, Negrete G, Phillips WT: Direct 99mTc labeling of pegylated liposomal doxorubicin (Doxil) for pharmacokinetic and non-invasive imaging studies. J Pharmacol Exp Ther 2004;308:419–425.
54 Koukourakis MI, Koukouraki S, Giatromanolaki A, et al: High intratumoral accumulation of stealth liposomal doxorubicin in sarcomas – rationale for combination with radiotherapy. Acta Oncol 2000;39:207–211.
55 Spinazzi A, Ceriati S, Pianezzola P, et al: Safety and pharmacokinetics of a new liposomal liver-specific contrast agent for CT: results of clinical testing in nonpatient volunteers. Invest Radiol 2000;35:1–7.
56 Leander P, Hoglund P, Borseth A, Kloster Y, Berg A: A new liposomal liver-specific contrast agent for CT: first human phase-I clinical trial assessing efficacy and safety. Eur Radiol 2001;11:698–704.
57 Dams ET, Oyen WJ, Boerman OC, et al: 99mTc-PEG liposomes for the scintigraphic detection of infection and inflammation: clinical evaluation. J Nucl Med 2000;41:622–630.
58 Peng G, Hakim M, Broza YY, et al: Detection of lung, breast, colorectal, and prostate cancers from exhaled breath using a single array of nanosensors. Br J Cancer 2010;103:542–551.
59 Kirchner C, Liedl T, Kudera S, et al: Cytotoxicity of colloidal CdSe and CdSe/ZnS nanoparticles. Nano Lett 2005;5:331–338.
60 Pons T, Pic E, Lequeux N, et al: Cadmium-free CuInS2/ZnS quantum dots for sentinel lymph node imaging with reduced toxicity. ACS Nano 2010;4:2531–2538.
61 National Cancer Institute: NCI Alliance for Nanotechnology in Cancer; Learn About Nanotechnology; Where it stands now. http://nano.cancer.gov/learn/now/clinical-trials.asp.

Mukesh Harisinghani, MD
Department of Radiology
Massachusetts General Hospital, 55 Fruit Street, White 270
Boston, MA 02114 (USA)
Tel. +1 617 726 8396, E-Mail mharisinghani@mgh.harvard.edu

Alexiou C (ed): Nanomedicine – Basic and Clinical Applications in Diagnostics and Therapy.
Else Kröner-Fresenius Symp. Basel, Karger, 2011, vol 2, pp 106–115

Medical Applications of Plasmonic Nanoparticles

Ramón A. Alvarez-Puebla · Luis M. Liz-Marzán

Departamento de Química-Física y Unidad Asociada CSIC, Universidade de Vigo, Vigo, España

Abstract

In this chapter, we present an overview of the main medical applications of metal nanoparticles displaying localized surface plasmon resonances (LSPRs). A description of diagnostic methods based on selective nanoparticle aggregation (plasmon coupling), LSPR shifts, and surface-enhanced spectroscopies [surface-enhanced Raman scattering (SERS) in particular] is presented and the main advantages and drawbacks of each method are briefly discussed. Additionally, a final section is devoted to applications in therapy, regarding the use of gold nanoparticles as drug delivery systems, as well as photothermal cell death.

Nanomedicine is the field of nanotechnology devoted to the development of nanoscale systems and devices with applications in medicine. In fact, this area could be defined as the science and technology involved in the rational design of complex nanosystems and nanodevices, with prospective applications in the diagnosis, detection, and/or therapy of diseases. Nanomedicine joins physics, chemistry, biology, and medicine and includes a wide variety of materials such as conducting metals, semiconductors, magnetic particles, polymers, and others, which are usually combined with biomolecules (lipids, sugars, proteins, and nucleic acids) and/or drugs to achieve the desired functionality.

The so-called plasmonic nanoparticles provide a particularly useful platform, demonstrating unique properties that can be used in various ways in diagnostic and therapeutic applications [1–3]. Plasmonic nanomaterials are characterized by the ability to generate a highly intense constrained electromagnetic field due to the collective excitation (oscillation) of the free-electron gas density – characteristic of conducting materials – upon excitation with an appropriate radiation. These electron oscillations are known as localized surface plasmon resonances (LSPRs) and can be observed at the interface between a dielectric and a nanoscaled metal (with dimensions much smaller than the wavelength of the excitation light) in which the real part of the dielectric function (a complex value) is negative in the spectral range of interest, and its magnitude is greater than that of the dielectric constant of the dielectric next to it. This condition is met in the visible and near-infrared regions for air/metal or water/metal interfaces (where the real dielectric function of the metal is negative and that of air or water is positive) for coinage metals, aluminum, titanium, chromium, or alkaline metals. However, most of these metals display a low chemical stability and thus gold and silver are, by far, the ones that have been most often studied and employed in practice.

In this chapter, we discuss general approaches toward the integration of plasmonic nanoparticles into functional materials for applications in nanomedicine. We have chosen 3 major areas: (1) diagnosis and biodetection, (2) bioimaging, and (3) phototherapy.

Plasmonic Nanoparticles for Diagnosis and Biodetection

The fast, accurate, sensitive and non- (or minimally) invasive detection of biological agents, disease markers, and toxic materials in the biofluids of patients is a key objective for biomedical diagnosis, forensic analysis, and environmental monitoring [4, 5]. The recognition event in sensors is always based on the change of at least one measurable property in the sensing element upon exposure to the sample of interest. The magnitude in which this property changes is then observed by a transduction element, which in this case contains plasmonic nanoparticles.

Sensors Based on Localized Surface Plasmon Resonance Shifts

One of the simplest ways to use plasmonic nanostructures as sensors simply comprises direct monitoring of their LSPRs. As previously discussed, LSPRs are confined to the interface between the nanoscaled metal (typically silver or gold) and the surrounding dielectric. These oscillations are thus extremely sensitive to any change occurring at the interface, such as the adsorption of molecules on the metal surface or other changes in the composition of the surrounding medium (fig. 1a). The reason is that the LSPR wavelength shifts toward higher values when the medium refractive index is increased. This approach is rather simple and a standard, inexpensive UV-Vis spectrometer can be used as the transduction element. In fact, LSPR shifts in silver and gold nanoparticles have been widely demonstrated to allow detection of changes in the composition of the surrounding fluid (air or solvent) [6] as well as direct recognition of the presence of a given biomarker adsorbing onto the surface of the nanoparticles. Unfortunately, not all of the interesting targets can spontaneously adsorb onto gold or silver surfaces. Another important drawback is that the transduction property, i.e. the LSPR shift, is not selective (cannot distinguish the reason behind the refractive index change), and therefore, when studying complex mixtures such as biological fluids, many different components can be the origin of this response. Consequently, the nanoparticle surfaces must be previously functionalized with highly selective molecular systems such as antibodies or nucleic acid fragments, among others (fig. 1b). However, while these sensors do allow us to specifically detect a target molecule within a complex fluid, the magnitude of the LSPR red shift is still small, which limits the sensitivity of these biosensors [7].

An alternative to the direct use of LSPR shifts as the sensing magnitude is the design of functionalized particles that form aggregates in the presence of the target analyte (fig. 2). The aggregation of plasmonic nanoparticles also leads to large changes (shift and broadening) in the LSPR modes, but these changes are principally due, in this case, to the interaction between nanoparticles (plasmon coupling) when they are located at a short distance from each other rather than to refractive index changes arising from variation in the composition of the surrounding medium. By using this approach, the LSPR changes induced by the presence of the analyte can be observed visually since the color of the colloidal dispersion changes dramatically (typically from red to blue) [8]. Therefore, the use of any instrumental technique as a transduction element can be complete-

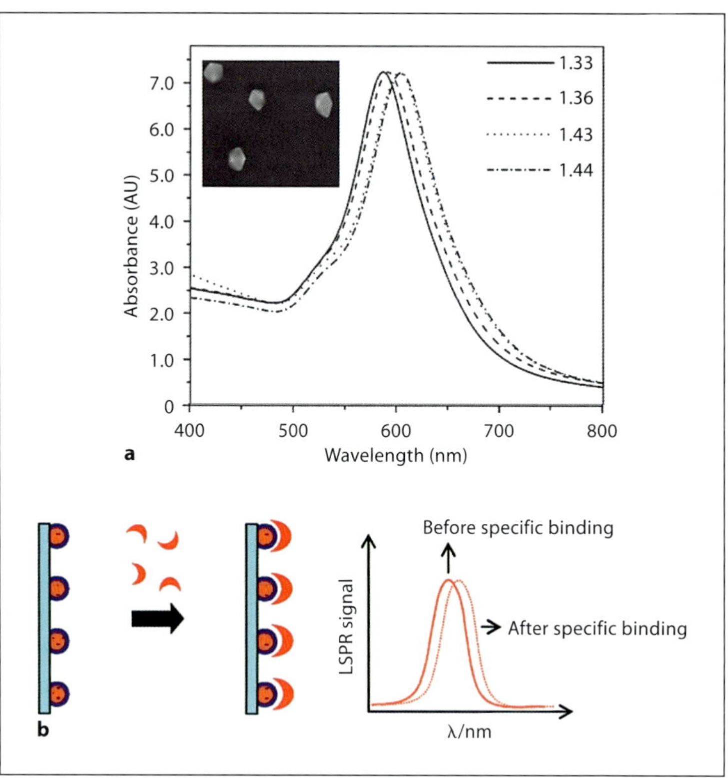

Fig. 1. a Experimental UV-Vis spectra of small (approximately 50 nm, see **inset**) decahedra in different solvents (water, ethanol, and mixtures of ethanol and toluene). **b** Schematic representation of the preparation and response of LSPR biosensors based on refractive index changes of functionalized nanoparticles. The metal particles are modified with a selective sensor moiety where the analyte attaches from the solution specifically onto the recognition function adsorbed onto the particles, causing a change in the refractive index around the particle which results in an LSPR red shift. **a** Adapted with permission from Pastoriza-Santos et al. [6]. Copyright 2007 Wiley-VCH. **b** Adapted with permission from Sepúlveda et al. [7]. Copyright 2009 Elsevier.

ly avoided, resulting in a much more sensitive detection technique. In fact, several of these sensors have been fabricated for different bioanalytes (DNA, proteins, etc.), with detection limits around the nanomolar regime [9, 10].

Sensors Based on Surface-Enhanced Spectroscopies

While the direct use of LSPR band shifts as the sensing property is still a research field with a great deal of activity, it presents some limitations such as the number of analytes that can be efficiently detected, lack of multiplex application, and insufficient detection limits. However, the large electromagnetic field generated when plasmon modes are excited in metal nanoparticles can be used to enhance the spectroscopic cross sections of a wide variety of molecules for different optical spectroscopic techniques, giving rise to the so-called surface-enhanced spectroscopies (SES) [11, 12]. SES include surface-enhanced fluorescence (SEF), surface-enhanced Raman scattering (SERS), and surface-enhanced infrared absorption (SEIRA) spectroscopies. Among these, the most important one by far is SERS [13], which brings together extraordinary detection limits down to a single molecule, rich structural information (and thus selectivity), and the opportunity to operate under biological conditions.

SERS can be utilized in 2 main operation modes: *direct SERS* refers to identification of the target analyte(s) through direct registering of their vibrational spectra, while the use of spectroscopically *encoded labels* allows detection of the target analyte through selective complexation to a plasmonic particle that carries a highly efficient SERS tag. Contrary to direct LSPR detection, in which the signal arises from changes

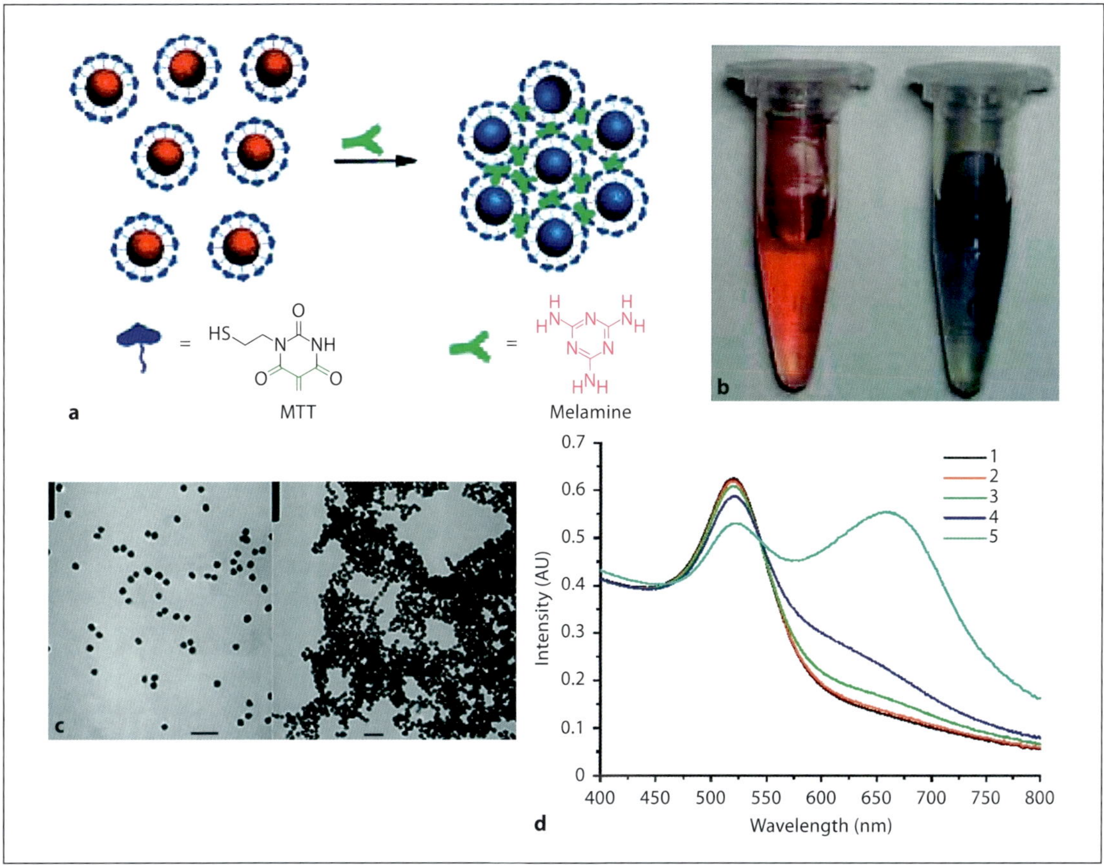

Fig. 2. a Schematic representation of the induced aggregation of gold nanoparticles functionalized with MTT before and after exposure to melamine. The change can be observed visually (**b**) and is consistent with the aggregation of the sample as demonstrated by TEM (**c**). **d** UV-Vis spectra (LSPR) of the system upon exposure to different melamine concentrations: 1 = without any addition; 2 = with the addition of the extract from blank raw milk; 3 = with addition of the extract containing 1 ppm (final concentration 8 ppb); 4 = with addition of the extract containing 2.5 ppm (final concentration 20 ppb), and 5 = with addition of the extract containing 5 ppm (final concentration 40 ppb). Adapted with permission from Ai et al. [8]. Copyright 2010 American Chemical Society. Scale bars = 50 nm.

in a broad band with a very simple structure, direct SERS offers the complete vibrational fingerprint of the molecule under study. Thus, this technique has been successfully exploited in biological sensing for the detection of pathogens, disease markers, or doping agents, as well as in vivo monitoring of health parameters. Notwithstanding, when dealing with very complex systems, such as biological fluids, and especially when the target analyte is diluted to minute concentrations and does not readily bind to gold or silver surfaces, the background vibrational signal of the sample can overlap and hide the signal of the desired target. In analogy to direct LSPR sensing, the nanoparticles can be functionalized to attract/concentrate the desired analyte onto the plasmonic surfaces and thereby solve this problem. Two different alternatives have been developed for this purpose. The first one consists of functionalizing the nanoparticle surface with

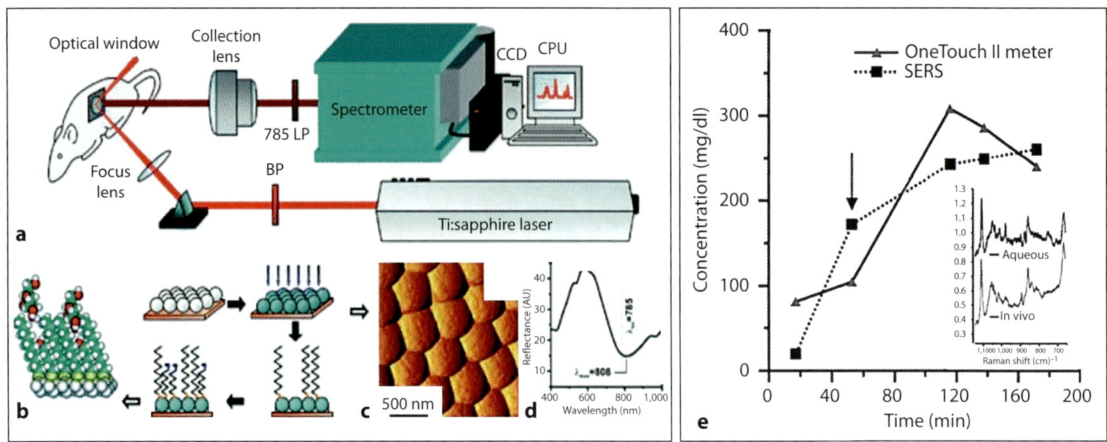

Fig. 3. Schematics of the instrumental apparatus (**a**), sensor preparation (**b**), morphology (**c**), and optical characterization (**d**). **a** A rat with a surgically implanted sensor and optical window was integrated into a conventional laboratory Raman spectroscopy system consisting of a Ti:sapphire laser (λ = 785 nm), a band-pass filter (BP), steering and collection optics, and a long-pass filter (LP) that rejects Rayleigh-scattered light. **b** Silver film over nanospheres (AgFONs) were prepared by depositing metal through a mask of self-assembled nanospheres. The AgFON was then functionalized by successive immersions in ethanolic solutions of decanethiol and mecaptohexanol. Glucose can partition into and out of the decanethiol/mecaptohexanol layer, as shown in the left of **b**. The resultant structure is shown in the atomic force micrograph (**c**). **d** After functionalization, a reflectance spectrum was collected to determine the position of the LSPR. **e** Time trace of the in vivo glucose measurement. Triangles correspond to measurements from a OneTouch II blood glucose meter, and squares correspond to measurements from the SERS sensor. Glucose infusion was started at 60 min as indicated by the arrow. The **inset** shows a typical in vivo spectrum compared to a typical ex vivo spectrum of the same surface (λ_{ex} = 785 nm, P = 50 mW, time = 2 min). Adapted with permission from Stuart et al. [14]. Copyright 2006 American Chemical Society.

molecules that display high affinity toward the target analyte but are characterized by a low SERS cross section, such as aliphatic thiol monolayers or stimuli responsive microgels. This has been successfully exploited in in vivo monitoring of glucose levels in blood (fig. 3) [14, 15]. The second method is based on the functionalization of nanoparticles with highly selective biomolecules such as antibodies (proteins) or aptamers (nucleic acids). In this case, the collected spectra are those of the ligand (antibody or aptamer) before and after the recognition event. This alternative permits quantitative ultradetection of the analyte in very complex samples such as blood, saliva, or urine and has been demonstrated for the detection of cocaine consumption [16], thrombin [17], or platelet-derived growth factor [18].

Simultaneous detection of several components (multiplexing) is difficult to achieve using direct SERS due to the inherent complexity of biological fluids. These limitations might, however, be overcome using the so-called encoded beads, which are labeled materials that can be directly bound to a biointerface, providing it with a specific signature for identifying a recognition event. These detection schemes can be termed *indirect methods* because the signal does not originate from the analyte but originates from the labeling material instead [19]. Encoded particles [20, 21] are cost-efficient platforms that address some of the limitations posed by more conventional substrates, such as: (1) amenability to high-throughput screening and multiplexing [22], (2) a larger surface area for receptor conjugation or solid-phase synthesis, (3) better accessibility of the analytes to the entire sample volume for interaction with bead-conjugated receptors, and (4) greater versatility in terms of sample analysis and data acquisition. Encoded nanoparticles can be used either in suspension assays or patterned into microchips. In a typical suspension assay, a mixture of different antibodies, each of them previously labeled with one particular encoded bead, is added to the fluid of interest. After reaction with their specific antigens, the beads are separated by centrifugation, washed, and exposed to a detection antibody (labeled with a fluorophore), which selectively binds to those particles that had been complexed to the corresponding capture antibody. After fluorophore binding, the particles are centrifuged and washed again, and a small portion is cast onto a glass slide and analyzed in a micro-Raman system. Positive diagnosis of the presence of a pathogen, disease marker, etc., requires the presence of both fluorescent labels and SERS tags on the same particle [23]. Although in principle this method does not require homogeneity in the size or shape of the beads, if automation by flow cytometry or microfluidics is to be implemented, size homogeneity is essential to avoid interference due to different Rayleigh scattering signals (which are morphology dependent) [24]. An alternative route to suspension assays is the generation of microchips containing encoded particles within well-defined spatial regions and fixed to a surface, which allows fast recognition and avoids the need for washing steps such as centrifugation. On the other hand, the readout from the chip can be performed automatically. However, preparation of these platforms requires using particles with a high homogeneity in terms of both size and shape [22].

Bioimaging

Encoded particles have not only paved the way toward the design of new and fast advanced sensor devices capable of monitoring multiple parameters in a single readout, but they can also be applied in bioimaging. In 1991, Nabiev et al. [25] reported the SERS spectra of the antitumor drug doxorubicin, recorded from treated cancer cells, incubated with silver colloids. These spectra were the first feasibility data demonstrating the promising prospectives of SERS spectroscopy in bioimaging, and they have currently become a key topic of research. SERS bioimaging is based on the functionalization of encoded particles with bioligands displaying a high affinity toward specific receptors of the cell membrane. This technique can be applied to in vivo imaging of cells [26], tissues [27], and organs [28] and has a great projection, e.g. in surgery. Surgery is one of the most effective and widely used procedures for treating human cancers, but a major problem is that the surgeon often fails to remove the entire tumor, leaving behind tumor-positive margins, metastatic lymph nodes, and/or satellite tumor nodules. Encoded nanoparticles with the appropriate functionalization can be selectively retained around the tumor. This, in conjunction with modern developments in optical fiber technology providing excellent optical gain far away from the measuring system, allows for precise detection, both preoperatively and intraoperatively, of the tumor contours and thus continuous monitoring of both positive and negative tumor margins around the surgical cavity, thereby raising new possibilities for real-time tumor detection and image-guided surgery (fig. 4) [29].

Phototherapy

The unique optical, chemical, and biological properties of gold nanoparticles make them attractive for applications in clinical therapy, including chemical (drug and gene delivery) and physical (photothermia) therapeutic methods. The attractive features of gold nanoparticles include their surface plasmon resonance, especial-

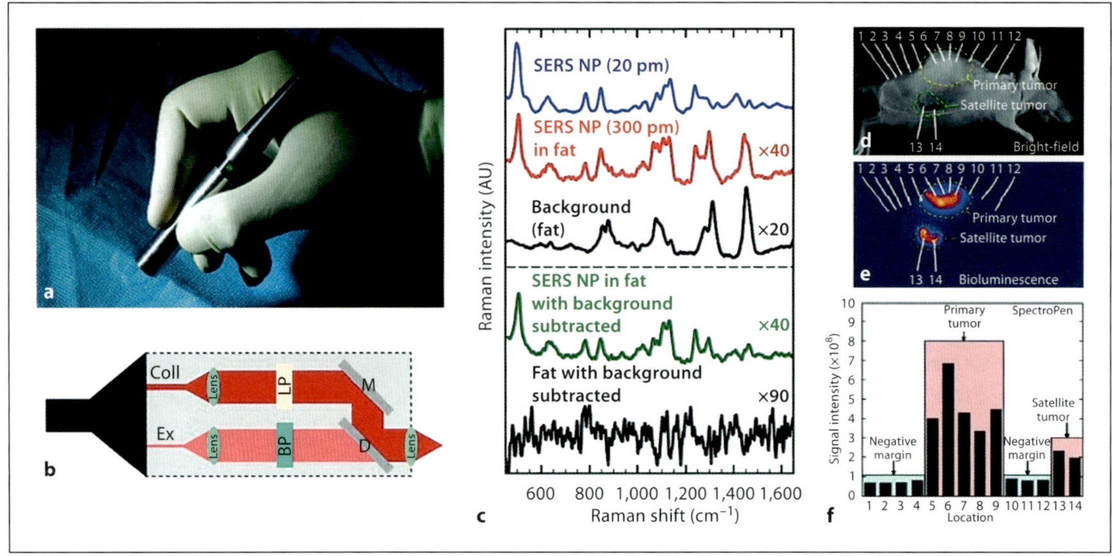

Fig. 4. a Photograph showing the SpectroPen held in the operator's hand in a surgical setting. **b** Optical beam paths of the SpectroPen. The excitation light is provided by a 785-nm diode laser (200-mW output); Ex = Excitation fiber; Coll = collection fiber; BP = band-pass filter; LP = long-pass filter; D = dichroic filter; M = reflective mirror. **c** Raman spectra of pure SERS nanoparticles, animal fat, and a mixture of SERS nanoparticles and animal fat before (upper) and after background subtraction (lower). All spectra were taken with the SpectroPen positioned 1 cm above the top layer of tissue. Spectra were acquired over 0.1–10 s. The background was obtained by averaging 4 different spectra from control tissues and was subtracted from the contrast-enhanced spectra or from single background measurements. Signal intensities relative to those of pure ICG or SERS samples are indicated by scaling factors. NP = Nanoparticles. The Raman reporter dye was diethylthiatricarbocyanine (DTTC). **d** Bright-field image showing the anatomical locations of a primary 4T1 breast tumor and 2 satellite nodules (dashed circles). The specific locations for SpectroPen measurement are indicated by numbers 1–12 for the primary tumor and 13 and 14 for the satellite nodules. **e** Bioluminescence image of the mouse showing the primary and satellite tumors (red signals). **f** Integrated ICG signal intensities as recorded by the SpectroPen at various locations corresponding to the numbers in **d** and **e**. Adapted with permission from Mohs et al. [29]. Copyright 2010 American Chemical Society.

ly in the near-infrared region (tissue transparency window), and their nontoxic nature. These attributes can be exploited to provide an effective and selective platform to target tissues and either releasing therapeutic drugs or selectively destroying malignant cells by light-induced heating. Both applications are mainly based on the same effect: localized heating of a gold nanoparticle when irradiated with a laser beam that overlaps the corresponding LSPR band due to the excitation of electron oscillations and thermal relaxation [30].

In general terms, the strategy comprises the preparation of gold particles with the appropriate morphological features so that they absorb radiation with wavelengths above 830 nm (the so-called biological window). This is essential to avoid damage to normal tissues and to ensure that light reaches the nanocarrier (or nanokiller) inside a living organism. Until very recently, this spectral window was almost restricted to gold nanorods and nanoshells. However, gold nanoparticles with other shapes have been developed which also display the LSPR frequencies of inter-

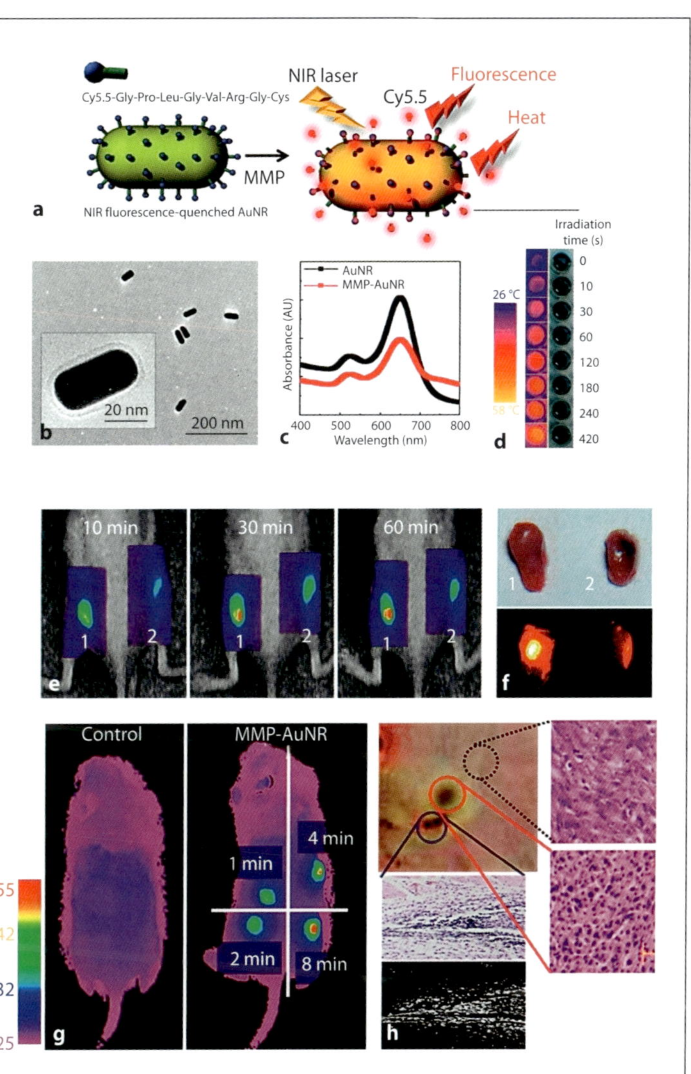

Fig. 5. a Schematic representation of the synthesis of matrix metalloprotease (MMP)-AuNR for simultaneous imaging and photothermal therapy. **b** TEM images of MMP-AuNR. **c** UV-Vis spectroscopy of bare AuNR and Cy-MMP peptide conjugated AuNR (MMP-AuNR). **d** Optical and infrared images showing the temperature increase of MMP-AuNR in a 96-well plate as a function of the laser irradiation time. **e** Near-infrared fluorescence tomographic images of SCC-7 tumor-bearing mice after intratumoral injection of the MMP-AuNR probe without (1) and with (2) inhibitor. **f** Optical and near-infrared fluorescence images of the excised tumor after injection of MMP-AuNR without and with MMP-2 inhibitor. **g** Infrared thermal images of tumor-bearing mice according to different laser irradiation times. AuNR was injected intratumorally prior to laser irradiation. **h** Optical images of the exterior of the SCC-7 tumor irradiated by laser. Histology (right) and dark-field images (lower) of the tumor after treatment are also presented. Adapted with permission from Yi et al. [34]. Copyright 2010 American Chemical Society.

est [31]. Some of these new morphologies additionally include nanoporosity, which may be advantageous for loading drugs within the plasmonic nanocarrier. The second essential ingredient (only for nanocarriers) is appropriate surface functionalization with a thermoresponsive material. This can be achieved by the use of a thermoresponsive coating shell that acts as the drug carrier and contains surface functional groups for functionalization with a directing agent, which is the third essential element. Alternatively, gold functionalization can be carried out with DNA. In this case, a thiolated DNA fragment is adsorbed onto the gold surface and subsequently, in a solution containing the drug to be delivered, the complementary DNA strand functionalized with the directing group is assembled [32]. Controlled release of the drug in the target area is achieved by denaturation of the DNA double helix, induced by thermal heating. Another alternative that also includes the use of DNA consists of trapping the drug at the gaps between nanoparticles which have been previously functionalized with appropriate DNA strands [33]. Nevertheless, the use of such aggregates leads to an increase in the size of the resulting nanocarrier, which is of relevance when the drug release is to be performed in vivo. Regardless of the nature of the photothermal therapy (chemical, physical, or both), the nanocarriers/nanokillers are to be injected intravenously or intratumorally into the individual. After some time, which may range from minutes to days depending of the way of injection and the anatomical disposition of the target, to allow the nanosytems to find their targets while the uncomplexed nanoparticles are eliminated, the selected area is illuminated with a near-infrared laser which penetrates the tissue and selectively increases the temperature of the gold core and the surrounding area, thereby triggering the release of the drug or killing the cell tissue where it has been selectively adsorbed. It is worth noting that phototherapy can be accompanied by bioimaging (fig. 5) by simply adding a fluorophore or a SERS label. In such cases, localization, monitoring, and elimination of the target can be achieved with the same material [34].

Conclusions and Outlook

Among the wide variety of techniques and materials that can be included within the field of nanomedicine, plasmonic metal nanoparticles display a number of important features that can be applied in several different ways to both diagnosis and therapy. In all cases, the operational principle is related to the collective oscillation of conduction electrons, in resonance with electromagnetic radiation (LSPR), within the visible or near-infrared ranges. Optimization of the nanoparticles' morphology and surface chemistry is a fundamental requisite to achieving a sufficient degree of performance since these properties define the operational wavelength range and the interaction between the nanoparticles and the biological samples (fluids, tissue, etc.).

Although applications in both fields of diagnosis and therapy have seen tremendous advances during the past decade, some of the proposed techniques seem to be significantly closer than others to practical application and eventual commercialization. In diagnosis, surface enhanced spectroscopies are particularly appealing since they offer a variety of application modes and the possibility of ultrasensitive and multiplexed analysis, integration in microfluidic systems (lab-on-a-chip), and bioimaging through highly specific labeling. Therapeutic techniques based on hyperthermia are advancing very fast and some of the most promising materials are currently at the clinical test stage.

Thus, an eye should be kept on the further development of these novel techniques since they are ready to compete with other popular nanomaterials, such as magnetic iron oxides or semiconductor quantum dots.

Acknowledgements

R.A.A.-P. acknowledges the RyC (MEC, Spain) program. This work was funded by the Spanish Ministerio de Ciencia e Innovación (grants MAT2010-15374 and MAT2008-05755, and Consolider Ingenio 2010-CSD2006-12) and the Xunta de Galicia (grants 09TMT011314PR and 08TMT008314PR).

Disclosure Statement

The authors have nothing to disclose.

References

1 Stewart ME, Anderton CR, Thompson LB, et al: Nanostructured plasmonic sensors. Chem Rev 2008;108:494–521.
2 Giljohann DA, Seferos DS, Daniel WL, Massich MD, Patel PC, Mirkin CA: Gold nanoparticles for biology and medicine. Angew Chem Int Ed Engl 2010;49:3280–3294.
3 De M, Ghosh PS, Rotello VM: Applications of nanoparticles in biology. Adv Mater 2008;20:4225–4241.
4 Alvarez-Puebla RA, Liz-Marzan LM: Environmental applications of plasmon assisted Raman scattering. Energy Environ Sci 2010;3:1011–1017.
5 Alvarez-Puebla RA, Liz-Marzan LM: SERS-based diagnosis and biodetection. Small 2010;6:604–610.
6 Pastoriza-Santos I, Sánchez-Iglesias A, García de Abajo FJ, Liz-Marzán LM: Environmental optical sensitivity of gold nanodecahedra. Adv Funct Mater 2007;17:1443–1450.
7 Sepúlveda B, Angelomé PC, Lechuga LM, Liz-Marzán LM: LSPR-based nanobiosensors. Nano Today 2009;4: 244–251.
8 Ai K, Liu Y, Lu L: Hydrogen-bonding recognition-induced color change of gold nanoparticles for visual detection of melamine in raw milk and infant formula. J Am Chem Soc 2009;131:9496–9497.
9 Lu W, Arumugam SR, Senapati D, et al: Multifunctional oval-shaped gold-nanoparticle-based selective detection of breast cancer cells using simple colorimetric and highly sensitive two-photon scattering assay. ACS Nano 2010;4: 1739–1749.
10 Hong S, Choi I, Lee S, Yang YI, Kang T, Yi J: Sensitive and colorimetric detection of the structural evolution of superoxide dismutase with gold nanoparticles. Anal Chem 2009;81:1378–1382.
11 Pieczonka NP, Goulet PJ, Aroca RF: Applications of the enhancement of resonance Raman scattering and fluorescence by strongly coupled metallic nanostructures. Top Appl Phys 2006; 113:197–216.
12 Aroca R: Surface-Enhanced Vibrational Spectroscopy. Chichester, Wiley, 2006.
13 Moskovits M: Spectroscopy: expanding versatility. Nature 2010;464:357.
14 Stuart DA, Yuen JM, Shah N, et al: In vivo glucose measurement by surface-enhanced Raman spectroscopy. Anal Chem 2006;78:7211–7315.
15 Yuen JM, Shah NC, Walsh JT, Glucksberg MR, Van Duyne RP: Transcutaneous glucose sensing by surface-enhanced spatially offset Raman spectroscopy in a rat model. Anal Chem 2010;82:8382–8385.
16 Sanles-Sobrido M, Rodriguez-Lorenzo L, Lorenzo-Abalde S, et al: Label-free SERS detection of relevant bioanalytes on silver-coated carbon nanotubes: the case of cocaine. Nanoscale 2009;1:153–158.
17 Hu J, Zheng PC, Jiang JH, Shen GL, Yu RQ, Liu GK: Electrostatic interaction based approach to thrombin detection by surface-enhanced Raman spectroscopy. Anal Chem 2008;81:87–93.
18 Neumann O, Zhang D, Tam F, Lal S, Wittung-Stafshede P, Halas NJ: Direct optical detection of aptamer conformational changes induced by target molecules. Anal Chem 2009;81:10002–10006.
19 Bake KD, Walt DR: Multiplexed spectroscopic detections. Ann Rev Anal Chem (Palo Alto Calif) 2008;1:515–547.
20 Fenniri H, Alvarez-Puebla R: High-throughput screening flows along. Nature Chem Biol 2007;3:247–249.
21 Sanles-Sobrido M, Exner W, Rodríguez-Lorenzo L, et al: Design of SERS-encoded, submicron, hollow particles through confined growth of encapsulated metal nanoparticles. J Am Chem Soc 2009;131:2699–2705.
22 Raez J, Blais DR, Zhang Y, et al: Spectroscopically encoded microspheres for antigen biosensing. Langmuir 2007;23: 6482–6485.
23 Cao YC, Jin R, Mirkin CA: Nanoparticles with Raman spectroscopic fingerprints for DNA and RNA detection. Science 2002;297:1536–1540.
24 Goddard G, Brown LO, Habbersett R, et al: High-resolution spectral analysis of individual SERS-active nanoparticles in flow. J Am Chem Soc 2010;132:6081–6090.
25 Nabiev IR, Morjani H, Manfait M: Selective analysis of antitumor drug interaction with living cancer cells as probed by surface-enhanced Raman spectroscopy. Eur Biophys J 1991;19: 311–316.
26 Chourpa I, Lei FH, Dubois P, Manfait M, Sockalingum GD: Intracellular applications of analytical SERS spectroscopy and multispectral imaging. Chem Soc Rev 2008;37:993–1000.
27 Hu Q, Tay LL, Noestheden M, Pezacki JP: Mammalian cell surface imaging with nitrile-functionalized nanoprobes: biophysical characterization of aggregation and polarization anisotropy in SERS imaging. J Am Chem Soc 2006;129:14–15.
28 Lutz BR, Dentinger CE, Nguyen LN, et al: Spectral analysis of multiplex Raman probe signatures. ACS Nano 2008; 2:2306–2314.
29 Mohs AM, Mancini MC, Singhal S, et al: Hand-held spectroscopic device for in vivo and intraoperative tumor detection: contrast enhancement, detection sensitivity, and tissue penetration. Anal Chem 2010;82:9058–9065.
30 Baffou G, Quidant R, García de Abajo FJ: Nanoscale control of optical heating in complex plasmonic systems. ACS Nano 2010;4:709–716.
31 Yang M, Alvarez-Puebla Rn, Kim HS, Aldeanueva-Potel P, Liz-Marzán LM, Kotov NA: SERS-active gold lace nanoshells with built-in hotspots. Nano Lett 2010;10:4013–4019.
32 Huschka R, Neumann O, Barhoumi A, Halas NJ: Visualizing light-triggered release of molecules inside living cells. Nano Lett 2010;10:4117–4122.
33 Tam JM, Tam JO, Murthy A, et al: Controlled assembly of biodegradable plasmonic nanoclusters for rear-infrared imaging and therapeutic applications. ACS Nano 2010;4:2178–2184.
34 Yi DK, Sun IC, Ryu JH, et al: Matrix metalloproteinase sensitive gold nanorod for simultaneous bioimaging and photothermal therapy of cancer. Bioconjugate Chem 2010, DOI: 10.1021/bc100308p.

Luis M. Liz-Marzán
Departamento de Química Física
Universidade de Vigo
ES–36310 Vigo (Spain)
Tel. +34 986 812 298, E-Mail lmarzan@uvigo.es

Alexiou C (ed): Nanomedicine – Basic and Clinical Applications in Diagnostics and Therapy.
Else Kröner-Fresenius Symp. Basel, Karger, 2011, vol 2, pp 116–134

Pancreatic Cancer Stem Cells as New Targets for Diagnostics and Therapy

Jenifer Clausell-Tormos · Christopher Heeschen

Stem Cells and Cancer Group, Clinical Research Program, Spanish National Cancer Research Centre (CNIO), Madrid, Spain

Abstract

According to the cancer progression model, multiple events are required for progression from normal epithelium to carcinoma. Due to their extended life span, stem cells represent the most likely target for the accumulation of these genetic events; however, this has not been proven for most solid cancers. Clinically, cancer stem cells (CSCs) seem to harbor mechanisms protecting them from standard cytotoxic therapy; this was recently demonstrated for pancreatic CSCs. Therefore, novel approaches targeting CSC-intrinsic and CSC-extrinsic properties might overcome these resistance mechanisms. To improve our understanding of in vivo CSC biology, novel imaging modalities in conjunction with the most clinically relevant CSC models need to be developed. Here, multimodal magnetic nanoparticles, due to their potential on CSC tracking, targeting and elimination, might help to elucidate the underlying regulatory mechanisms of CSCs and develop platforms for targeted theragnostics. Eventually, these studies should allow improvement of the prognosis of patients suffering from these deadly diseases.

Solid tumors represent a major cancer burden. In particular, epithelial cancers arising in tissues that include breast, lung, colon, prostate, and ovary constitute approximately 80% of all cancers while other, less frequent tumor entities such as glioblastoma multiforme and pancreatic ductal adenocarcinoma are characterized by extraordinarily high mortality rates [1]. Tumors are generally assessed clinically at the gross level by histology and by expression of specific markers. In combination with gene expression analysis, this has lead to the definition of distinct tumor subtypes. However, the cellular origin of most solid tumors remains unknown in most cases. It is hypothesized that different subtypes correspond to distinct cells of origin at the time of tumor initiation. Indeed, the cell of origin, ranging from somatic stem cells to their postmitotic progenies, may differ between patients and may determine the properties of the evolving cancers. Thereby, in different tumor subtypes, cells within the tumor population frequently exhibit functional diversity termed tumor heterogeneity [2];

some cells exhibit high proliferative and differentiative capacities while others show poor differentiation.

The fundamental cellular mechanisms determining this tumor heterogeneity have been the subject of intense research efforts. As early as 1961, rather exceptional studies by Southam and Brunschwig [3] provided the first evidence of heterogeneity also in tumorigenicity by autologous transplantation of malignant cells from patients with different carcinomas into subcutaneous tissue. Intriguingly, the smallest number of cells capable of resulting in transplant growth was 10^6, indicating that a large number of viable cells is necessary to promote tumor growth. While it is possible that there are growth-inhibiting factors that may need to be overcome by larger populations of cells or that the provision of a peculiar local milieu suitable to the growth of the transplanted cells is critical for cell engraftment, these data are also consistent with a hierarchical organization of tumor cells. This concept would imply that only subpopulations of cells are capable of engrafting consistently.

However, at the time of these experiments, technologies to detect cellular heterogeneity were lacking. Therefore, despite these intriguing early data, it was not until 1997 that pioneering studies from John Dick's laboratory identified for the first time leukemia-initiating stem cells [4, 5]; this was followed by landmark studies in breast cancer [6] and then rapidly emerging investigations on tumor-initiating stem cells, also termed cancer stem cells (CSCs), in numerous other solid tumors. While this review will focus on pancreatic cancer [7, 8] and the potential use of nanoparticles for in vivo imaging and drug delivery, evidence for CSCs using several different (surface) markers has been found and functionally investigated in most other tumors such as colon [9] and breast [6] cancer, glioblastoma [10], prostate cancer [11, 12], melanoma [13, 14], lung cancer [15, 16], head and neck squamous cell carcinoma [17], and liver cancer [18, 19].

Pancreatic Cancer in 2011 – Still a Devastating Diagnosis

Pancreatic adenocarcinoma is the deadliest solid cancer and currently the fourth most frequent cause for cancer-related deaths. The disease is characterized by late diagnosis due to lack of early symptoms, extensive metastasis, and high resistance to both chemotherapy and radiation. The only currently available effective treatment modality for pancreatic cancer requires a very invasive and complex surgical procedure, also known as the Whipple procedure, for which only a limited number of patients (about 20%) with local disease are eligible [20].

Despite increasing research, there has hardly been any substantial progress leading to new therapies capable of affecting clinical endpoints over the past decades. The latest achievement in improving the treatment of patients with pancreatic cancer is the introduction of the nucleoside analog and chemotherapeutic agent gemcitabine. This drug not only improved the clinical response in terms of pain reduction and loss of weight but it also lead to a significant increase in median survival as compared to 5-fluouracil (fig. 1a) [21]. Subsequently, gemcitabine became the first-line chemotherapeutic agent in pancreatic cancer utilized in patients with locally advanced and metastasized disease who cannot undergo surgery. Overall, however, with a 5-year survival rate of 1–4% and a median survival period of 4–6 months, the prognosis of these patients has remained extremely poor [20, 22–26]. The most recent addition to our armory for fighting pancreatic cancer is the EGF receptor inhibitor erlotinib; it is the only other approved agent apart from gemcitabine and has not resulted in markedly improved survival (fig. 1b) [27]. Therefore, new ways to target the complex biology of pancreatic cancer are desperately needed to pave the way for the development of more effective treatments for these patients.

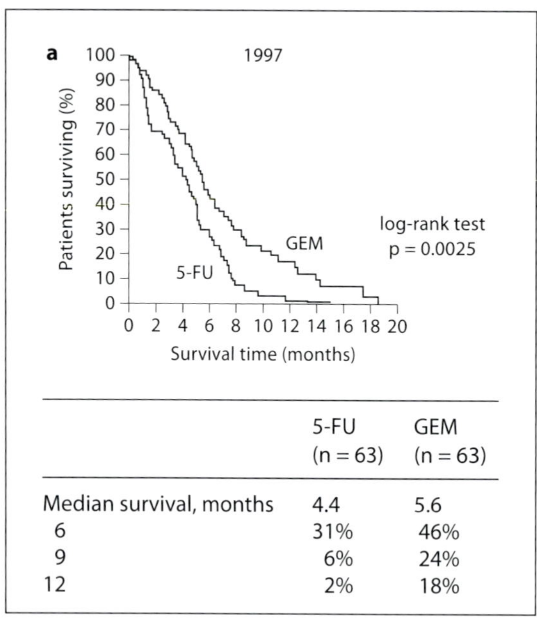

	5-FU (n = 63)	GEM (n = 63)
Median survival, months	4.4	5.6
6	31%	46%
9	6%	24%
12	2%	18%

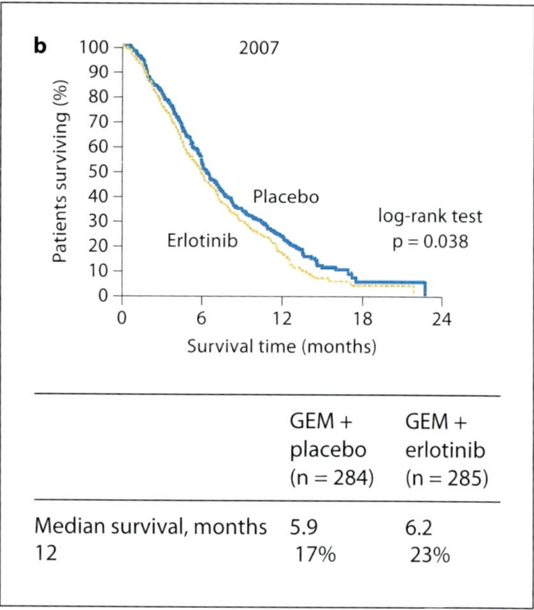

	GEM + placebo (n = 284)	GEM + erlotinib (n = 285)
Median survival, months	5.9	6.2
12	17%	23%

Fig. 1. Gemcitabine as the current best chemotherapeutic agent in pancreatic cancer. **a** In 1997, gemcitabine was shown to significantly improve the median survival in patients with advanced pancreatic cancer from 4.4 months to 5.6 months [21]. **b** Ten years later, the only other approved targeted treatment modality using the EGF receptor inhibitor erlotinib enhanced survival by no more than 10 days [27]. GEM = Gemcitabine; 5-FU = 5-fluoracil.

The Cancer Stem Cell Concept

Over the past decades, at least 2 not necessarily exclusive models have emerged trying to explain the heterogeneity and inherent differences in tumor-regenerating capacity: the clonal model, whereby mutant tumor cells with a growth advantage are selected and expanded during tumorigenesis [28], and the hierarchical model, in which primary tumors and metastatic cancer are initiated by a subpopulation of CSCs [29]. The acquisition of genetic alterations underpins the clonal model, but epigenetic modifications and the influence of the microenvironment are also likely to play important roles during later progression. Clonal evolution may involve a stochastic component whereby a distinct population of tumor cells acquires an appropriate set of somatic mutations and develops metastatic capability [5]. However, the CSC hypothesis could also be a relevant model to account for the functional heterogeneity that is observed in solid tumors [30]. It proposes a hierarchical organization of cells within the tumor, in which a subpopulation of cells with stem-like features is responsible for sustaining tumor growth, driving metastasis, and resistance to standard therapy. One important implication of this organization is that CSCs are placed at the top of this hierarchy and may represent the source of tumor relapse as these cells bear exclusive tumorigenicity [4]. It is important to note, however, that the two models are not mutually exclusive as CSCs themselves undergo clonal evolution, as has been shown for leukemia stem cells [31].

Specifically, the CSC hypothesis implies that a distinct population of cells with stem cell properties is essential for the development and perpetuation of various human cancers, including pancreatic cancer [6, 8]. CSCs are a population of cancer-initiating cells that usually constitute a

Table 1. Diverse markers have been used for the identification of cancer stem cells in solid tumors

Tumor entity	Markers	First author (year)	Reference
Breast cancer	$CD44^{+}CD24^{-/low}$	Al-Hajj (2003)	[6]
	$CD133^{+}$	Wright (2008)	[37]
	$CD133^{+}CXCR4^{+}$	Hwang-Verslues (2009)	[38]
	ALDH-1^{+}	Ginestier (2007)	[39]
	$CD49F^{+}DLL1^{high}DNER^{high}$	Pece (2010)	[40]
Glioblastoma	$CD133^{+}$	Singh (2004)	[10]
	SSEA-1^{+}	Son (2009)	[41]
Prostate cancer	$CD44^{+}$alpha2beta$1^{high}CD133^{+}$	Collins (2005)	[11]
	$CD133^{+}CXCR4^{+}$	Miki (2007)	[12]
Melanoma	$CD20^{+}$	Fang (2005)	[13]
	$ABCB5^{+}$	Schatton (2008)	[14]
	CD271	Boiko (2010)	[42]
Lung cancer	Sca-$1^{+}CD45^{-}PECAM^{-}CD34^{+}$	Kim (2005)	[15]
	$CD133^{+}CXCR4^{+}$	Bertolini (2009)	[16]
Colon cancer	$CD133^{+}$	O'Brien (2007), Ricci-Vitiani (2007)	[43, 44]
	$EpCAM^{+}CD44^{+}CD166^{+}$	Dalerba (2007)	[45]
Pancreatic cancer	$EpCAM^{+}CD44^{+}CD24^{+}$	Li (2007)	[8]
	$CD133^{+}$	Hermann (2007)	[46]
	$CD133^{+}CXCR4^{+}$	Hermann (2007)	[46]
	ALDH-1^{+}	Feldmann (2007), Jimeno (2009), Rasheed (2010)	[47–49]
Liver cancer	$CD133^{+}$	Ma (2007)	[18]
	$CD90^{+}$	Yang (2008)	[19]

small but variable percentage of the total tumor mass and show 3 defining features: their capability to self-renew, generation of all of the heterogeneous cell types a tumor contains, and exclusive tumorigenicity in secondary recipients. In general, CSCs should have characteristic traits like a distinct surface marker expression profile (although none of the currently available markers specifically labels CSCs) and the capacity for asymmetric/symmetric cell division that allows the CSC population to maintain/expand itself while generating the more differentiated progeny of tumor cells [32]. Most importantly, CSCs are often resistant to standard therapy [8, 33, 34]. The concept of CSCs has been under debate in the past few years, but while the first evidence was shown early for leukemia and myeloma [35, 36] their existence has now been validated in several kinds of solid tumors, such as breast [6, 37–40], glioblastoma [10, 41], prostate [11, 12], melanoma [13, 14, 42], lung [15, 16], colon [43–45], pancreas [8, 46–49], and liver [18, 19] tumors (table 1).

Apparently, several questions remain to be addressed including the interpretation of data from human tumors generated with different xenograft models [50]. In each of the studies listed in table 1, CSCs were identified by a set of specific criteria in order to demonstrate their role in tumor initiation and progression [29]: the expression of surface markers depending on the tumor of origin, the capacity for self-renewal and differentiation, and the ability to recapitulate the phenotype of the original tumor using in vitro and in vivo tumorigenicity assays. The CSC theory is an intriguing model to explain both the wide heterogeneity observed in an originally monoclonal tumor and tumor relapse after treatment due to the presence of a therapy-resistant population.

While the latter study suggests a low frequency of CSCs, it is important to note that only early-stage tumor samples were used for these investigations. Indeed, the frequency of CSCs in solid tumors appears to vary considerably between tumors of the same entity from almost undetectable in some tumors to highly abundant in other tumors. It is currently not clear whether this increase in CSC numbers is indeed related to tumor progression or to limitations of the available set of markers for their identification, emphasizing the urgent need for more definitive CSC markers. This must be resolved by clonality studies and more extensive studies in genetically engineered mouse models covering the full spectrum of tumor progression.

While the application of this theory to the development and progression of pancreatic cancer and the identification, quantification, and clinical relevance of pancreatic CSCs could have profound clinical implications, the cell origin of pancreatic CSCs remains to be determined [51]. In the pancreas, CSCs may arise from somatic stem or progenitor cells with genetic alterations that lead to malignant behavior. Another possibility is that CSCs originate from the dedifferentiation of a lineage-committed cell that has (re)acquired stem cell characteristics through mutation. This issue is further complicated by the current lack of convincing evidence for a stable stem cell population in the normal pancreas. The identification and precise investigation of a putative pancreatic stem cell including its niche using genetically engineered mouse models will hopefully lend further insights into the origin of CSCs.

Pancreatic Cancer Stem Cell Markers and the Heterogeneity of Pancreatic Cancer Cells

Putative pancreatic CSCs, defined for the first time by the simultaneous expression of CD44, CD24, and EpCAM [7], are highly tumorigenic and possess the ability to both self-renew and produce differentiated progeny in secondary recipients that reflects the heterogeneity of the patient's primary tumor. However, it should be noted that in this first study putative CSCs were compared to a population of cells that were negative for all 3 markers. Since EpCAM identifies epithelial cells within the tumor, the confinement of the study to EpCAM-negative cells as the control population may have been too restrictive as these cells should primarily represent nonepithelial inflammatory, stromal, and vascular cells [7]. Using a different cell surface marker, Hermann et al. [8] showed that $CD133^+$ cells in primary pancreatic cancers and pancreatic cancer cell lines also discriminate cells with enhanced proliferative capacity which also show the defining CSC traits. Interestingly, while $CD133^+$ cells are inherently contaminated by small numbers of endothelial and hematopoietic progenitor cells, they also showed that this $CD133^+$ population also contains but does not exclusively consists of the previously $CD44^+CD24^+EpCAM^+$ cells. Interestingly, CD133 has also been used to distinguish different types of pancreatic cancers [52]. Other distinctive markers have also been used for the characterization of CSC; aldehyde dehydrogenase-1 (ALDH-1) has been shown to be associated with tumorigenic cells in pancreatic cancer [47–49], although more recent comprehensive investigations suggest abundant expression of ALDH-1 in normal pancreas tissue [53], which would disqualify ALDH-1 as a suitable marker for CSC in humans. Moreover, side population cells, which exclude the DNA dye Hoechst 33342, proved to be cancer-initiating cells in several tumors [54], but these data require further validation as the use of side population cells in gastrointestinal cancers has generated conflicting data [55].

Apparently, a number of studies have published conflicting data on the expression of these markers for the identification of pancreatic CSCs, suggesting that the analyzed CSC populations are by no means pure and/or technical obstacles remain. Importantly, it was reported that the use of different CD133 antibodies can translate into significantly different findings [56]. Two studies using different CD133 antibodies for histological analysis of pancreatic cancer tissue led to opposing results with respect to the CD133 expression patterns [57, 58]. Moreover, the differentiation of colon CSCs did not coincide with a change in CD133 promoter activity, mRNA, splice variants, protein expression, or even cell surface expression of CD133. In contrast, a change occurred in CD133 glycosylation, suggesting that CD133 is expressed on both CSCs and differentiated tumor cells but is probably differentially folded as a result of differential glycosylation to mask specific epitopes [59]. Therefore, CSCs can be reliably identified by AC133, which only binds to this modified form of CD133, but use of this antibody should still be interpreted with great caution as handling of the cells prior to and during the staining procedure may affect the results.

Although none of these markers appears to selectively characterize a pure population of CSCs, their use not only increases our knowledge about the biology of these cells but also produces consistent data for a strong enrichment of CSCs. Moreover, another important aspect to consider before the evaluation of data is the origin of the investigated sample since the results obtained from fresh patient-derived samples could be significantly different from those obtained from in vivo expanded tumors and, of course, those derived from established cancer cell lines. Indeed, the analysis of clinical samples will certainly provide clinically more relevant data as established cell lines have adapted to in vitro culture conditions, and potentially no longer resemble their primary counterparts, especially after long-term passaging. However, well-characterized primary cancer cell lines derived from a diverse set of patients will still be important for some mechanistic work since primary cells are usually much more difficult to handle (e.g. low transfection efficiency and reduced viability in culture conditions) [8].

Nevertheless, the most important approach for testing the CSC hypothesis and potential targeted therapeutics is still the use of optimized preclinical model systems. The current gold standard for this is the investigation of primary patient-derived xenografts since these xenografts closely resemble the heterogeneity of mutations and cellular composition of the original primary tumors. Recent data provide further experimental proof of a close relationship between primary tumors and their corresponding xenografts [60]. Interestingly, this study also demonstrated that xenografts bear a signature very similar to metastatic lesions, suggesting that these secondary tumors may be formed by a subpopulation of migratory cells within the primary tumor that is more potent in engrafting in immunocompromised mice.

A strong interindividual heterogeneity concerning the expression patterns of markers that have been used for the enrichment of CSCs has been reported that may be related to differences in tumor stages [51, 56] but may also be related to the digestion and subsequent culturing of the primary tissue. It may indeed also indicate that the CSC hypothesis is not a universal model for all individual patient samples. Whether an individual tumor follows the CSC model or not may depend on whether the initializing mutation occurred in the stem cell compartment or in more differentiated progenitor cells, but it may switch to follow the clonal evolution model during tumor progression and metastasis [61].

Cancer Stem Cell Heterogeneity – Change of Phenotype during Progression and Treatment

CSCs are defined by their self-renewal and differentiation capacities. Normal stem cells accomplish these two tasks by asymmetric cell division,

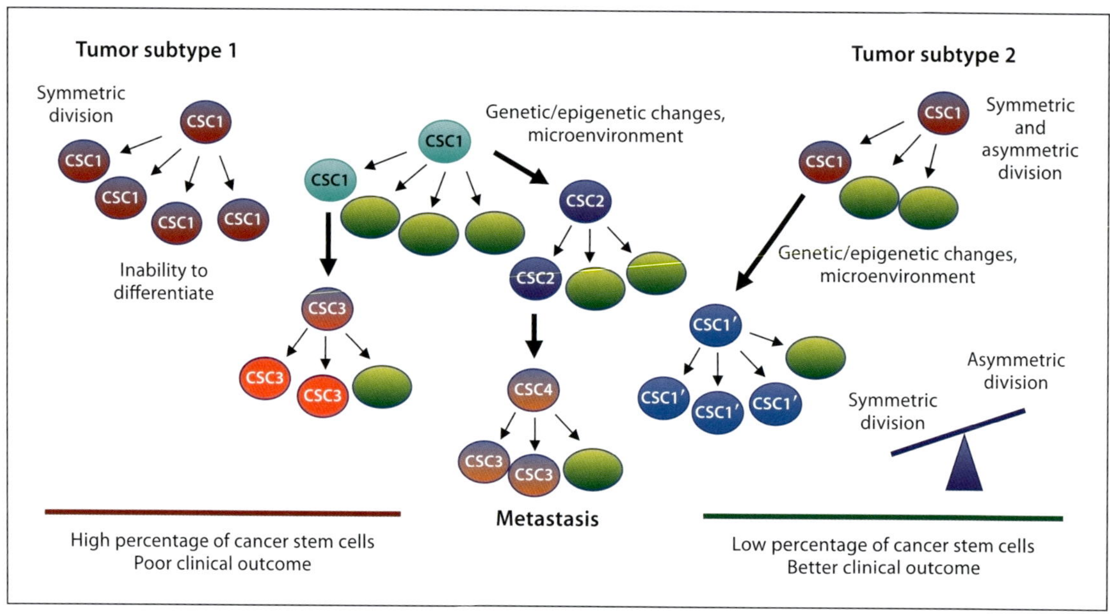

Fig. 2. Frequency and genetic heterogeneity of CSCs. The number of CSCs has been linked to prognosis, but this may not represent a fixed stage as the frequency of CSCs may increase during progression due to genetic/epigenetic as well as environmental changes.

a process where the stem cell divides to generate one stem cell bearing self-renewal capacity and one daughter cell that subsequently differentiates (fig. 2, middle and right). This modus maintains a stable number of stem cells. However, it has been shown that CSC numbers can increase markedly during progression, requiring a more preferential symmetric division modus [40]. Cell division can become symmetric in several ways: (1) the dividing cell resides within a polar environment and exposure to different environments induces alternative fates and (2) genetic/epigenetic changes drive symmetric division (e.g. aberrant DNA methylation, abnormal RNA interference, and chromatin remodeling) (fig. 2, middle and right).

CSCs must bear distinct strategies for controlling self-renewal and leading to dynamic modulation of their numbers, cell proliferation, and tumor relapse. It is currently unknown whether CSC homeostasis, which is readily established after in vivo implantation of highly purified CSCs, is maintained by asymmetric divisions or by a strategy that uses symmetric divisions to balance CSCs and more differentiated progeny. CSCs use symmetric divisions at a much higher rate as compared to their normal counterparts to constantly expand their pool or to generate more differentiated progeny resulting in tumor growth [40]. Symmetric divisions are defined as the generation of daughter cells that are destined to acquire the same fate (fig. 2, left). Indeed, Pece et al. [40] recently reported that a different ratio of asymmetric versus symmetric division as compared to normal stem cells is causal for the increasing numbers of CSCs in G3 versus G1 tumors, which could explain at least in part the biological and clinical heterogeneity of breast cancers at different stages. CSCs are defined by their 'potential' to generate CSCs and differentiated daughters. Therefore, a pool of CSCs with equivalent developmental potential may produce only new CSCs in some divisions and only more differentiated progeny in others. Furthermore, symmetric division may confer developmental plasticity, increased growth, and enhanced regenerative capacity as well as an inherent risk of cancer [62]. CSCs may be capable of using either symmetric division only or a combination of symmetric and asymmetric division (fig. 2, right). The preferential modus may be determined by the sustained activation of different developmental cascades such as certain hormones, growth factors (fibroblast growth factor and epidermal growth factor), and signaling pathways [sonic hedgehog (Shh), Wnt/β-catenin, and/or Notch], which are involved in the strict control of self-renewal and differentiation of CSCs.

Cancer Stem Cell Signaling as New Target for Novel Treatment Modalities

Despite increasing research efforts over the past decades, pancreatic cancer is still one of the deadliest cancer-related diseases in the world [20]. Therefore, the development of novel therapeutic strategies is an issue of outstanding significance for finally improving patients' prognosis. In the case of pancreatic cancer, cell cycle analyses of $CD133^+$ CSCs have proven that while these cells stop proliferating under the influence of the cytotoxic agent gemcitabine they do not undergo apoptosis, and as soon as gemcitabine is withdrawn these cells immediately start to repopulate the cancer (stem) cell pool [8]. In contrast, the more differentiated CD133-negative cells representing the vast majority of the tumor cell population became apoptotic after the application of gemcitabine. Thus, evidence is accumulating that only more differentiated tumor cells can be targeted with standard therapy resulting in a selection process for undifferentiated tumor-initiating and metastasis-propagating CSCs. These data also emphasize that the mere monitoring of tumor size does not represent a suitable endpoint for monitoring the efficacy of CSC-targeting therapies.

Since these CSCs have been shown to exclusively generate tumors in secondary recipients [7, 8], the question arises of whether these undifferentiated cancer cells represent an important new target for more efficient therapeutic approaches, eventually allowing targeting of the root of pancreatic cancer. There are mainly 2 possible ways of depleting cancers for CSCs: (1) finding therapeutic agents that selectively kill these cells or (2) identifying treatment modalities that drive CSCs into differentiation, thus making them susceptible to standard (chemo)therapy.

Stem cells in general, but CSCs in particular, possess distinctive traits that provide them with higher resistance levels against classic cytotoxic agents. Stem cells strongly express ABC membrane transporters that can exclude toxic substances from the cell [63], have an extraordinarily high capacity to repair DNA damage, display a reduced immunogenicity, and have inherent antiapoptotic properties [30]. Even more importantly, however, the tumor compartment may contain a quiescent subpopulation of CSCs, which could evade the effects of most cytotoxic drugs at least in part due to lack of proliferation. Indeed, quiescence protects the stem cell compartment from most injuries and ensures its functionality during their long life span. As a consequence, quiescent stem cells have been shown to survive conventional cancer chemotherapy and radiation [64]. Moreover, Pece et al. [40] isolated stem cells on the basis of their ability to retain the lipophilic dye PKH26 as a consequence of their quiescent nature [6, 37–39].

Therefore, if CSCs represent the key target for new therapeutic approaches, all of these traits and defense mechanisms need to be overcome. Apparently, one of the currently most extensively studied approaches for targeting stem cells is the inhibition of developmental and/or stem cell-associated pathways [e.g. Shh, mammalian target of rapamycin (mTOR), notch, BMI, and BMP]. Further therapeutic targets may represent specific enzymes (e.g. telomerase), membrane transporters (e.g. ABC transporters), or RNA translation [65]. A publication by Feldmann et al. [47] recently described increased Shh activity in pancreatic cancers, which spurred interest in this pathway in the context of CSCs. Shh signals via inhibition of the transmembrane receptor Patched, which again inhibits Smoothened in the absence of Shh. Patched inactivation after binding of Shh leads to activation of Smoothened, which in turn leads to transcription of the Gli protein family target genes.

Although several genetically engineered mouse models have been established to investigate a causative role of Hh signaling in pancreatic tumorigenesis, none of these distinguish paracrine versus autocrine canonical Hh signaling in pancreatic cancer. Yauch et al. [66] recently demonstrated a paracrine requirement for the Hh pathway in xenograft models of pancreatic cancer, where Hh ligand is produced by tumor cells and the pathway is activated by the adjacent stroma. To address whether a paracrine Hh signal is present in autochthonous mouse pancreatic tumors, and to test if epithelial cancer cells are competent to transduce the Hh signal, Tian et al. [67] used an oncogenic form of Smoothened to activate the pathway cell autonomously. These data indicate that Hh signaling is restricted to tumor stroma, contradicting previous reports that suggest a key role for ligand-driven epithelial Hh signaling in tumor cell growth. However, these studies are indeed consistent with the observation that subpopulations of CSC, which do not express markers of epithelial differentiation, rely on Hh signaling [7, 47].

Indeed, after ex vivo pretreatment of pancreatic cancer cells with the naturally occurring Hh pathway inhibitor cyclopamine, a decline but not complete abrogation of the CSC content was observed. Therefore, it was not surprising that this effect did not translate into significantly reduced tumorigenic activity of pancreatic CSCs in a single-agent therapy [56]. Interestingly, however, cyclopamine alone significantly decreased the metastatic activity of the treated cells as compared to treatment with gemcitabine alone. This is compatible with earlier publications indicating an antimetastatic effect of cyclopamine in an orthotopic mouse model of pancreatic cancer, using different pancreatic cancer cell lines [47]. Interestingly, combined treatment with simultaneous application of cyclopamine and gemcitabine completely eliminated the $CD133^{+}CXCR4^{+}$ migrating CSC population, which has been demonstrated to be exclusively responsible for the metastatic spread of pancreatic cancers [56].

Since Hh pathway inhibition alone was not capable of eliminating the entire CSC population, additional targets needed to be investigated. Mueller et al. [56] were able to demonstrate that $CD133^{+}$ cells in pancreatic cancers show particularly high activity for mTOR signaling. mTOR is a serine/threonine kinase which belongs to the phosphatidylinositol 3-kinase (PI3K) super family and is the target of a widely branched signaling pathway that activates mTOR among other downstream effectors [68]. Interestingly, it was recently demonstrated in the hematopoietic system that deletion of the upstream signaling molecule Pten leads to depletion of normal stem cells while also resulting in an expansion of leukemia-initiating cells [69]. For pancreatic CSCs, single-agent therapy with rapamycin alone also resulted in a significant decrease in $CD133^{+}$ CSCs but did not completely abrogated the in vivo tumorigenicity of the remaining cells. Only inhibition of both the Hh pathway and mTOR by the combined use of cyclopamine and rapamycin, together with gemcitabine, resulted in the desired complete depletion of the pancreatic CSC pool. Implantation of cells that were pretreated ex vivo demonstrated that tumorigenic activity was completely abrogated. In a clinically more relevant setting, the authors then investigated the effects of the triple therapy on established pancreatic cancers utilizing patient-derived pancreatic cancer tissues [56]. For the first time, the authors were thus able to show that a multimodal therapy involving the inhibition of 2 relevant stem cell pathways and additional chemotherapy represents a very promising approach, resulting in virtually complete elimination of CSC, significantly reduced tumorigenic and metastatic activity, and long-term event-free survival (fig. 3).

In vivo Imaging and Drug Delivery to Target Cancer Stem Cells

These recent breakthroughs in the CSC field also highlight the need to find novel strategies for the combined in vivo tracking and subsequent elimination of CSCs. Current techniques for CSC tracking mainly utilize in vitro approaches, such as fluorescence-activated cell sorting (FACS) which is capable of efficiently isolating viable populations of rare cells from the bulk of tumor cells based on their expression of several cell surface markers, and/or histological analysis of the generated tumors. However, in vivo molecular imaging may allow noninvasive tracking of CSCs in their natural in vivo environment without prior disruption of the microenvironmental context. Therefore, in vivo imaging should enable us to understand molecular pathways as well as physiologic or anatomical changes in living organisms in a noninvasive or minimally invasive manner. In this context, the selection of a specific imaging modality and a reporter gene/probe should be based on the intended use (animal or human) and spatial requirements (organs versus cellular resolution and depth) [70].

Molecular imaging enables visualization, characterization, and quantification of biological processes at the cellular and subcellular level [71]. This technique, which is mostly optical based, encompasses various distinct fields including genetic, proteomic, and metabolic imaging and is therefore collectively called molecular-genetic imaging for its role in imaging reporter genes [71]. The developments in the field of optical imaging, such as fluorescent and bioluminescent techniques, although still far from being used in the clinical scenario, have gained considerable momentum for small animal imaging in the context of visualization of the movement of specific cells, monitoring of promoter activity, or distribution of cellular factors using reporters emitting photons in the near-infrared region. Fortunately, molecular imaging is a rather affordable technology that also bears the distinct advantage of being simple, convenient to use, and more user friendly than most other imaging technologies.

In this respect, bioluminescence has emerged as a particularly potential imaging modality for detecting low levels of gene expression. Using bioluminescence imaging, Tiede et al. [72] was able to noninvasively and longitudinally track mammary stem cells in genetically engineered mouse models by quantifying luciferase expression in vivo. Bioluminescent imaging has also been used for investigating mesenchymal stem cell tropism in tumor and wound microenvironments [73, 74]. Finally, this technology has also been used for imaging human embryonic stem cells transplanted into chick and murine models [75]. On the other hand, when physiologic or biochemical changes are of interest, single-photon emission tomography (SPECT) [76] or positron emission tomography (PET) [77] are of great use. However, for simultaneous recording of anatomic morphologic changes, conventional methods such as radiography, ultrasound, computed X-ray tomography (CT), and magnetic resonance imaging (MRI) are very valuable technologies [78]. Indeed, MRI has evolved as a very reliable and robust imaging modality owing it to its high spatial resolution which allows the extraction of physiologic and anatomic information simultaneously. MRI has been extensively used for detecting single cells in small animals for a wide range of applications extending from stem cell to cancer cell tracking [79]. By labeling human

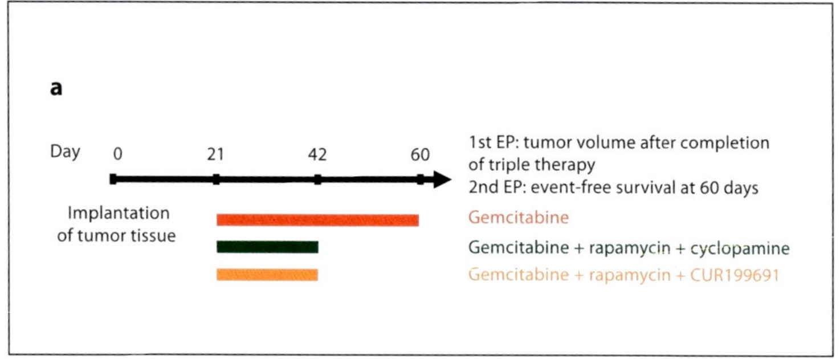

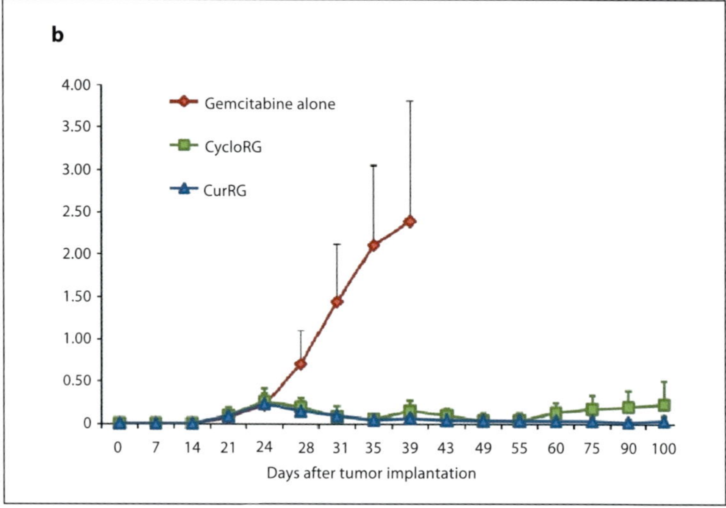

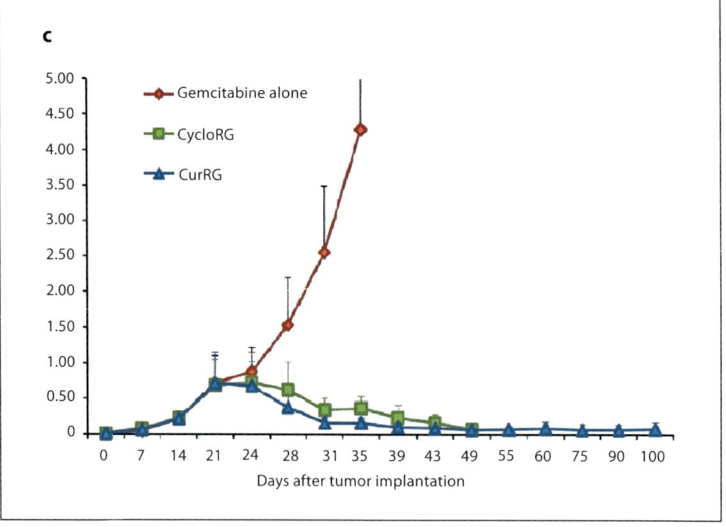

Fig. 3. Study design of combination therapy for the treatment of mice bearing primary pancreatic cancers. **a** Mice were followed for 100 days for relapse of cancers after treatment with only a chemotherapeutic drug (gemcitabine) and the combination of this drug with other molecules involved in the Hh and mTOR pathway (cyclopamine, rapamycin, and CUR199691). 1st EP = Primary endpoint; 2nd EP = secondary endpoint. **b**, **c** The experiment was performed for 2 tumors, i.e. No. 185 (**b**) and 265 (**c**) [56].

Table 2. Characteristics of currently available imaging modalities

Modality	Temporal resolution	Spatial resolution[1]	Penetration depth	Sensitivity	Best performance	Safety	Cost
SPECT	minutes	8–12 mm	whole body	medium	physiologic/ biochemical assays	medium	high
CT	milliseconds	20–300 μm	whole body	medium	anatomic assays	low	medium
MRI	milliseconds	10–100 μm	whole body	low	anatomic/ physiologic assays	high	high
PET	minutes	3–4 mm	whole body	high	physiologic/ biochemical assays	medium	high
US	milliseconds	50 μm	whole body	high	anatomic assays	high	medium
Optical	milliseconds	1–10 mm	cellular level	high	genetic assays	high	low

US = Ultrasound.
[1] Usually depends on the size of the imaged object; therefore, a range is provided.

(LNCaP, DU145, and PC3) and rodent prostate cancer cells (TRAMPC1 and YPEN-1), human breast cancer cells (MDA-MB-231), and mouse mammary cancer cells (MYC/VEGF) with fluorescent superparamagnetic iron oxide nanoparticles (SPION), Rodriguez et al. [80] demonstrated the feasibility of in vivo tracking using fluorescence stereomicroscopy and 3-D MRI.

The advantages and disadvantages of each imaging modality with respect to their spatial and temporal resolution, sensitivity, safety, and cost have been previously discussed in detail [81] but are also summarized in table 2. Importantly, none of these imaging techniques on its own possess all of the required capabilities for comprehensive imaging. Therefore, multimodal imaging is quickly gaining momentum in state-of-the-art biomedical research, clinical diagnostics, and therapeutics. In particular, using multimodal techniques, expression levels of exogenous and endogenous genes can be visualized just as well as signal transduction pathways, protein-protein interactions, and nuclear receptor activities as well as intracellular phenomena. This enables a prudent approach towards monitoring tumor mass and composition including visualization of the in vivo distribution of target cells, such as immune cells and stem cells. This is useful when screening the body for primary-to-metastatic cancer [82, 83], neuroassessment for gliomas [84], and functional neuroimaging examinations [85]. The multimodal systems combine 2 or more imaging modalities, preferably in a single examination, e.g. dual- or triple-labeled optical or nuclear medicine reporter agents or by performing ultrasound or optical studies together with PET or fluorescent imaging of reporter genes with MRI, SPECT, or CT.

Excellent examples of potential multimodal probes are magnetic nanoparticles. They can be used as diagnostic agents as well as drug nanocarriers for therapeutic purposes. In addition to functioning as MRI contrast-enhancing probes due to their magnetic core, they can also be used as a platform for the addition of several functional moieties including fluorescence tags, radionuclides, and other biomolecules for cellular trafficking, cell-specific targeting, and/or gene or

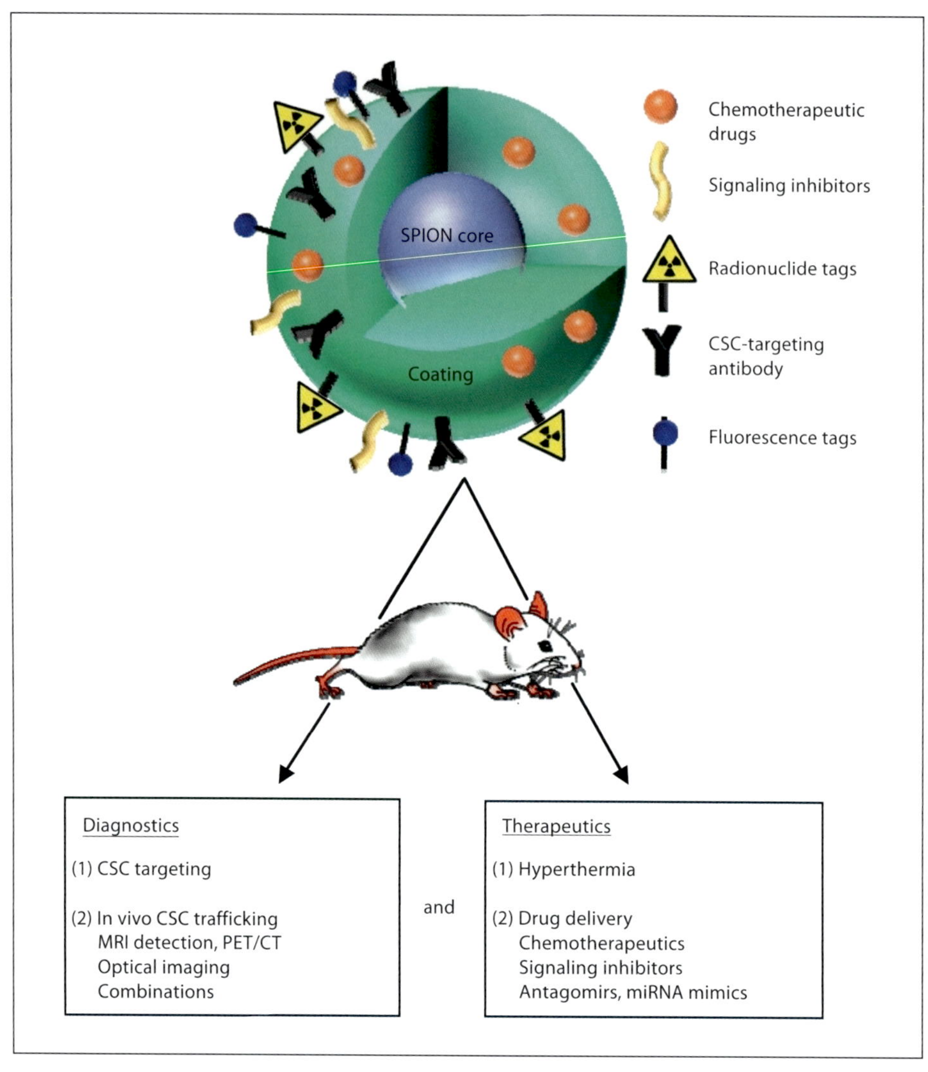

Fig. 4. Overview of the multimodal imaging platform for CSC tracking and targeted elimination. Schematic design of a multimodal SPION functionalized with CSC-specific antibodies, radionuclide, and fluorescence tags for cancer diagnostic purposes and conjugated with signaling inhibitors and chemotherapeutic drugs for cancer therapy purposes.

drug delivery (fig. 4). The efficiency of nanoparticles for medical applications depends on their cellular specificity, organ distribution (which is also size related), cellular uptake, the ability to enhance their visualization, drug delivery potential (therapeutic potential), and most importantly their toxicity to the human organism [86–89]. For this purpose, in order to improve each of these aspects, different methods of nanoparticle synthesis (e.g. coprecipitation, thermal decomposition, or laser pyrolysis), different coatings of the nanoparticle core [anionic (e.g. dimercaptosuccinic acid; DMSA], and cationic (e.g. aminodextran) and neutral (e.g. dextran) conjugation

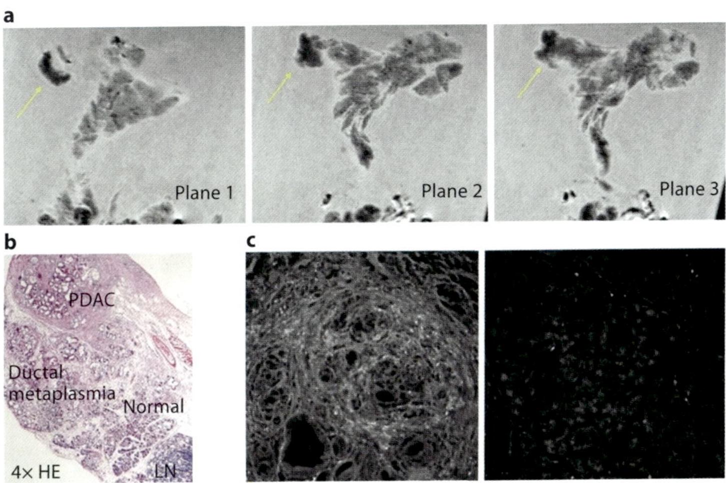

Fig. 5. Nanoparticle uptake in pancreatic ductal adenocarcinoma (PDAC). **a** Three consecutives slices from an ex vivo MRI of a mouse pancreas show NP-Cy5.5-PTP (PTP = plectin-1 targeted peptide) uptake (yellow arrow). **b** Pancreas presenting regions of pancreatic ductal adenocarcinoma and ductal metaplasia and a normal region. **c** Fluorescence microscopy of the sections demonstrated nanoparticle accumulation in areas of pancreatic ductal adenocarcinoma (left) but not in areas of normal pancreas (right) [91]. LN = Lymph node.

with distinct functional moieties are currently being developed and tested [86].

SPION have also been used as MRI contrast agents for imaging different organs including the liver, spleen, brain, and pancreas [87, 90]. They have also been used as tumor diagnostics based on their selective accumulation in tumors. For example, detection of pancreas cancer has been enhanced by the use of SPION targeting plectin-1 (fig. 5) [91] or urokinase plasminogen activator (uPAR, a cellular receptor that is highly expressed in pancreatic cancer and tumor stromal cells) (fig. 6) [87] in combination with optical imaging (e.g. Cy5.5, a near-infrared molecule). On the other hand, SPION can also be used for therapeutic purposes when utilized as drug nanocarriers or for generation of hyperthermia using oscillating magnetic fields [92]. In case of drug nanocarriers, therapeutic agents can be delivered into the cell either in a temperature- [93] (usually using nanoparticle made of natural materials or polymers) or a pH-dependent manner [94]. For example, magnetic nanoparticles conjugated with the cytokine interferon gamma (IFN-γ) as a tumor suppressor have delivered in vitro in a pH-dependent manner to murine macrophages for their activation leading to increased production and secretion of several cytokines [94].

Other approaches that have been explored are the utility of metallic gold nanoparticles. They have been used for colorimetric sensing and ultrasensitive surface-enhanced Raman detection of biomolecules such as DNA and cancer biomarkers [95, 96] as well as for drug delivery [88]. Furthermore, quantum dots have been used as probes for optically imaging many biological systems ranging from DNA, small organelles, and tumors to biological activities including cell-to-cell interactions and cell signaling processes [97, 98].

Taken together, the field of (cancer) stem cell tracking using nanotechnology-based approaches is gaining momentum in the field of oncology and cancer therapeutics. However, this technology is still lacking approaches aimed at the development of multifunctional nanoparticles. We envisage that the coming decade will see major advancements in the combination of the described novel imaging agents, new specific (cancer) stem cell markers, novel therapeutic molecules, and multimodal imaging in order to track (cancer) stem cells in tissues and circulating CSCs.

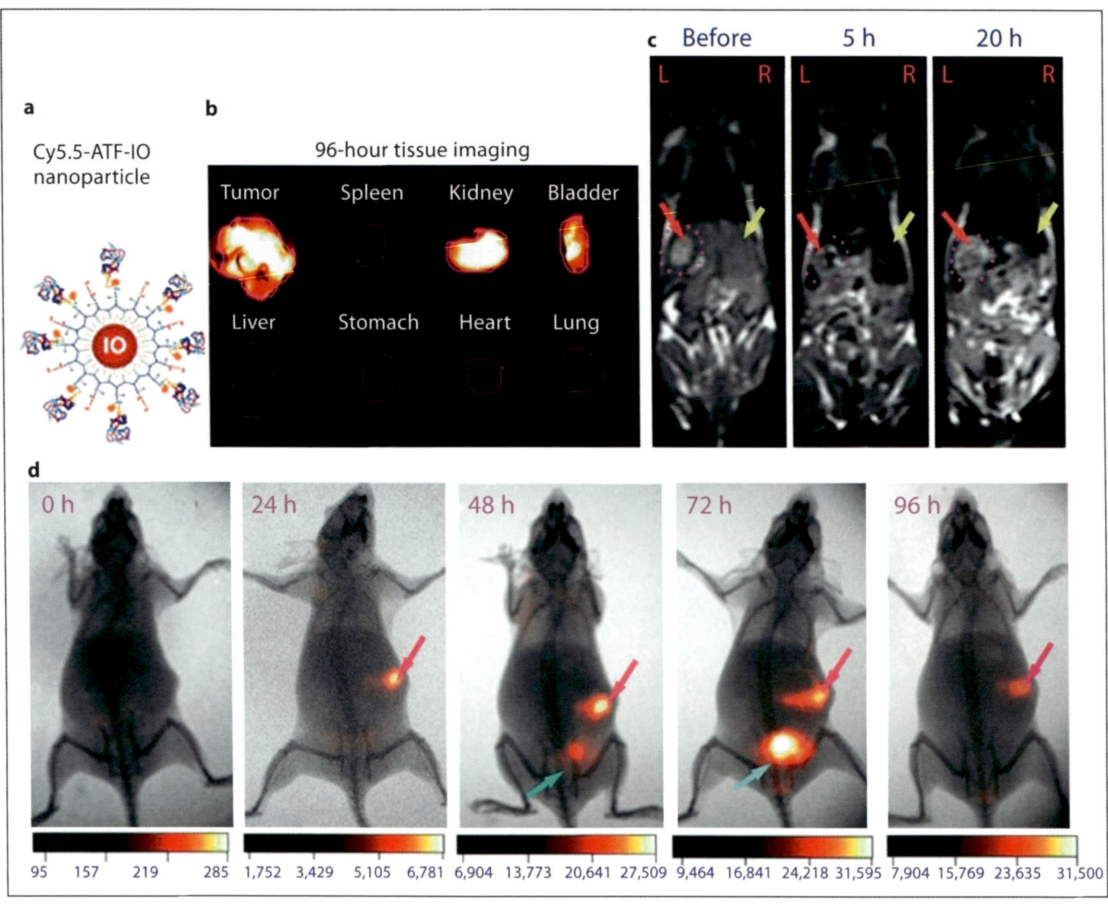

Fig. 6. Iron oxide nanoparticles for in vivo imaging of a pancreatic tumor. **a** Iron oxide nanoparticle conjugated with Cy5.5 (for its fluorescent detection) and further functionalized with an amino-terminal fragment [ATF; sequence of aminoacids with high affinity for urokinase plasminogen activator receptor (uPAR)] for specific targeting of pancreatic tumor cells. **b** Organ distribution of Cy5.5-ATF-IO nanoparticles after 96 h of their injection into a mouse. **c** MRI of a tumor-bearing mouse before and after injection (5 and 20 h) of ATF-IO nanoparticles. **d** NIR images of the mouse after Cy5.5-ATF-IO nanoparticle injection for a period of 96 h [87].

Conclusions and Perspectives

According to the cancer progression model postulated by Fearon and Vogelstein [99] in 1990, at least 4–5 genetic events are required for progression from normal epithelium to carcinoma. Due to their extended life span, stem cells represent a rather likely target for the accumulation of these genetic events. However, irrespective of their cell of origin, CSCs seem to harbor mechanisms which protect them from standard therapy [8, 56, 100]. To further foster our understanding of CSC biology, synergy between the development of novel probes such as nanoparticles and the corresponding imaging modalities will be of paramount importance in building strategies for the robust and efficient tracking of CSCs and their respective niches under both in vitro and in vivo conditions. These studies will pave the way to a better understanding of the underlying regulatory mechanisms of CSCs and develop platforms for targeted theragnostics. Evidence is accumulating for putative therapeutic approaches to overcome resistance mechanisms in CSCs, thus promoting the search for new and better clinical therapies based on the CSC concept, which may eventually help improving the prognosis of patients suffering from this deadly disease.

Disclosure Statement

The authors have nothing to disclose.

References

1 Jemal A, Siegel R, Ward E, Hao Y, Xu J, Murray T, Thun MJ: Cancer statistics 2008. J Clin 2008;58:71–96.

2 Heppner GH, Miller BE: Tumor heterogeneity: biological implications and therapeutic consequences. Cancer Metastasis Rev 1983;2:5–23.

3 Southam CM, Brunschwig A: Quantitative studies of autotranplantation of human cancer. Cancer 1961;14:971–978.

4 Bonnet D, Dick JE: Human acute myeloid leukemia is organized as a hierarchy that originates from a primitive hematopoietic cell. Nat Med 1997;3:730–737.

5 Reya T, Morrison SJ, Clarke MF, Weissman IL: Stem cells, cancer, and cancer stem cells. Nature 2001;414:105–111.

6 Al-Hajj M, Wicha MS, Benito-Hernandez A, Morrison S, Clarke MF: Prospective identification of tumorigenic breast cancer cells. Proc Natl Acad Sci USA 2003;100:3983–3988.

7 Li C, Heidt DG, Dalerba P, Burant CF, Zhang L, Adsay V, Wicha M, Clarke MF, Simeone DM: Identification of pancreatic cancer stem cells. Cancer Res 2007;67:1030–1037.

8 Hermann P, Huber S, Herrler T, Aicher A, Ellwart J, Guba M, Bruns C, Heeschen C: Distinct populations of cancer stem cells determine tumor growth and metastatic activity in human pancreatic cancer. Cell Stem Cell 2007;1:313–323.

9 Ricci-Vitiani L, Lombardi DG, Pilozzi E, Biffoni M, Todaro M, Peschle C, De Maria R: Identification and expansion of human colon-cancer-initiating cells. Nature 2007;445:111–115.

10 Singh SK, Hawkins C, Clarke ID, Squire JA, Bayani J, Hide T, Henkelman RM, Cusimano MD, Dirks PB: Identification of human brain tumour initiating cells. Nature 2004;432:396–401.

11 Collins AT, Berry PA, Hyde C, Stower MJ, Maitland NJ: Prospective identification of tumorigenic prostate cancer stem cells. Cancer Res 2005;65:10946–10951.

12 Miki J, Furusato B, Li H, Gu Y, Takahashi H, Egawa S, Sesterhenn IA, McLeod DG, Srivastava S, Rhim JS: Identification of putative stem cell markers, CD133 and CXCR4, in hTERT-immortalized primary nonmalignant and malignant tumor-derived human prostate epithelial cell lines and in prostate cancer specimens. Cancer Res 2007;67:3153–3161.

13 Fang D, Nguyen TK, Leishear K, Finko R, Kulp AN, Hotz S, Van Belle PA, Xu X, Elder DE, Herlyn M: A tumorigenic subpopulation with stem cell properties in melanomas. Cancer Res 2005;65:9328–9337.

14 Schatton T, Murphy G, Frank N, Yamaura K, Waaga-Gasser A, Gasser M, Zhan Q, Jordan S, Duncan L, Weishaupt C, Fuhlbrigge R, Kupper T, Sayegh M, Frank M: Identification of cells initiating human melanomas. Nature 2008;451:345–349.

15 Kim C, Jackson E, Woolfenden A, Lawrence S, Babar I, Vogel S, Crowley D, Bronson R, Jacks T: Identification of bronchioalveolar stem cells in normal lung and lung cancer. Cell 2005;121:823–835.

16 Bertolini G, Roz L, Perego P, Tortoreto M, Fontanella E, Gatti L, Pratesi G, Fabbri A, Andriani F, Tinelli S, Roz E, Caserini R, Lo Vullo S, Camerini T, Mariani L, Delia D, Calabro E, Pastorino U, Sozzi G: Highly tumorigenic lung cancer $CD133^+$ cells display stem-like features and are spared by cisplatin treatment. Proc Natl Acad Sci USA 2009;106:16281–16286.

17 Prince M, Sivanandan R, Kaczorowski A, Wolf G, Kaplan M, Dalerba P, Weissman I, Clarke M, Ailles L: Identification of a subpopulation of cells with cancer stem cell properties in head and neck squamous cell carcinoma. Proc Natl Acad Sci USA 2007;104:973–978.

18 Significance of $CD90^+$ cancer stem cells in human liver cancer. Cancer Cell 2008;13:153–166.

19 Ma S, Chan K, Hu L, Lee T, Wo J, Ng I, Zheng B, Guan X: Identification and characterization of tumorigenic liver cancer stem/progenitor cells. Gastroenterology 2007;132:2542–2556.

20 Philip PA, Mooney M, Jaffe D, Eckhardt G, Moore M, Meropol N, Emens L, O'Reilly E, Korc M, Ellis L, Benedetti J, Rothenberg M, Willett C, Tempero M, Lowy A, Abbruzzese J, Simeone D, Hingorani S, Berlin J, Tepper J: Consensus report of the National Cancer Institute clinical trials planning meeting on pancreas cancer treatment. J Clin Oncol 2009;27:5660–5669.

21 Burris HA 3rd, Moore MJ, Andersen J, Green MR, Rothenberg ML, Modiano MR, Cripps MC, Portenoy RK, Storniolo AM, Tarassoff P, Nelson R, Dorr FA, Stephens CD, Von Hoff DD: Improvements in survival and clinical benefit with gemcitabine as first-line therapy for patients with advanced pancreas cancer: a randomized trial. J Clin Oncol 1997;15:2403–2413.

22 Ahlgren JD: Chemotherapy for pancreatic carcinoma. Cancer 1996;78:654–663.

23 Jemal A, Tiwari RC, Murray T, Ghafoor A, Samuels A, Ward E, Feuer EJ, Thun MJ: Cancer statistics, 2004. CA Cancer J Clin 2004;54:8–29.

24 Rosenberg L: Treatment of pancreatic cancer: promises and problems of tamoxifen, somatostatin analogs, and gemcitabine. Int J Pancreatol 1997;22: 81–93.
25 Rothenberg ML, Moore MJ, Cripps MC, Andersen JS, Portenoy RK, Burris HA 3rd, Green MR, Tarassoff PG, Brown TD, Casper ES, Storniolo AM, Von Hoff DD: A phase II trial of gemcitabine in patients with 5-FU-refractory pancreas cancer. Ann Oncol 1996;7:347–353.
26 Warshaw AL, Fernandez-del Castillo C: Pancreatic carcinoma. N Engl J Med 1992;326:455–465.
27 Moore MJ, Goldstein D, Hamm J, Figer A, Hecht JR, Gallinger S, Au HJ, Murawa P, Walde D, Wolff RA, Campos D, Lim R, Ding K, Clark G, Voskoglou-Nomikos T, Ptasynski M, Parulekar W: Erlotinib plus gemcitabine compared with gemcitabine alone in patients with advanced pancreatic cancer: a phase III trial of the National Cancer Institute of Canada Clinical Trials Group. J Clin Oncol 2007;25:1960–1966.
28 Nowell PC: The clonal evolution of tumor cell populations. Science 1976;194: 23–28.
29 Clarke MF, Dick JE, Dirks PB, Eaves CJ, Jamieson CH, Jones DL, Visvader J, Weissman IL, Wahl GM: Cancer stem cells – perspectives on current status and future directions: AACR Workshop on cancer stem cells. Cancer Res 2006; 66:9339–9344.
30 Visvader JE, Lindeman GJ: Cancer stem cells in solid tumours: accumulating evidence and unresolved questions. Nat Rev Cancer 2008;8:755–768.
31 Barabe F, Kennedy JA, Hope KJ, Dick JE: Modeling the initiation and progression of human acute leukemia in mice. Science 2007;316:600–604.
32 Wicha MS: Cancer stem cells: an old idea – a paradigm shift. Cancer Res 2006;66: 1883–1890, discussion 1895–1896.
33 Bar EE, Chaudhry A, Lin A, Fan X, Schreck K, Matsui W, Piccirillo S, Vescovi AL, DiMeco F, Olivi A, Eberhart CG: Cyclopamine-mediated hedgehog pathway inhibition depletes stem-like cancer cells in glioblastoma. Stem Cells 2007;25:2524–2533.
34 Hermann PC, Huber SL, Heeschen C: Metastatic cancer stem cells: a new target for anti-cancer therapy? Cell Cycle 2008;7:188–193.
35 Bruce WR, Van Der Gaag H: A quantitative assay for the number of murine lymphoma cells capable of proliferation in vivo. Nature 1963;199:79–80.
36 Park CH, Bergsagel DE, McCulloch EA: Mouse myeloma tumor stem cells: a primary cell culture assay. J Natl Cancer Inst 1971;46:411–422.
37 Wright MH, Calcagno AM, Salcido CD, Carlson MD, Ambudkar SV, Varticovski L: Brca1 breast tumors contain distinct $CD44^+/CD24^-$ and $CD133^+$ cells with cancer stem cell characteristics. Breast Cancer Res 2008;10:R10.
38 Hwang-Verslues WW, Kuo WH, Chang PH, Pan CC, Wang HH, Tsai ST, Jeng YM, Shew JY, Kung JT, Chen CH, Lee EY, Chang KJ, Lee WH: Multiple lineages of human breast cancer stem/progenitor cells identified by profiling with stem cell markers. PLoS One 2009;4:e8377.
39 Ginestier C, Hur MH, Charafe-Jauffret E, Monville F, Dutcher J, Brown M, Jacquemier J, Viens P, Kleer CG, Liu S, Schott A, Hayes D, Birnbaum D, Wicha MS, Dontu G: ALDH1 is a marker of normal and malignant human mammary stem cells and a predictor of poor clinical outcome. Cell Stem Cell 2007;1: 555–567.
40 Pece S, Tosoni D, Confalonieri S, Mazzarol G, Vecchi M, Ronzoni S, Bernard L, Viale G, Pelicci PG, Di Fiore PP: Biological and molecular heterogeneity of breast cancers correlates with their cancer stem cell content. Cell 2010;140: 62–73.
41 Son MJ, Woolard K, Nam DH, Lee J, Fine HA: SSEA-1 is an enrichment marker for tumor-initiating cells in human glioblastoma. Cell Stem Cell 2009;4:440–452.
42 Boiko AD, Razorenova OV, van de Rijn M, Swetter SM, Johnson DL, Ly DP, Butler PD, Yang GP, Joshua B, Kaplan MJ, Longaker MT, Weissman IL: Human melanoma-initiating cells express neural crest nerve growth factor receptor CD271. Nature 2010;466:133–137.
43 O'Brien CA, Pollett A, Gallinger S, Dick JE: A human colon cancer cell capable of initiating tumour growth in immunodeficient mice. Nature 2007;445:106–110.
44 Ricci-Vitiani L, Lombardi D, Pilozzi E, Biffoni M, Todaro M, Peschle C, De Maria R: Identification and expansion of human colon-cancer-initiating cells. Nature 2007;445:111–115.
45 Dalerba P, Dylla SJ, Park IK, Liu R, Wang X, Cho RW, Hoey T, Gurney A, Huang EH, Simeone DM, Shelton AA, Parmiani G, Castelli C, Clarke MF: Phenotypic characterization of human colorectal cancer stem cells. Proc Natl Acad Sci USA 2007;104:10158–10163.
46 Li C, Heidt D, Dalerba P, Burant C, Zhang L, Adsay V, Wicha M, Clarke M, Simeone D: Identification of pancreatic cancer stem cells. Cancer Research 2007;67:1030–1037.
47 Feldmann G, Dhara S, Fendrich V, Bedja D, Beaty R, Mullendore M, Karikari C, Alvarez H, Iacobuzio-Donahue C, Jimeno A, Gabrielson KL, Matsui W, Maitra A: Blockade of hedgehog signaling inhibits pancreatic cancer invasion and metastases: a new paradigm for combination therapy in solid cancers. Cancer Res 2007;67:2187–2196.
48 Jimeno A, Feldmann G, Suarez-Gauthier A, Rasheed Z, Solomon A, Zou GM, Rubio-Viqueira B, Garcia-Garcia E, Lopez-Rios F, Matsui W, Maitra A, Hidalgo M: A direct pancreatic cancer xenograft model as a platform for cancer stem cell therapeutic development. Mol Cancer Ther 2009;8:310–314.
49 Rasheed ZA, Yang J, Wang Q, Kowalski J, Freed I, Murter C, Hong SM, Koorstra JB, Rajeshkumar NV, He X, Goggins M, Iacobuzio-Donahue C, Berman DM, Laheru D, Jimeno A, Hidalgo M, Maitra A, Matsui W: Prognostic significance of tumorigenic cells with mesenchymal features in pancreatic adenocarcinoma. J Natl Cancer Inst 2010; 102:340–351.
50 Quintana E, Shackleton M, Sabel MS, Fullen DR, Johnson TM, Morrison SJ: Efficient tumour formation by single human melanoma cells. Nature 2008; 456:593–598.
51 Hermann PC, Mueller MT, Heeschen C: Pancreatic cancer stem cells – insights and perspectives. Exp Opin Biol Ther 2009;9:1271–1278.

52 Shimizu K, Itoh T, Shimizu M, Ku Y, Hori Y: CD133 expression pattern distinguishes intraductal papillary mucinous neoplasms from ductal adenocarcinomas of the pancreas. Pancreas 2009;38:e207–e214.

53 Deng S, Yang X, Lassus H, Liang S, Kaur S, Ye Q, Li C, Wang LP, Roby KF, Orsulic S, Connolly DC, Zhang Y, Montone K, Butzow R, Coukos G, Zhang L: Distinct expression levels and patterns of stem cell marker, aldehyde dehydrogenase isoform 1 (ALDH1), in human epithelial cancers. PLoS One 2010; 5:e10277.

54 Hirschmann-Jax C, Foster AE, Wulf GG, Nuchtern JG, Jax TW, Gobel U, Goodell MA, Brenner MK: A distinct 'side population' of cells with high drug efflux capacity in human tumor cells. Proc Natl Acad Sci USA 2004;101: 14228–14233.

55 Burkert J, Otto W, Wright N: Side populations of gastrointestinal cancers are not enriched in stem cells. J Pathol 2008;214:564–573.

56 Mueller MT, Hermann PC, Witthauer J, Rubio-Viqueira B, Leicht SF, Huber S, Ellwart JW, Mustafa M, Bartenstein P, D'Haese JG, Schoenberg MH, Berger F, Hidalgo M, Heeschen C: Combined targeted treatment to eliminate tumorigenic cancer stem cells in human pancreatic cancer. Gastroenterology 2009; 137:1102–1113.

57 Immervoll H, Hoem D, Sakariassen P, Steffensen O, Molven A: Expression of the 'stem cell marker' CD133 in pancreas and pancreatic ductal adenocarcinomas. BMC Cancer 2008;8:48.

58 Maeda S, Shinchi H, Kurahara H, Mataki Y, Maemura K, Sato M, Natsugoe S, Aikou T, Takao S: CD133 expression is correlated with lymph node metastasis and vascular endothelial growth factor-C expression in pancreatic cancer. Br J Cancer 2008;98:1389–1397.

59 Kemper K, Sprick MR, de Bree M, Scopelliti A, Vermeulen L, Hoek M, Zeilstra J, Pals ST, Mehmet H, Stassi G, Medema JP: The AC133 epitope, but not the CD133 protein, is lost upon cancer stem cell differentiation. Cancer Res 2010;70:719–729.

60 Ding L, Ellis MJ, Li S, Larson DE, Chen K, Wallis JW, Harris CC, McLellan MD, Fulton RS, Fulton LL, Abbott RM, Hoog J, Dooling DJ, Koboldt DC, Schmidt H, Kalicki J, Zhang Q, Chen L, Lin L, Wendl MC, McMichael JF, Magrini VJ, Cook L, McGrath SD, Vickery TL, Appelbaum E, Deschryver K, Davies S, Guintoli T, Crowder R, Tao Y, Snider JE, Smith SM, Dukes AF, Sanderson GE, Pohl CS, Delehaunty KD, Fronick CC, Pape KA, Reed JS, Robinson JS, Hodges JS, Schierding W, Dees ND, Shen D, Locke DP, Wiechert ME, Eldred JM, Peck JB, Oberkfell BJ, Lolofie JT, Du F, Hawkins AE, O'Laughlin MD, Bernard KE, Cunningham M, Elliott G, Mason MD, Thompson DM Jr, Ivanovich JL, Goodfellow PJ, Perou CM, Weinstock GM, Aft R, Watson M, Ley TJ, Wilson RK, Mardis ER: Genome remodelling in a basal-like breast cancer metastasis and xenograft. Nature 2010;464:999–1005.

61 Shmelkov S, Butler J, Hooper A, Hormigo A, Kushner J, Milde T, St Clair R, Baljevic M, White I, Jin D, Chadburn A, Murphy A, Valenzuela D, Gale N, Thurston G, Yancopoulos G, D'Angelica M, Kemeny N, Lyden D, Rafii S: CD133 expression is not restricted to stem cells, and both $CD133^+$ and $CD133^-$ metastatic colon cancer cells initiate tumors. J Clin Invest 2008:118:2111–2120.

62 Morrison SJ, Kimble J: Asymmetric and symmetric stem-cell divisions in development and cancer. Nature 2006;441: 1068–1074.

63 Goodell MA, Brose K, Paradis G, Conner AS, Mulligan RC: Isolation and functional properties of murine hematopoietic stem cells that are replicating in vivo. J Exp Med 1996;183:1797–1806.

64 Wilson A, Laurenti E, Oser G, van der Wath RC, Blanco-Bose W, Jaworski M, Offner S, Dunant CF, Eshkind L, Bockamp E, Lio P, Macdonald HR, Trumpp A: Hematopoietic stem cells reversibly switch from dormancy to self-renewal during homeostasis and repair. Cell 2008;135:1118–1129.

65 Ji Q, Hao X, Zhang M, Tang W, Yang M, Li L, Xiang D, Desano JT, Bommer GT, Fan D, Fearon ER, Lawrence TS, Xu L: MicroRNA miR-34 inhibits human pancreatic cancer tumor-initiating cells. PLoS One 2009;4:e6816.

66 Yauch RL, Gould SE, Scales SJ, Tang T, Tian H, Ahn CP, Marshall D, Fu L, Januario T, Kallop D, Nannini-Pepe M, Kotkow K, Marsters JC, Rubin LL, de Sauvage FJ: A paracrine requirement for hedgehog signalling in cancer. Nature 2008;455:406–410.

67 Tian H, Callahan CA, DuPree KJ, Darbonne WC, Ahn CP, Scales SJ, de Sauvage FJ: Hedgehog signaling is restricted to the stromal compartment during pancreatic carcinogenesis. Proc Natl Acad Sci USA 2009;106:4254–4259.

68 Inoki K, Corradetti MN, Guan KL: Dysregulation of the TSC-mTOR pathway in human disease. Nat Genet 2005;37: 19–24.

69 Yilmaz Ö, Valdez R, Theisen B, Guo W, Ferguson D, Wu H, Morrison S: Pten dependence distinguishes haematopoietic stem cells from leukaemia-initiating cells. Nature 2006;441:475–482.

70 Shah K, Jacobs A, Breakefield XO, Weissleder R: Molecular imaging of gene therapy for cancer. Gene Ther 2004;11:1175–1187.

71 Kang JH, Chung JK: Molecular-genetic imaging based on reporter gene expression. J Nucl Med 2008;49(suppl 2):164S–179S.

72 Tiede BJ, Owens LA, Li F, DeCoste C, Kang Y: A novel mouse model for noninvasive single marker tracking of mammary stem cells in vivo reveals stem cell dynamics throughout pregnancy. PLoS One 2009;4:e8035.

73 Kidd S, Spaeth E, Dembinski JL, Dietrich M, Watson K, Klopp A, Battula VL, Weil M, Andreeff M, Marini FC: Direct evidence of mesenchymal stem cell tropism for tumor and wounding microenvironments using in vivo bioluminescent imaging. Stem Cells 2009; 27:2614–2623.

74 Tolar J, Nauta AJ, Osborn MJ, Panoskaltsis Mortari A, McElmurry RT, Bell S, Xia L, Zhou N, Riddle M, Schroeder TM, Westendorf JJ, McIvor RS, Hogendoorn PC, Szuhai K, Oseth L, Hirsch B, Yant SR, Kay MA, Peister A, Prockop DJ, Fibbe WE, Blazar BR: Sarcoma derived from cultured mesenchymal stem cells. Stem Cells 2007;25:371–379.

75 Priddle H, Grabowska A, Morris T, Clarke PA, McKenzie AJ, Sottile V, Denning C, Young L, Watson S: Bioluminescence imaging of human embryonic stem cells transplanted in vivo in murine and chick models. Cloning Stem Cells 2009;11:259–267.

76 Karacalioglu AO, Jata B, Kilic S, Arslan N, Ilgan S, Ozguven MA: A physiologic approach to decreasing upward creep of the heart during myocardial perfusion imaging. J Nucl Med Technol 2006; 34:215–219.
77 Galban CJ, Bhojani MS, Lee KC, Meyer CR, Van Dort ME, Kuszpit KK, Koeppe RA, Ranga R, Moffat BA, Johnson TD, Chenevert TL, Rehemtulla A, Ross BD: Evaluation of treatment-associated inflammatory response on diffusion-weighted magnetic resonance imaging and 2-[18F]-fluoro-2-deoxy-D-glucose-positron emission tomography imaging biomarkers. Clin Cancer Res 2010;16: 1542–1552.
78 Choueiri NE, Balci NC, Alkaade S, Burton FR: Advanced imaging of chronic pancreatitis. Curr Gastroenterol Rep 2010;12:114–120.
79 Heyn C, Ronald JA, Mackenzie LT, MacDonald IC, Chambers AF, Rutt BK, Foster PJ: In vivo magnetic resonance imaging of single cells in mouse brain with optical validation. Magn Reson Med 2006;55:23–29.
80 Rodriguez O, Fricke S, Chien C, Dettin L, VanMeter J, Shapiro E, Dai HN, Casimiro M, Ileva L, Dagata J, Johnson MD, Lisanti MP, Koretsky A, Albanese C: Contrast-enhanced in vivo imaging of breast and prostate cancer cells by MRI. Cell Cycle 2006;5:113–119.
81 Miller JC, Thrall JH: Clinical molecular imaging. J Am Coll Radiol 2004;1:4–23.
82 Schirner M, Menrad A, Stephens A, Frenzel T, Hauff P, Licha K: Molecular imaging of tumor angiogenesis. Ann NY Acad Sci 2004;1014:67–75.
83 Scheidhauer K, Walter C, Seemann MD: FDG PET and other imaging modalities in the primary diagnosis of suspicious breast lesions. Eur J Nucl Med Mol Imaging 2004;31(suppl 1):S70–S79.
84 Jacobs AH, Dittmar C, Winkeler A, Garlip G, Heiss WD: Molecular imaging of gliomas. Mol Imaging 2002;1: 309–335.
85 Ward NS, Frackowiak RS: Towards a new mapping of brain cortex function. Cerebrovasc Dis 2004;17(suppl 3):35–38.
86 Villanueva A, Canete M, Roca AG, Calero M, Veintemillas-Verdaguer S, Serna CJ, Morales Mdel P, Miranda R: The influence of surface functionalization on the enhanced internalization of magnetic nanoparticles in cancer cells. Nanotechnology 2009;20:115103.
87 Yang L, Mao H, Cao Z, Wang YA, Peng X, Wang X, Sajja HK, Wang L, Duan H, Ni C, Staley CA, Wood WC, Gao X, Nie S: Molecular imaging of pancreatic cancer in an animal model using targeted multifunctional nanoparticles. Gastroenterology 2009;136:1514–1525.e2.
88 De Jong WH, Borm PJ: Drug delivery and nanoparticles: applications and hazards. Int J Nanomedicine 2008;3: 133–149.
89 Giouroudi I, Kosel J: Recent progress in biomedical applications of magnetic nanoparticles. Recent Pat Nanotechnol 2010;4:111–118.
90 Mejias R, Perez-Yague S, Roca AG, Perez N, Villanueva A, Canete M, Manes S, Ruiz-Cabello J, Benito M, Labarta A, Batlle X, Veintemillas-Verdaguer S, Morales MP, Barber DF, Serna CJ: Liver and brain imaging through dimercaptosuccinic acid-coated iron oxide nanoparticles. Nanomedicine (Lond) 2010;5:397–408.
91 Kelly KA, Bardeesy N, Anbazhagan R, Gurumurthy S, Berger J, Alencar H, Depinho RA, Mahmood U, Weissleder R: Targeted nanoparticles for imaging incipient pancreatic ductal adenocarcinoma. PLoS Med 2008;5:e85.
92 Park H, Park HJ, Kim JA, Lee SH, Kim JH, Yoon J, Park TH: Inactivation of *Pseudomonas aeruginosa* PA01 biofilms by hyperthermia using superparamagnetic nanoparticles. J Microbiol Methods 2011;84:41–45.
93 Yu MK, Park J, Jeong YY, Moon WK, Jon S: Integrin-targeting thermally cross-linked superparamagnetic iron oxide nanoparticles for combined cancer imaging and drug delivery. Nanotechnology 2010;21:415102.
94 Mejias R, Costo R, Roca AG, Arias CF, Veintemillas-Verdaguer S, Gonzalez-Carreno T, del Puerto Morales M, Serna CJ, Manes S, Barber DF: Cytokine adsorption/release on uniform magnetic nanoparticles for localized drug delivery. J Control Release 2008;130:168–174.
95 Elghanian R, Storhoff JJ, Mucic RC, Letsinger RL, Mirkin CA: Selective colorimetric detection of polynucleotides based on the distance-dependent optical properties of gold nanoparticles. Science 1997;277:1078–1081.
96 Qian X, Peng XH, Ansari DO, Yin-Goen Q, Chen GZ, Shin DM, Yang L, Young AN, Wang MD, Nie S: In vivo tumor targeting and spectroscopic detection with surface-enhanced Raman nanoparticle tags. Nat Biotechnol 2008; 26:83–90.
97 Michalet X, Pinaud FF, Bentolila LA, Tsay JM, Doose S, Li JJ, Sundaresan G, Wu AM, Gambhir SS, Weiss S: Quantum dots for live cells, in vivo imaging, and diagnostics. Science 2005;307:538–544.
98 Medintz IL, Uyeda HT, Goldman ER, Mattoussi H: Quantum dot bioconjugates for imaging, labelling and sensing. Nat Mater 2005;4:435–446.
99 Fearon ER, Vogelstein B: A genetic model for colorectal tumorigenesis. Cell 1990;61:759–767.
100 Bao S, Wu Q, Mclendon R, Hao Y, Shi Q, Hjelmeland A, Dewhirst M, Bigner D, Rich J: Glioma stem cells promote radioresistance by preferential activation of the DNA damage response. Nature 2006;444:756–760.

Dr. Christopher Heeschen
Stem Cells and Cancer Group, Clinical Research Program
Spanish National Cancer Research Centre (CNIO)
Melchor Fernández Almagro 3, ES–28029 Madrid (Spain)
Tel. +34 91 732 8000, ext. 2911, E-Mail christopher.heeschen@cnio.es

Alexiou C (ed): Nanomedicine – Basic and Clinical Applications in Diagnostics and Therapy.
Else Kröner-Fresenius Symp. Basel, Karger, 2011, vol 2, pp 135–144

Nanomedicine Approaches for Cancer Stem Cell Targeting and Personalized Cancer Treatment

Ines Block · Steffen Schmidt · Pernille Lund Hansen · Angela Riedel · Helle Christiansen · Jan Mollenhauer

Lundbeckfonden Center of Excellence NanoCAN and Molecular Oncology, Institute of Molecular Medicine, University of Southern Denmark, Odense, Denmark

Abstract

Nanotechnology is on the route to providing novel breakthroughs in medicine. While there are already examples of the success of the new field of nanomedicine in cancer therapy, it remains open how nanomedicine can appropriately address future demands for advanced cancer therapy, such as personalization of treatment and elimination of cancer stem cells. This article introduces the particular challenges associated with these goals, which relate to the discussion about the definition of nanodrugs. Interdisciplinary sites of intersection, by which nanomedicine, i.e. research at the small nanoscale, can profit from the most recent developments in 'omics' technologies, i.e. research at the very large scale, are discussed. We propose strategies for the systematic identification and assembly of advanced nanodrugs, to which functional genomics and synthetic biology approaches could make important future contributions. Combining these disciplines can potentially lead to the design of advanced nanodrugs that may meet the challenges of the future.

Nanomedicine – A New Discipline Still Searching for Definitions

As always occurs when a new discipline is launched, the new field of nanomedicine is in search of defining itself. Being an interdisciplinary area, nanomedicine has broad roots, including diverse branches such as diagnostic and therapeutic ones, imaging in clinical and basic research, i.e. at the organism and molecular scale, nanotechnology, molecular biology, pharmacology, toxicology, physics, and chemistry. With the nanoscale aspect as the unifying element, nanomedicine crosses borders between different diseases, and present research covers all main areas, including cardiovascular and neurological diseases as well as infection, inflammation, cancer, and regenerative medicine – to mention only a few examples. In general, a broad definition is the appropriate expression of an interdisciplinary character, while narrow definitions are helpful to visualize the uniqueness of a new field. This creates a certain dilemma which has precipitated even down to the question of what a nanodrug actually is.

The baseline knowledge is that particles with a size of 50 nm or less efficiently enter cells which have diameters in the range of 10,000–20,000 nm. Particles with a size of 20 nm or less are capable of efficiently passing through blood vessels. Hence, an often used definition for a nanodrug is a drug that at least in one dimension has a size of 100 nm or less, which approximately corresponds to the size of a virus. This broad definition is able to cover the prototype nanodrugs that are presently in clinical use, such as doxil, which is often referred to as the first nanodrug. With a diameter of about 100 nm, doxil resides precisely at the maximal limit of the presently used definition.

Why care about a definition for nanodrugs? Such a definition can have far-reaching consequences as it may include or exclude entire research fields from nanomedicine, not least also with respect to funding programs. Furthermore, a definition is a valuable and necessary instrument to visualize novel areas for stakeholders and the public. Finally, the definition of a nanodrug may have a practical impact on analyses of efficacy and toxicity and thereby in the long term eventually even on drug approval processes.

Applying the size criterion in a consequent manner, small molecule drugs and short interfering RNAs (siRNAs) with dimensions in the range of approximately 1–10 nm represent nanodrugs, as do therapeutic proteins and antibodies with dimensions of about 5–20 nm in monomeric form. It may be obvious that this general definition has the shortcoming that it is necessary but not sufficient to emphasize the aspects of novelty, which would discern nanodrugs from conventional therapeutics. In fact, there are ongoing discussions as to whether, for example, antibodies represent nanodrugs or not. Asking a chemist, a physicist, and a molecular biologist, will probably yield at least three different answers about what a nanodrug is. A stricter containment to underscore the novel aspects of nanomedicine could in principle be made by excluding small molecules and biologicals from the definition of a nanodrug. That this is not reasonable is evidenced by the fact that the paradigm of a nanodrug, i.e. doxil, would in this case not score as a nanodrug anymore. It would thus cut off nanomedicine from the entire spectrum of therapeutic molecules we can nowadays make use of, leaving a narrow spectrum of instruments probably not suited to match the future requirements for advanced drugs.

Hence, this article will operate with a pragmatic definition of the term nanodrug; in simple terms: a nanodrug is where size matters. This means a drug whose pharmacological properties can be or have been changed in a beneficial manner by variations in size or structure at the nanoscale. Within this definition, the total dimensions of a nanodrug could in theory be substantially larger than 100 nm if, for example, its positive therapeutic effect is increased or its toxicity is decreased by a nanostructured surface. Accordingly, an antibody, a therapeutic polypeptide, a small molecule, or an siRNA alone would not be sufficient to meet the definition of a nanodrug unless it is, for example, assembled into nanoscaled complexes or combined with nanostructured devices in order to alter its properties in a beneficial way. This definition shall also relate to what the actual goal of nanomedicine is: to provide the patient with improved options for diagnosis and treatment. With this pragmatic definition of a nanodrug it can now be discussed what the actual challenges and demands for future cancer therapies are and what the general design of nanodrugs meeting these requirements could look like.

Cancer Stem Cells and Associated Challenges

The term cancer stem cells (CSCs), also sometimes designated as cancer- (or tumor-) initiating cells, refers to a small population of cancer cells that occurs within most tumors and has specific properties, such as self-renewal, asymmetric division, and pluripotency. This closely resembles the properties of normal embryonic and adult stem cells, which are responsible for tissue and organ development and regeneration, respectively. These analogies have actually led to designating this tumor cell subpopulation as CSCs.

In fact, numerous links between stem cells and cancer have meanwhile been uncovered as reviewed in detail elsewhere [1–3]. For example, mesenchymal stem cells, which maintain the capacity to differentiate into bone-forming cells when immortalized with telomerase [4], have been demonstrated to spontaneously convert in vitro into cells that give rise to tumors. This includes loss-of-function mutations in the tumor suppressor p16 [5]. Stem cells in the intestinal crypts have been shown to give rise to colorectal cancer in vivo [6, 7]. Single bona fide CSCs sorted from melanoma are capable of creating a full-blown tumor when implanted into mice [8], which resembles the capacity of single adult stem cells to recreate a complete organ, as recently shown for the mammary gland [9].

Similar to the situation in nanomedicine, there are also intense discussions going on in the field of CSC research with regard to definitions and conceptual aspects, where part of the scientific community still regards the concept of CSCs with hesitation or criticizes the term 'cancer stem cells'. However, also in this case a pragmatic perspective could help. To this end, it has been shown that most tumors contain a subpopulation of cells that share a number of markers and properties with normal adult stem cells, e.g. positive status for ALDH1. These cells can be isolated from primary tumors as well as established cancer cell lines and are particularly tumorigenic when implanted into mice, highly metastatic, and resistant to most conventional therapies due to specific DNA repair mechanisms and transporters that shuttle drugs to the outside of the cells [1–3]. Estimations of the percentage of this cancer cell population in tumors cover a broad range from less than 1% to up to 25% or even more, depending on the source, the markers used for isolation, and the experimental model systems employed for counting the frequency of these cells [8, 10, 11]. This phenomenological perspective harbors two aspects of major importance for future cancer nanomedicine. Firstly, it would be the bona fide CSCs that need to be addressed by future diagnostics and therapies to achieve further progress, because their elimination bears the promise to potentially overcome cancer recurrence after therapy, therapy resistance, and mortality due to incurable metastatic cancer (fig. 1). Secondly, the circumstance that a special minor population of cells within a tumor needs to be addressed has profound consequences for the selection of the starting points that may be most suitable to design the next generation of nanodrugs. The latter aspect is tightly linked to the goal of personalized cancer nanomedicine, as discussed below.

Personalized Cancer Treatment and Associated Challenges

The ultimate goal of personalized treatment is to identify the therapy that would match the individual patient's tumor in an optimal manner, thus increasing therapy success and saving patients from inappropriate or unnecessary treatment with all of the commonly associated adverse side effects [for a detailed review see 12, 13]. A further predicted beneficial effect in societal terms would be cost savings for health care systems. Two major strategies are nowadays pursued to come closer to this ideal goal. The first strategy comprises attempts to retrospectively personalize non-personalized treatment strategies and basically aims at identifying markers for treatment response prediction to existing conventional drugs, which so far has provided limited success. The second strategy is to design new personalized drugs which target a scenario specific for a cancer cell and for which a responder identification can be performed in advance based on the rational design of the drug. Two recent examples are the discovery of PARP inhibitors for breast cancer therapy and of BRAF inhibitors for melanoma therapy. PARP inhibitors block a protein that is involved in DNA repair and selectively kills cells deficient for the tumor suppressors BRCA1 or BRCA2, which likewise function in DNA repair [14, 15]. Inactivating mutations in BRCA1/2 compromise the correction of DNA errors and thereby can cause breast cancer. Breast cancer cells with such mutations are now addicted to PARP to maintain a basal level of DNA repair for survival, while the normal cells within the same patient do not depend on PARP because they still have functional BRCA1/2. This phenomenon is referred to as pathway addiction [16]. Thus, PARP inhibitors selectively kill breast cancer cells deficient for BRCA1/2, but not normal cells, providing a broad therapeutic window. The specific killing of cells with a molecular defect is referred to as synthetic lethality [16]. BRAF is a kinase involved in the positive regulation of Ras/Raf-MEK-ERK signaling and was discovered by a systematic screen, showing hyperactivating mutations in about 50–60% of the melanoma [17]. BRAF inhibitors have demonstrated promising effects already in clinical phase I studies in patients with otherwise incurable metastatic melanoma [18]. The striking advantage in both examples is that the prospective responders can be identified before starting the treatment based on analyzing BRCA1/2 and BRAF mutations, respectively.

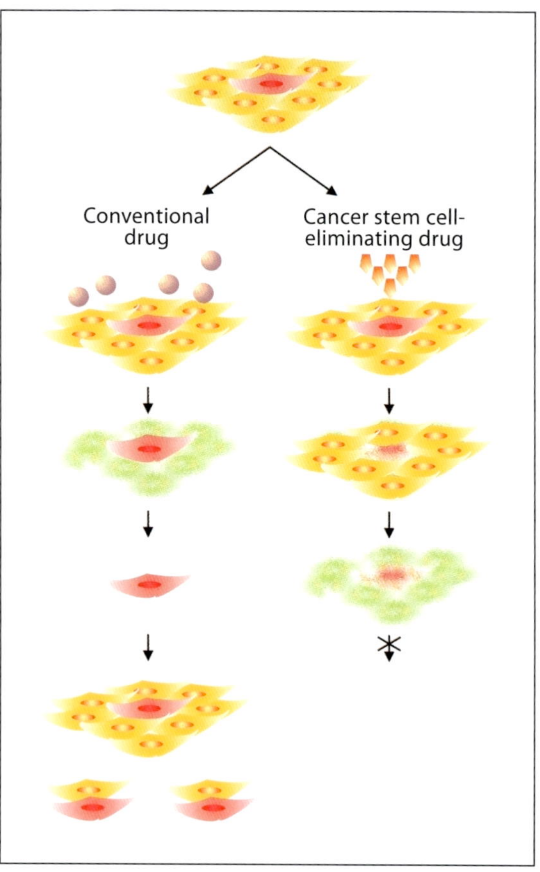

Fig. 1. CSC-eliminating therapy versus conventional cancer therapy. Conventional anticancer drugs (left) may eliminate conventional tumor cells (orange cells), leading to an apparent reduction of the tumor. CSCs (red cells), however, may stay alive, giving rise to a recurrent tumor with a high likelihood of therapy resistance and metastatic spread, resulting in incurable cancer. CSC-eliminating drugs (right) would target and eradicate the CSCs, which would result in a cutoff of the supply to the tumor. Lacking cells that steadily give rise to fresh cancer cells, the tumor would be predicted to starve with minimized risk for cancer recurrence, development of resistance, and metastatic spread.

The concept of CSCs bears some novel particular challenges for personalized cancer treatment. Personalized nanodrugs should in principle be directed against molecular traits or pathways to which the CSCs are addicted. However, the vast majority of the nowadays available molecular profiles have been generated using unfractionated cancer cells as the source and thus do not reflect the profiles of the CSCs which would be needed to design drugs that, for example, eliminate CSCs by causing synthetic lethality. As far as has been investigated, CSCs have been found to display quite different molecular profiles compared to bulk cancer cells [10].

Possibilities of Modern Genomics and Related Technologies

The postgenomics era has given cancer research a number of valuable tools at hand. Determination of the activity patterns of all 20,000 human genes, also called mRNA expression profiling, is meanwhile routinely performed in tumors. It has recently become possible to also analyze the so-called micro-RNAs (miRNAs), short cellular RNA molecules that function as key regulators of entire groups of genes, via chip-based methods. Chip-based methods for comprehensively mapping epigenetic changes, e.g. DNA methylation at a genome-wide scale, are available to analyze cancer cells. Advances in proteomics enable determination of the levels and modifications of several thousands of proteins in a single experiment. Next-generation sequencing has recently begun to add to this powerful portfolio, which allows sequencing of all 20,000 human genes within a tumor within a couple of days to systematically map mutations. The first genome-wide sequencing approach in a number of breast cancer cases identified a set of 139 genes with possibly important mutations [19].

In principle, all of these profiles comprise rich resources for personalized cancer nanomedicine. However, if one intends to target CSCs, the past data generated by most of these techniques unfortunately cannot be used. As mentioned above, the molecular profiles of CSCs are different from the profiles of bulk cancer cells, and the vast majority of the nowadays available profiles have been determined without sorting for CSCs. This essentially narrows down the starting points for CSC-targeting personalized cancer medicine to mutation profiles from next-generation sequencing. Mutations would represent the only molecular changes that are inherited by the CSC offspring and thus would be expected to be informative for CSCs even when using unfractionated tumor material as the starting point (fig. 2). It is, however, predictable that within the next few years an increasing number of molecular profiles which characterize CSCs will become available. Nonetheless, it is remarkable that some of the most promising starting points for personalized cancer treatment have in fact been developed based on mutations – e.g. therapies based on PARP and BRAF inhibition as depicted above.

Conceptual Approaches towards Novel Nanodrugs

The goal of therapeutically addressing CSCs and applying this kind of therapy on a personalized basis poses substantial challenges with regard to the mode the drugs need to operate with. The next generation of drugs will have to exert selective effects in CSCs without harming normal adult stem cells, which have virtually identical cell surface marker profiles and very similar active biological pathways. As a consequence, especially in conjunction with personalization, the most optimal drug target(s) have to be addressed in a highly selective manner. Accordingly, it is predictable that future nanodrugs will contain small molecules and/or biologicals as the active drug and the targeting components. It is difficult to imagine that without such components, i.e. by pure variation of the physical or chemical properties through nanotechnology, the necessary specificity and selectivity can be created. Unless strategies have been developed that enable rational mimicking of the huge diversity of small molecules and biologicals, the valuable role of nanotechnology will be based on its potential to convert less efficient or toxic drugs into more efficient and/or less toxic drugs.

There are arguments for why certain biologicals may break the present dominance of small molecules in cancer therapy. RNA interference is a process which is part of the mechanisms that defend the cell against invading viruses and regulates also other major cellular processes via RNA molecules (miRNAs) produced by the host

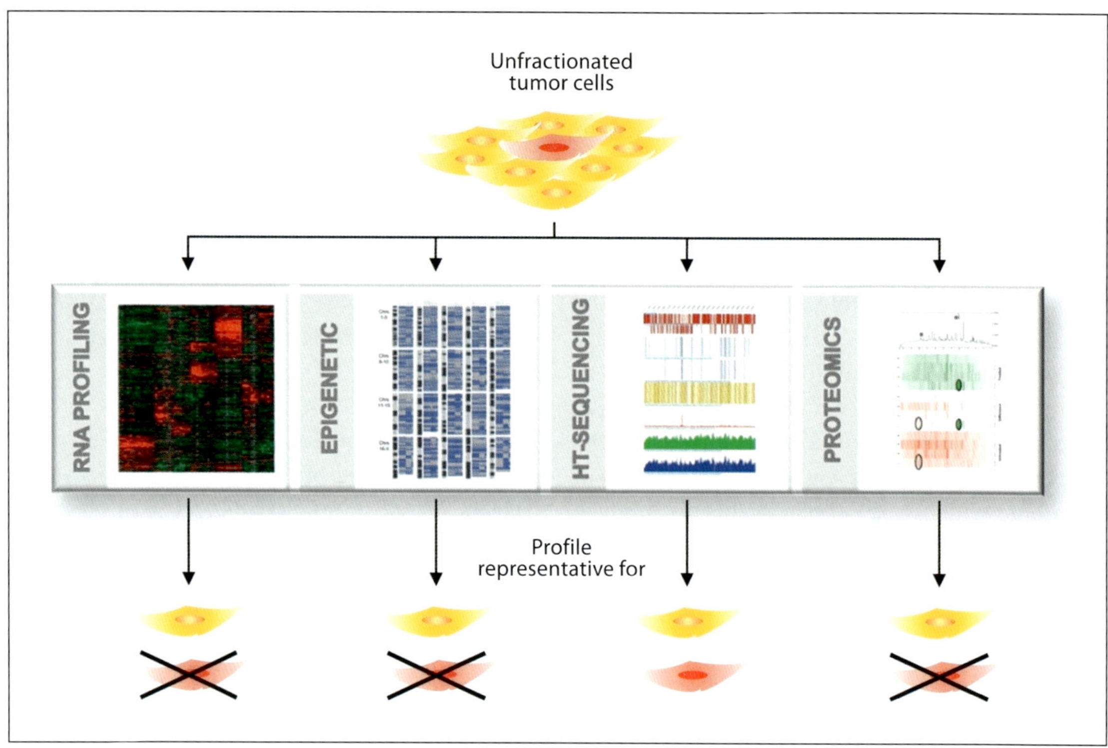

Fig. 2. Impact of CSCs on the applicability of molecular profiles for the development of personalized nanodrugs against cancer. The vast majority of the nowadays available profiles are derived from tumor samples that contain mixtures of cancer cells with a low percentage of CSCs and a high percentage of 'bulk' cancer cells. The mRNA, miRNA, and epigenetic and proteomics profiles are likely different between these two types of cancer cells, because they are expected to change when 'bulk' cancer cells develop from CSCs, similar to what happens to normal differentiated cells that emerge from embryonic or adult stem cells. Thus, these profiles reflect the patterns of the abundant 'bulk' cancer cells. By contrast, important mutations that contributed to tumorigenesis are expected to be inherited by the 'bulk' cancer cells from the CSCs. Hence, mutation profiles determined by next-generation sequencing are expected to be informative for the CSCs despite using unfractionated cancer cells as the starting point. This predicts that mutation profiles represent the most promising starting point for the development of personalized nanodrugs that could eliminate CSCs.

cells. Back in 2001, Elbashir et al. [20] successfully converted this principle into a technique that allows for the silencing (knockdown) of genes in mammalian cells by transfer of 21-mer RNA oligonucleotides (siRNAs) with a match to the target gene sequence. Consecutively, siRNA libraries with genome-wide coverage have been generated which allow for the inactivation of each of the individual 20,000 human genes. In conjunction with advanced robotic screening devices, genome-wide siRNA screens have successfully been employed to systematically identify new components of relevant signaling pathways, developmental and differentiation processes, or novel starting points to understand and intervene in human disease [21–24]. Genome-wide or smaller scaled siRNAs screens have also been successfully applied to find synthetic lethal drug targets in cancer [25–27]. These novel drug targets have the potential to give rise to novel personalized treatment options in the future.

The two most remarkable advantages of using siRNAs as a screening tool and as the prospective active drug component in nanodrugs are that all 20,000 human genes can be targeted, while small molecules commonly target only about 33% of human genes, and the uniform chemistry of the siRNAs compared to the diverse formulations of chemical compounds.

In conclusion, future nanodrugs aimed at eliminating CSCs and at operating in a personalized manner will most probably have to include three major elements: one or more biological targeting device(s) which allow(s) for addressing the target cell type, one or more active compound(s) like small molecules or siRNAs, which represent the 'warheads' exerting the therapeutic effect, and a nanocarrier, which combines these devices in a manner that is most favorable for its pharmacological properties. As discussed below, this would allow for a modular drug design, which could substantially accelerate the drug development process. The continuing need to make use of small molecules and biologicals as nanodrug components supports the view that the definition of nanodrugs should be pragmatic and goal-oriented rather than too restrictive, which may exclude important components from future developments.

A Possible Roadmap to Cancer Stem Cell Targeting and Personalized Cancer Nanomedicine

Developing a feasible roadmap from these considerations, which represents the strategy pursued in the Lundbeckfonden Center of Excellence NanoCAN and the German-Danish High Technology Platform HiT-ID, would mean that the most promising starting point is the mutational profiles of cancers generated by next-generation sequencing (fig. 3), because there is a high likelihood that the relevant mutations are already present in the CSCs. As this technique advances, it may even become a standard diagnostic approach in which the DNA of an individual patient's tumor is completely sequenced in order to provide a comprehensive instrument for therapy decisions on a personalized basis. A possible endpoint is the assembly of nanodrugs with targeting devices that guide the nanoparticles to the CSCs and with active compounds selectively causing synthetic lethality to the CSCs only, because they inhibit pathways to which only the CSCs are addicted due to a causal mutation in one (or more) cancer gene(s). A synthetic lethal active compound is also desirable from the perspective that all CSC markers known to date occur also on normal cells or normal adult stem cells. Making use of the principles of pathway addiction and synthetic lethality will greatly assist in avoiding collateral damage, i.e. toxicity caused by the nanodrugs.

If the active drug and the targeting components can be designed with a uniform chemistry, for example by using RNA to design siRNAs as drug components and RNA- or DNA-aptamers as targeting components, the preclinical drug development process can be parallelized in a favorable fashion and nanocarrier development can focus on the optimal way of delivering RNA (fig. 3). The respective siRNAs can systematically be recovered by comprehensive genome-wide screens and be directly forwarded to optimization processes. Personalization of the nanodrug is possible via the search strategy as well as through exchange of the drug and the targeting components, giving rise to the necessary diversity.

There is still one gap to be closed, i.e. that the relevant gene mutations remain to be found. The recent sequencing of all genes in about 40 breast cancer cases delivered an initial set of 139 genes with potentially important mutations [19]. For more than 90% of these genes either the function or the role of their mutations in cancer is still unknown. The number of such genes can be expected to increase dramatically because 500 cases of each of the most prevalent cancer types are pres-

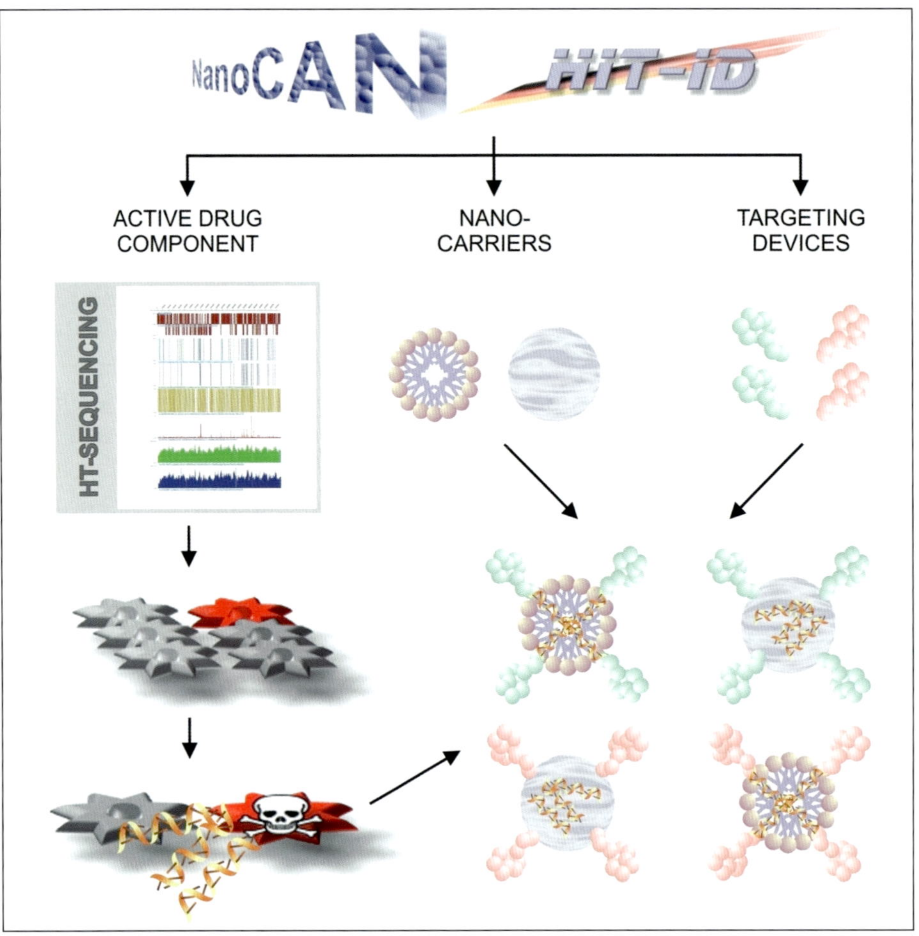

Fig. 3. Strategies for the development of CSC-targeting and personalized nanodrugs. A conceivable strategy, also pursued at the NanoCAN Center and the German-Danish High Technology Platform HiT-ID, is to parallelize the development of the three nanodrug components. The active drug component can be identified by genome-wide synthetic lethal siRNA screens, which identify siRNAs that eliminate only cells with a cancer-specific mutation. Mutation profiles determined by next-generation sequencing can serve as a starting point, which requires that serial functional studies are included to identify the relevant genes and mutations from the comprehensive profiles. Chemical modifications meanwhile allow for stabilization and optimized encapsulation of nucleic acids such as siRNAs so that these can be used directly as the active drug component. Thus, efforts can be focused on developing nanocarriers for nucleic acid delivery. A third branch, i.e. devices for targeting CSC surface markers, can be developed in parallel. Using a uniform chemistry also for the targeting devices, e.g. RNA or DNA aptamers, would allow for combination of each nanocarrier with any of the siRNA drugs and any of the targeting devices so that permutations of nanodrugs for testing can easily be generated. A further advantage of the modular drug design is that exchangeable drug and targeting components allow for personalization.

ently sequenced for mutations in all 20,000 human genes within the International Cancer Genome Consortium (ICGC; www.icgc.org). To design the most promising scenarios for synthetic lethal screens, novel tools are required allowing for systematic studies of gene function at a large scale and in a standardized manner. To meet this challenge, nanomedicine can profit from interdisciplinary approaches with functional genomics and synthetic biology. Such approaches comprise part of the strategies of the Lundbeckfonden Center of Excellence NanoCAN and the German-Danish High Technology Platform HiT-ID.

Conclusions

Cancer nanomedicine as the discipline of performing biomedical research at the nanoscale can greatly profit from incorporating 'omics' approaches at a very large scale that have the capacity to scan for cancer-specific alterations and selective possibilities for intervention at the scale of all 20,000 human genes. Such an alliance has the potential to assemble modular nanodrugs, meeting the high demands for the next generation of anticancer drugs, i.e. to efficiently and selectively eradicate CSCs in a personalized fashion. The potential of such approaches is indicated by the success of PARP- and BRAF-inhibitors, which are conventional but rationally designed drugs presently under evaluation in clinical trials. Functional genomics and synthetic biology will prospectively provide solutions for closing the still existent gaps in cancer nanomedicine, emphasizing its highly interdisciplinary character.

Acknowledgments

The authors would like to thank the Lundbeck Foundation, Kræftens Bekæmpelse, Gangstedfonden, the A.P. Møller Foundation for the Advancement of Medical Science, Fru Astrid Thaysens Legat for Lægensvidenskabelig Grundforskning, the Aase og Ejnar Danielsens Fond, the Grosserer M. Brogaard og Hustrus Mindefond, the Augustinus Fonden, the Lippmann Fonden, the Broedrene Hartmanns Fonden, the Familien Hede Nielsens Fond, the INTERREG4A Syddanmark-Schleswig-K.E.R.N. program of the EU, the Region Syddanmark, and the Christians den Tiendes Fond for their support of our research.

Disclosure Statement

The authors have nothing to disclose.

References

1 Clarke MF, Fuller M: Stem cells and cancer: two faces of Eve. Cell 2006;124: 1111–1115.
2 Wicha MS, Liu S, Dontu G: Cancer stem cells: an old idea – a paradigm shift. Cancer Res 2006;66:1883–1890.
3 Visvader JE, Lindeman GJ: Cancer stem cells in solid tumours: accumulating evidence and unresolved questions. Nat Rev Cancer 2008;8:755–768.
4 Simonsen JL, Rosada C, Serakinci N, Justesen J, Stenderup K, Rattan SI, Jensen TG, Kassem M: Telomerase expression extends the proliferative life-span and maintains the osteogenic potential of human bone marrow stromal cells. Nat Biotechnol 2002;20:592–596.
5 Serakinci N, Guldberg P, Burns JS, Abdallah B, Schroder H, Jensen T, Kassem M: Adult human mesenchymal stem cell as a target for neoplastic transformation. Oncogene 2004;23:5095–5098.
6 Zhu L, Gibson P, Currle DS, Tong Y, Richardson RJ, Bayazitov IT, Poppleton H, Zakharenko S, Ellison DW, Gilbertson RJ: Prominin 1 marks intestinal stem cells that are susceptible to neoplastic transformation. Nature 2009; 457:603–607.
7 Barker N, Ridgway RA, van Es JH, van de Wetering M, Begthel H, van den Born M, Danenberg E, Clarke AR, Sansom OJ, Clevers H: Crypt stem cells as the cells-of-origin of intestinal cancer. Nature 2009;457:608–611.
8 Quintana E, Shackleton M, Sabel MS, Fullen DR, Johnson TM, Morrison SJ: Efficient tumour formation by single human melanoma cells. Nature 2008; 456:593–598.
9 Shackleton M, Vaillant F, Simpson KJ, Stingl J, Smyth GK, Asselin-Labat ML, Wu L, Lindeman GJ, Visvader JE: Generation of a functional mammary gland from a single stem cell. Nature 2006; 439:84–88.

10 Charafe-Jauffret E, Ginestier C, Iovino F, Wicinski J, Cervera N, Finetti P, Hur MH, Diebel ME, Monville F, Dutcher J, Brown M, Viens P, Xerri L, Bertucci F, Stassi G, Dontu G, Birnbaum D, Wicha MS: Breast cancer cell lines contain functional cancer stem cells with metastatic capacity and a distinct molecular signature. Cancer Res 2009;69:1302–1313.

11 Ginestier C, Liu S, Diebel ME, Korkaya H, Luo M, Brown M, Wicinski J, Cabaud O, Charafe-Jauffret E, Birnbaum D, Guan JL, Dontu G, Wicha MS: CXCR1 blockade selectively targets human breast cancer stem cells in vitro and in xenografts. J Clin Invest 2010;120:485–497.

12 Olopade OI, Grushko TA, Nanda R, Huo D: Advances in breast cancer: pathways to personalized medicine. Clin Cancer Res 2008;14:7988–7999.

13 Overdevest JB, Theodorescu D, Lee JK: Utilizing the molecular gateway: the path to personalized cancer management. Clin Chem 2009;55: 684–697.

14 Farmer H, McCabe N, Lord CJ, Tutt AN, Johnson DA, Richardson TB, Santarosa M, Dillon KJ, Hickson I, Knights C, Martin NM, Jackson SP, Smith GC, Ashworth A: Targeting the DNA repair defect in BRCA mutant cells as a therapeutic strategy. Nature 2005;434:917–921.

15 Fong PC, Boss DS, Yap TA, Tutt A, Wu P, Mergui-Roelvink M, Mortimer P, Swaisland H, Lau A, O'Connor MJ, Ashworth A, Carmichael J, Kaye SB, Schellens JH, de Bono JS: Inhibition of poly(ADP-ribose) polymerase in tumors from BRCA mutation carriers. N Engl J Med 2009;361:123–134.

16 Kaelin WG Jr: The concept of synthetic lethality in the context of anticancer therapy. Nat Rev Cancer 2005;5:689–698.

17 Davies H, Bignell GR, Cox C, Stephens P, Edkins S, Clegg S, Teague J, Woffendin H, Garnett MJ, Bottomley W, Davis N, Dicks E, Ewing R, Floyd Y, Gray K, Hall S, Hawes R, Hughes J, Kosmidou V, Menzies A, Mould C, Parker A, Stevens C, Watt S, Hooper S, Wilson R, Jayatilake H, Gusterson BA, Cooper C, Shipley J, Hargrave D, Pritchard-Jones K, Maitland N, Chenevix-Trench G, Riggins GJ, Bigner DD, Palmieri G, Cossu A, Flanagan A, Nicholson A, Ho JW, Leung SY, Yuen ST, Weber BL, Seigler HF, Darrow TL, Paterson H, Marais R, Marshall CJ, Wooster R, Stratton MR, Futreal PA: Mutations of the BRAF gene in human cancer. Nature 2002; 417:949–954.

18 Flaherty KT, Puzanov I, Kim KB, Ribas A, McArthur GA, Sosman JA, O'Dwyer PJ, Lee RJ, Grippo JF, Nolop K, Chapman PB: Inhibition of mutated, activated BRAF in metastatic melanoma. N Engl J Med 2010;363:809–819.

19 Wood LD, Parsons DW, Jones S, Lin J, Sjöblom T, Leary RJ, Shen D, Boca SM, Barber T, Ptak J, Silliman N, Szabo S, Dezso Z, Ustyanksky V, Nikolskaya T, Nikolsky Y, Karchin R, Wilson PA, Kaminker JS, Zhang Z, Croshaw R, Willis J, Dawson D, Shipitsin M, Willson JK, Sukumar S, Polyak K, Park BH, Pethiyagoda CL, Pant PV, Ballinger DG, Sparks AB, Hartigan J, Smith DR, Suh E, Papadopoulos N, Buckhaults P, Markowitz SD, Parmigiani G, Kinzler KW, Velculescu VE, Vogelstein B: The genomic landscapes of human breast and colorectal cancers. Science 2007;318: 1108–1113.

20 Elbashir SM, Harborth J, Lendeckel W, Yalcin A, Weber K, Tuschl T: Duplexes of 21-nucleotide RNAs mediate RNA interference in cultured mammalian cells. Nature 2001;411:494–498.

21 Kittler R, Pelletier L, Heninger AK, Slabicki M, Theis M, Miroslaw L, Poser I, Lawo S, Grabner H, Kozak K, Wagner J, Surendranath V, Richter C, Bowen W, Jackson AL, Habermann B, Hyman AA, Buchholz F: Genome-scale RNAi profiling of cell division in human tissue culture cells. Nature Cell Biology 2007;9:1401–1412.

22 Müller P, Kuttenkeuler D, Gesellchen V, Zeidler MP, Boutros M: Identification of JAK/STAT signalling components by genome-wide RNA interference. Nature 2005;436:871–875.

23 Krishnan MN, Ng A, Sukumaran B, Gilfoy FD, Uchil PD, Sultana H, Brass AL, Adametz R, Tsui M, Qian F, Montgomery RR, Lev S, Mason PW, Koski RA, Elledge SJ, Xavier RJ, Agaisse H, Fikrig E: RNA interference screen for human genes associated with West Nile virus infection. Nature 2008;455:242–245.

24 Sieburth D, Ch'ng QL, Dybbs M, Tavazoie M, Kennedy S, Wang D, Dupuy D, Rual JF, Hill DE, Vidal M, Ruvkun G, Kaplan JM: Systematic analysis of genes required for synapse structure and function. Nature 2005;436:510–517.

25 Whitehurst AW, Bodemann BO, Cardenas J, Ferguson D, Girard L, Peyton M, Minna JD, Michnoff C, Hao W, Roth MG, Xie XJ, White MA: Synthetic lethal screen identification of chemosensitizer loci in cancer cells. Nature 2007;446: 815–819.

26 Sarthy AV, Morgan-Lappe SE, Zakula D, Vernetti L, Schurdak M, Packer JC, Anderson MG, Shirasawa S, Sasazuki T, Fesik SW: Survivin depletion preferentially reduces the survival of activated K-Ras-transformed cells. Mol Cancer Ther 2007;6:269–276.

27 Rottmann S, Wang Y, Nasoff M, Deveraux QL, Quon KC: A TRAIL receptor-dependent synthetic lethal relationship between MYC activation and GSK3beta/FBW7 loss of function. Proc Natl Acad Sci USA 2005;102: 15195–15200.

Prof. Dr. Jan Mollenhauer
Lundbeckfonden Center of Excellence NanoCAN and
Molecular Oncology, Institute of Molecular Medicine
University of Southern Denmark
JB Winsloews Vej 25, DK–5000 Odense (Denmark)
Tel. +45 6550 3970, E-Mail jmollenhauer@health.sdu.dk

Alexiou C (ed): Nanomedicine – Basic and Clinical Applications in Diagnostics and Therapy.
Else Kröner-Fresenius Symp. Basel, Karger, 2011, vol 2, pp 145–153

Targeted Iron Oxide Nanocomplex as a Theranostic Agent for Cancer

Esther H. Chang

Department of Oncology, Lombardi Comprehensive Cancer Center, Georgetown University Medical Center, Washington, D.C., USA

Abstract

Currently there are no tumor-specific superparamagnetic iron oxide (SPIO) nanocomplexes on the market for use as either diagnostic or hyperthermic agents. Moreover, none of the current hyperthermia approaches can efficiently treat disseminated (metastatic) cancer. Our anti-transferrin receptor scFv immunoliposome (scL) is a systemically administered, tumor-targeting nanocomplex for molecular medicines. It efficiently and specifically delivers various payloads to tumor cells in vivo and has successfully completed a phase I clinical trial for p53 gene therapy. We modified this scL complex to encapsulate SPIO. We have demonstrated that the scL nanocomplex can deliver SPIO specifically and efficiently into tumor cells, resulting in tumor growth inhibition in vivo. Thus, scL-SPIO should be capable of delivering sufficient SPIO to generate therapeutic temperatures only in the tumor when the patient is exposed to an alternating magnetic field. As SPIO is also used in MRI, use of scL-SPIO would allow simultaneous imaging and hyperthermic treatment. This scL-SPIO nanocomplex theranostic agent has the potential to simultaneously diagnose metastatic tumors, act as a therapeutic, and assess the treatment effect while minimizing heat effects on normal cells. Thus, it could have a significant clinical impact on the treatment of cancer, including metastatic lung cancer.

Lung cancer, with an estimated ~222,000 new cases and more than 157,000 deaths, is the current leading cause of cancer deaths for both men and women in the USA, accounting for 28% of all cancer deaths [1]. Worldwide, it is estimated that lung cancer causes nearly 1 million deaths annually. The 5-year survival rate is only 16%. One significant challenge is the accurate detection of small lung cancers. Current methods used for early identification of primary lung carcinoma include chest X-ray and computed tomography (CT). Although both methods are somewhat effective in identifying curable lung cancer, they possess a major drawback in that they commonly result in a high percentage of false positives, i.e. nodular areas that could indicate lung cancer but are in fact due to scars or focal inflammatory/infectious processes [2, 3]. For small lung nodules this is a frequent occurrence and a serious problem. While there are various diagnostic methods to distinguish between true malignancies and false positives, for small lung nodules the primary method is to obtain serial images over time looking for growth, a very inefficient and costly process. In some cases, contrast-enhanced CT or magnetic resonance imaging (MRI) is used to determine the vascularity of the suspect area, but increased vascularity is not necessarily specific for cancer. Moreover, a report by Bach et al. [4] on the results of a multicenter trial concluded that lung CT may not meaningfully reduce the risk of advanced lung cancer or death. Thus, there is a critical need for more effective means of early diagnosis and treatment for lung cancer.

A theranostic agent may best be described as one that integrates both therapeutics and diagnostics. The ideal delivery vehicle for such a theranostic agent would be one that could be systemically administered and then home specifically to tumor cells wherever they occur in the body. We have developed an antitransferrin receptor scFv-antibody fragment (TfRscFv) immunoliposome nanoscale delivery system platform technology for gene medicine: the scL complex. In this complex the payload is encapsulated within a cationic liposome, the surface of which is decorated with an antitransferrin receptor single-chain antibody fragment (TfRscFv) moiety. The tumor-specific targeting nature of this complex has been well established both in vitro and in vivo (in mice), demonstrating high selectivity for malignant tumors. We have shown that such a TfRscFv molecule can efficiently target the intravenously administered cationic liposome nanocomplex preferentially to tumors, both primary and metastatic (even in the brain), in animal models, delivering plasmid DNA (wt p53 gene and reporter genes) [5–7], AS ODN (HER-2) [8, 9], si/miRNA [10, 11], small molecules [12], and gadolinium compounds for MRI [13, 14]. Systemic administration of this targeting moiety-liposome-DNA nanocomplex targeted and sensitized various types of cancer to conventional radiation/chemotherapy and has just successfully completed a phase I clinical trial for p53 gene therapy. The modular nature of this platform technology suggested that it could also be employed for tumor-specific delivery of other molecules such as superparamagnetic iron oxide (SPIO). Thus, we have adapted this scL nanocomplex for encapsulation and systemic, tumor-specific delivery of SPIO particles to tumors as a theranostic agent, i.e. for use in both diagnosis (MRI) and treatment (hyperthermia) of lung cancer.

As with other examples of this platform technology, we have established that the scL-SPIO complex is indeed a nanoparticle [15]. Using dynamic light scattering, we found that the size of the scL-SPIO complex was 157.4 ± 12.5 nm (by intensity) as compared to 132.0 ± 25.1 nm for the scL delivery system without the SPIO payload. The polydispersity indexes, a measure of the size distribution within the sample, were 0.210 ± 0.02 and 0.282 ± 0.02, respectively, confirming the uniform size of the complex. The zeta potentials were positive, with values of 32.4 ± 2.2 and 29.4 ± 1.2 mV for the scL and scL-SPIO, respectively. Thus, encapsulation of the SPIO did not appreciably alter the size or charge of the scL complex.

Our previous studies have shown that the various payloads (plasmid DNA and siRNA) are encapsulated within the liposome of the scL nanocomplex [7, 11]. To confirm that this is also the case when SPIO is used as the payload, scanning probe microscopy (SPM), including fluid SPM, phase contrast, and magnetic force microscopy (MFM), was performed [15]. When examined using fluid SPM, intact liposomes encapsulating the SPIO exhibited a comma-like deformation in the topographical image due to the interaction of the complex with the SPM probe tip. However, when adsorbed onto a freshly cleaved mica substrate the cationic liposomes ruptured on the strongly negatively charged mica. The phase image shown in figure 1a reveals the presence of an aggregate surrounded by a lipid patch. The lipid appears dark since the interaction of the negatively charged Si_3N_4 SPM probe tip and cationic lipid is attractive, causing the rupture of the liposome. Encapsulation is further confirmed by examination of the scL-SPIO complex by MFM after drying of the samples on the substrate. Comparison of the simultaneously obtained topographical

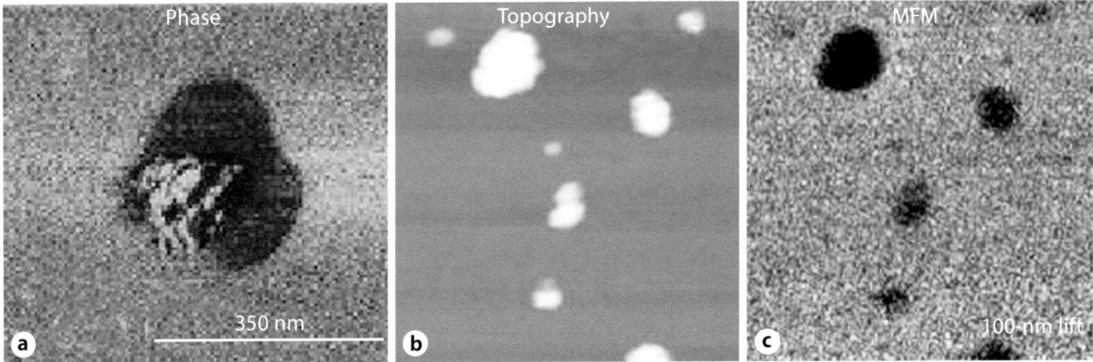

Fig. 1. SPM of the scL-SPIO nanocomplex. **a** Phase contrast image obtained by fluid SPM imaging of the scL-SPIO nanocomplex adsorbed onto freshly cleaved mica substrate. **b** Topographic image of the scL-SPIO nanocomplex after drying on an aged poly-L-lysine mica substrate. **c** Simultaneously obtained MFM image of the same sample as in **b**.

(fig. 1b) and MFM images (fig. 1c) demonstrates that a magnetic field due to the SPIO corresponds to the particles seen on the topographical image. These findings confirm the encapsulation of SPIO within the tumor-targeting nanocomplex delivery system, opening up the potential for use of scL-SPIO as a diagnostic/therapeutic agent.

Hyperthermia is the raising of the temperature of a part of the body or of the entire body to supraphysiological temperatures of 41–45°C for a defined period of time, resulting in cell death [16, 17]. The main molecular events triggered by this increase in heat are protein and DNA denaturation resulting in disruption of cellular structure and function through protein aggregation, chromosomal damage and misrepair, as well as disruption of DNA replication [18]. The presence of the denatured, aggregated proteins due to heat exposure also triggers transient expression of heat shock proteins.

The same inherent tumor characteristics (hypoxia, acidic pH, and poor nutrition) that result in the reduced response of tumors to radiation and chemotherapy actually render tumor cells more susceptible to hyperthermia [16]. Because of the connection between fever and immunological reactions, it has also been suggested that hyperthermia treatment may activate the immune system against the tumor [19]. Moreover, in tumor but not normal cells, the heat triggered heat shock proteins translocate to the cell membrane where they act as receptors for natural killer cells, macrophages, and antigen-presenting cells, thus contributing to the observed cytotoxic effect of hyperthermia on cancer cells [20]. Animal studies have demonstrated the anticancer efficacy of hyperthermia as a single-agent modality. Furthermore, hyperthermia has also been shown to be a potent enhancer of the tumor cell response to chemotherapy and radiation treatment, e.g. in combination with brachytherapy [17, 21, 22]. There have also been preclinical and clinical studies of a trimodal approach using both chemo- and radiotherapy in conjunction with hyperthermia [23].

Although over the last 2 decades there have been a large number of phase I, II, and III clinical trials of hyperthermia for various types of cancer including melanoma, breast, head and neck, bladder, and even glioblastomas [24], it is still considered by many to be experimental with no realistic future in clinical cancer therapy [17]. In addition to the technical challenges of equipment and standardization, this is primarily related to the fact that most current uses of hyperthermia (including whole-body and regional hyperthermia) are not tumor tissue specific. This limits the attainable tumor temperature due to the risk of normal tissue damage as some tissues, including normal brain, eye, and testes, are relatively sensitive to the effects of heating [25]. Other hyperthermia approaches also have limitations. Local hyperthermia is limited to a therapeutic depth of 3–4 cm [16]. While interstitial local hyperthermia can be used to treat deep tumors such as brain, it has the disadvantages of being invasive, with limited applicable sites [16]. Most significantly, none of these approaches are able to treat disseminated (metastatic) disease.

To overcome these current limitations, methods to heat only the tumor cells must be developed. One approach is the use of magnetic nanoparticles. Due to their unique properties, SPIO nanoparticles have found applications in increasingly diverse areas of biotechnology and biomedical sciences including MRI and hyperthermia [26]. SPIO particles have an overall magnetic moment which undergoes fluctuations due to the orientation of the molecules when exposed to an external alternating magnetic field (AMF). The induction heating of the SPIO that results (hysteresis) is due to the interaction of the magnetic moment of the SPIO particles with the AMF, and the magnetic energy is subsequently converted to thermal energy.

Various groups of investigators [27–31] have reported that such magnetically induced hyperthermia can kill cancer cells after intratumoral injection of the SPIO particles and exposure of the tumor to a localized AMF. Although intratumoral injection of the magnetic nanoparticles is useful in preclinical animal studies, this approach is also limited in its potential as an effective anticancer therapy as it does not affect disseminated disease. The ability to specifically and efficiently systemically deliver the SPIO particles to the tumors, both primary and metastatic, would limit the exposure of normal tissues, increasing efficacy and reducing toxicity, and would significantly enhance the utility of this therapeutic approach.

The recent advances in the field of nanoparticle delivery, coupled with the leakiness of tumor blood vessels (the enhanced permeability and retention effect), have resulted in improved delivery of molecular medicines to tumor cells.

Attempts have been made to target magnetic nanoparticles for hyperthermia, but with limited success. Natarajan et al. [32], produced a radio-conjugated SPIO nanoparticle that used a di-scFv fragment directed against the mucin-1 transmembrane molecule to target glandular epithelial cells. Pharmacokinetics and whole-body autoradiography demonstrated only 5% of the injected dose in the tumor after 24 h, with most accumulation in the liver, kidney, and spleen of the animal [32]. Ito and colleagues have published several papers [reviewed in 33] using magnetite (SPIO) encapsulated in either neutral or cationic liposomes, some including MAb for targeting with hyperthermia. In addition, Sato et al. [31] have used a melanogenesis substrate conjugated to magnetic nanoparticles for hyperthermic treatment of mice bearing subcutaneous mouse melanoma tumors. However, with both of these latter two groups, the complexes have only been used for intratumoral, but not systemic, delivery.

In contrast, as described above, our targeted delivery system is designed for systemic delivery of the payload, in this case SPIO, specifically and efficiently to tumor cells, even metastases (including those in the brain), with no uptake in normal cells such as liver. Initially we demonstrated in vitro that use of the scL delivery system would result in an increased uptake of SPIO by tumor cells when compared with free (unencapsulated) SPIO. We quantitated the in vitro uptake of scL-SPIO, L-SPIO (the nanocomplex minus the targeting moiety), and free SPIO in both human breast (MDA-MB-231) and human pancreatic cancer (PANC-1) cells using the Prussian

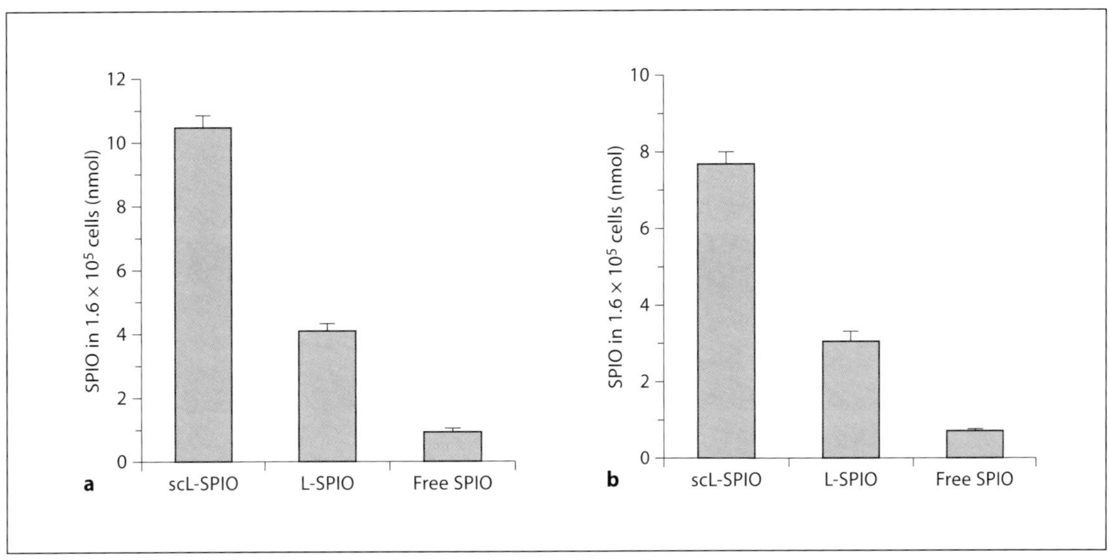

Fig. 2. In vitro uptake of SPIO in 2 human cancer cell lines via Prussian blue reaction. Comparison of iron content between scL-SPIO, L-SPIO (complex minus TfRscFv targeting moiety), and free SPIO in MDA-MB-231 breast cancer cells (**a**) and PANC-1 pancreatic cancer cells (**b**).

blue reaction [15]. As shown in figure 2a, b, the level of iron detected in both the breast and pancreatic cell lines after incubation with the complete complex was at least 2.5-fold greater than that obtained with the complex minus the TfRscFv targeting moiety. Moreover, delivery of SPIO by the scL nanocomplex resulted in an ~11-fold increase in SPIO uptake as compared to free SPIO in both cancer cell lines. The differences between scL-SPIO and either L-SPIO or free SPIO were statistically significant ($p \leq 0.001$). These studies demonstrated that encapsulation of SPIO in the scL nanocomplex can lead to enhanced uptake of SPIO by tumor cells. Moreover, this efficient uptake is in fact mediated by the TfRscFv targeting moiety on the surface of the liposome. Similar results were observed using human prostate cell line PC-3 and mouse renal cell carcinoma cell line RenCa, demonstrating that this approach is applicable to various types of cancer [15]. The enhanced in vitro uptake of fluorescently labeled SPIO compared to free SPIO or unliganded complex was also shown by confocal microscopy (fig. 3) [15].

The potential of the systemically administered scL-SPIO nanocomplex for use as a tumor-specific anticancer hyperthermia agent for metastatic disease was demonstrated using the syngeneic mouse renal cell carcinoma (RenCa) lung metastasis model in immune competent BALB/c mice [34]. This is a well-established model of metastatic disease to the lung [34]. The scL-SPIO complex, or free SPIO, at a dose of 345 μg SPIO/injection, was injected intravenously into the tail vein of the mice. One mouse received neither SPIO nor hyperthermia (untreated; UT). The mice were injected 12 times over 22 days. Four and one half hours after each injection, they were exposed to an AMF for 30 min. One day after the last treatment, the lungs were examined for the presence of metastases. There was no visible difference between the UT and free SPIO mice (fig. 4). In contrast, there were no visible nodules in the lungs from the animal that received the

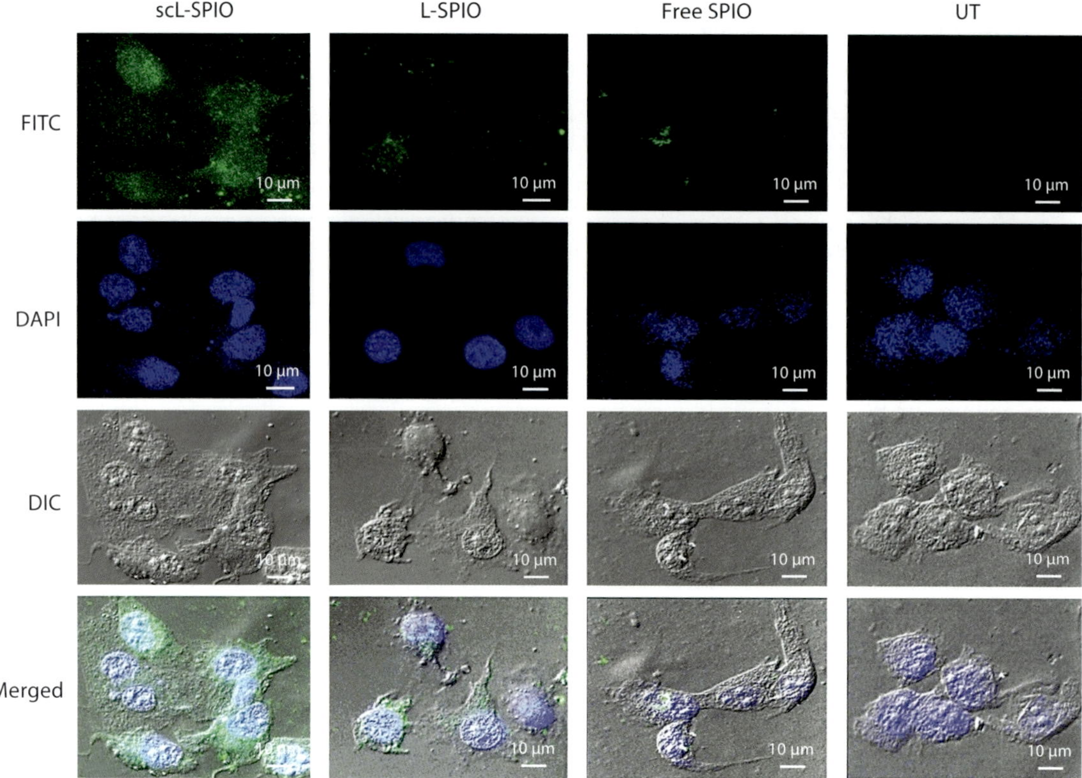

Fig. 3. In vitro localization of scL-complexed fluorescein-labeled SPIO nanoparticles in MDA-MB-231 cells by confocal microscopy. Confocal microscopy shows the localization of fluorescein-labeled SPIO nanoparticles in MDA-MB-231 cells after incubation with scL-complexed SPIO (scL-SPIO), the complex without the targeting moiety (Lip-SPIO), and uncomplexed SPIO (free SPIO). UT cells were used as a control. Intercalation of DAPI in the chromosomal DNA imparts a blue color to identify the nucleus. FITC = Fluorescence signal in the cells; DIC = differential interference contrast images.

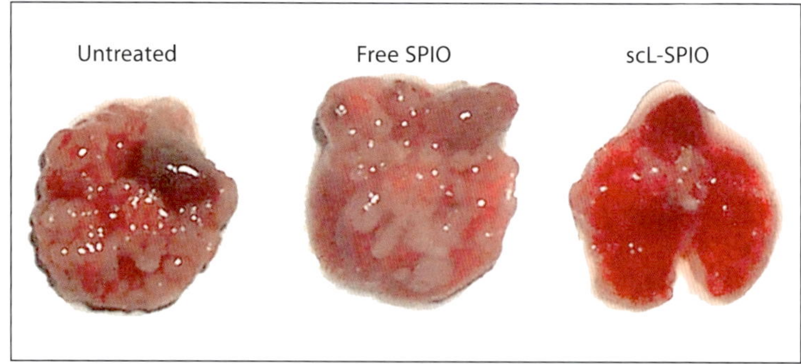

Fig. 4. Tumor cell growth inhibition in vivo by systemically administered scL-SPIO-mediated hyperthermia therapy. Lung metastases in BALB/c mice were induced by intravenous tail vein injection of 1.0×10^5 RenCa cells. Two days postinoculation, intravenous injection and hyperthermia treatment were initiated. After 11 treatments over 22 days, the lungs were examined for the presence of tumor nodules.

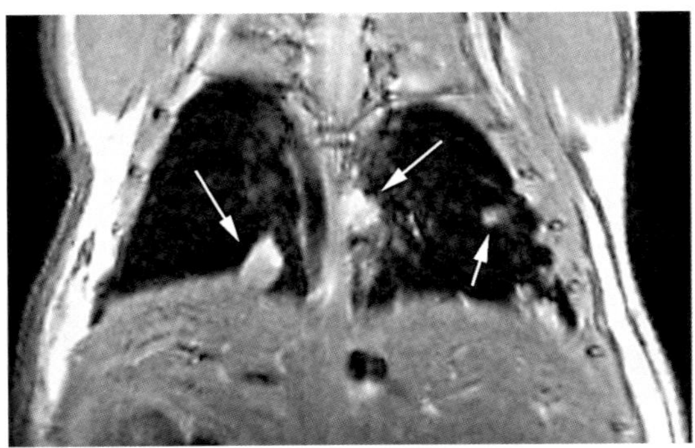

Fig. 5. scL-SPIO imaging of RenCa lung metastases. In this image from a T1-W respiratory gated experiment of a mouse with RenCa lung metastases, multiple nodules are seen (arrows). These images were obtained 6 h after injection and show the long persistence of SPIO delivered within the scL-SPIO complex. The peak values seen here in various nodules with scL-SPIO (measured via Image J software) are 14,954, 17,087, and 21,202.

scL-SPIO. These findings demonstrate that this approach has the potential to be developed into an effective anticancer hyperthermic agent.

In addition to tumor targeting, efficient accumulation of the payload in the target cell is critical for a therapeutic to be an effective theranostic agent. The scL-SPIO nanocomplex can overcome this challenge as well, as based upon our studies we conservatively estimate that approximately 20–71% of the injected SPIO accumulates in the tumor after systemic administration.

The encapsulation of SPIO within the scL nanocomplex results in a particle with multiple properties. In addition to its potential use in hyperthermic therapy discussed above, SPIO is being used as an MRI contrast agent both clinically and for research [35]. However, there are drawbacks to the use of SPIO as a contrast agent. Because of their size, when nontumor-targeting SPIO particles are injected into the blood, they are taken up by the reticuloendothelial system, passively accumulating in organs and tissues high in macrophages, such as the liver, lymph nodes, and bone marrow. Because of this dependence on uptake by the reticuloendothelial system, SPIO contrast agents are currently used for clinical imaging of these organs/tissues. However, since tumors do not generally accumulate macrophages or macrophage-like cells, the lack of tumor cell uptake places limitations on the usefulness of SPIO as a diagnostic agent.

SPIO can have both T1 (increases intensity) and T2* shortening (decreased intensity). Thus, if targeted to tumors, SPIO could be used for T1 imaging in areas not currently feasible, such as the lungs where the background is dark and the lung nodules would have increased intensity. The means to deliver SPIO specifically into the tumor cells themselves via the scL platform technology would significantly increase the utility of SPIO for MRI.

As the ultimate goal of this theranostic approach is to use the scL-SPIO complex for simultaneous MRI and hyperthermic therapy, we performed a study to examine the effect of scL-SPIO on MRI of lung metastases. Mice bearing RenCa lung metastases were intravenously tail vein injected with scL-SPIO complex. Six hours later the lungs were imaged using a T1-weighted imaging sequence.

As shown in figure 5, a very high pixel intensity is evident in the tumor nodules in the lung, while normal liver does not enhance. Thus, it appears that when imaged using T1-weighted conditions, scL-delivered SPIO can positively enhance imaging contrast in lung tumors, demonstrating the potential of scL-SPIO as a contrast agent for MRI with broader application than current forms of SPIO.

It is currently possible for hyperthermia systems to interface with MRI systems. This would allow not only simultaneous imaging and hyperthermic treatment of the tumors but also postassessment of tissue damage as well as noninvasive temperature monitoring of the treatment as described by MacFall and Soher [36]. This latter property is significant as accurate measurement of the temperature distribution within the tumor during hyperthermia is important in successful therapy. Thus, a multifunctional nanocomplex such as scL-SPIO which can simultaneously minimize heat effects on normal cells, monitor temperature, act as a therapeutic, and assess the therapeutic effect in a tumor-specific way would have a significant clinical impact on lung cancer.

Disclosure Statement

Esther H. Chang is one of the inventors of the described technology, for which several patents owned by Georgetown University have been issued. The patents have been licensed to SynerGene Therapeutics Inc. for commercial development. Dr. Chang owns equity interests in SynerGene. In addition, Dr. Chang serves as a non-paid scientific consultant to SynerGene.

References

1 American Cancer Society: Cancer facts and figures 2010. 2010. http://www.cancer.org.

2 Swensen SJ, Jett JR, Hartman TE, Midthun DE, Sloan JA, Sykes AM, Aughenbaugh GL, Clemens MA: Lung cancer screening with CT: Mayo Clinic experience. Radiology 2003;226:756–761.

3 Swensen SJ, Jett JR, Hartman TE, Midthun DE, Mandrekar SJ, Hillman SL, Sykes AM, Aughenbaugh GL, Bungum AO, Allen KL: CT screening for lung cancer: five-year prospective experience. Radiology 2005;235:259–265.

4 Bach PB, Jett JR, Pastorino U, Tockman MS, Swensen SJ, Begg CB: Computed tomography screening and lung cancer outcomes. JAMA 2007;297:953–961.

5 Xu L, Tang WH, Huang CC, Pirollo KF, Rait A, Chang EH: Systemic p53 gene therapy of cancer with immunolipoplexes targeted by anti-transferrin receptor scFv. Mol Med 2001;7:723–734.

6 Xu L, Huang C-C, Huang W-Q, Tang W-H, Rait A, Pirollo K, Chang EH: Systemic tumor-targeted gene delivery by anti-transferrin receptor scFv-immunoliposomes. Mol Cancer Ther 2002;1:337–346.

7 Xu L, Frederick P, Pirollo K, Rait A, Chang EH: Self-assembly of a virus-mimicking nanostructure system for efficient tumor-targeted gene delivery. Hum Gene Ther 2002;13:469–481.

8 Rait AS, Pirollo KF, Xiang L, Chang EH: Tumor-targeting, systemically delivered ASHER-2 chemosensitizes human breast cancer xenografts irrespective of HER-2 levels. Mol Med 2002;8:475–486.

9 Rait AS, Pirollo KF, Cullen K, Chang EH: HER-2-targeted AS oligonucleotide results in sensitization of head and neck cancer cells to chemotherapeutic agents. Ann NY Acad Sci 2003;1002:78–89.

10 Pirollo KF, Zon G, Rait A, Zhou Q, Yu W, Hogrefe R, Chang EH: Tumor-targeting nanoimmunoliposome complex for short interfering RNA delivery. Hum Gene Ther 2006;17:117–124.

11 Pirollo KF, Rait A, Dagata JA, Zon G, Hogrefe RI, Palchik G, Chang EH: Materializing the potential of siRNA via a tumor-targeting nanodelivery system. Cancer Res 2007;67:2932–2937.

12 Hwang SH, Rait A, Pirollo KF, Chinigo GM, Brown ML, Chang EH: Tumor-targeting nanodelivery enhances the anticancer activity of novel quinazolinone analog. Mol Cancer Ther 2008;7:1–10.

13 Pirollo KF, Dagata J, Wang P, Freedman M, Vladar A, Fricke S, Zhou Q, Chang EH: A tumor-targeted nanodelivery system to improve early MRI detection of cancer. Mol Imaging 2006;5:41–52.

14 Freedman M, Chang EH, Zhou Q, Pirollo KF: Nanodelivery of MRI contrast agent enhances sensitivity of detection of lung cancer metastases. Acad Radiol 16:627–637.

15 Yang C, Rait A, Pirollo K, Dagata JA, Farkas N, Chang E: Nanoimmunoliposome delivery of superparamagnetic iron oxide markedly enhances targeting/uptake in human cancer cells in vitro and in vivo. Nanomedicine 2008;4:318–329.

16 Habash RW, Bansal R, Krewski D, Alhafid HT: Thermal therapy, part 2: hyperthermia techniques. Crit Rev Biomed Eng 2006;34:491–542.

17 Horsman MR, Overgaard J: Hyperthermia: a potent enhancer of radiotherapy. Clin Oncol (R Coll Radiol) 2007;19:418–426.

18 Moyer HR, Delman KA: The role of hyperthermia in optimizing tumor response to regional therapy. Int J Hyperthermia 2008;24:251–261.

19 Dieing A, Ahlers O, Hildebrandt B, Kerner T, Tamm I, Possinger K, Wust P: The effect of induced hyperthermia on the immune system. Prog Brain Res 2007;162:137–152.

20 Dayanc BE, Beachy SH, Ostberg JR, Repasky EA: Dissecting the role of hyperthermia in natural killer cell mediated anti-tumor responses. Int J Hyperthermia 2008;24:41–56.

21 Kampinga HH: Cell biological effects of hyperthermia alone or combined with radiation or drugs: a short introduction to newcomers in the field. Int J Hyperthermia 2006;22:191–196.

22 Sugarbaker PH: Laboratory and clinical basis for hyperthermia as a component of intracavitary chemotherapy. Int J Hyperthermia 2007;23:431–442.

23 Bergs JW, Franken NA, Haveman J, Geijsen ED, Crezee J, van Bree C: Hyperthermia, cisplatin and radiation trimodality treatment: a promising cancer treatment? A review from preclinical studies to clinical application. Int J Hyperthermia 2007;23:329–341.

24 Falk MH, Issels RD: Hyperthermia in oncology. Int J Hyperthermia 2001;17:1–18.

25 Goldstein LS, Dewhirst MW, Repacholi M, Kheifets L: Summary, conclusions and recommendations: adverse temperature levels in the human body. Int J Hyperthermia 2003;19:373–384.

26 Gupta AK, Gupta M: Synthesis and surface engineering of iron oxide nanoparticles for biomedical applications. Biomaterials 2005;26:3995–4021.
27 Jordan A, Scholz R, Wust P, Fahling H, Krause J, Wlodarczyk W, Sander B, Vogl T, Felix R: Effects of magnetic fluid hyperthermia (MFH) on C3H mammary carcinoma in vivo. Int J Hyperthermia 1997:13:587–605.
28 Ito A, Honda H, Kobayashi T: Cancer immunotherapy based on intracellular hyperthermia using magnetite nanoparticles: a novel concept of 'heat-controlled necrosis' with heat shock protein expression. Cancer Immunol Immunother 2006;55:320–328.
29 Saito H, Mitobe K, Ito A, Sugawara Y, Maruyama K, Minamiya Y, Motoyama S, Yoshimura N, Ogawa JI: Self-regulation hyperthermia induced using thermosensitive ferromagnetic material with a low Curie temperature. Cancer Sci 2008;99:805–809.
30 Ito A, Saito H, Mitobe K, Minamiya Y, Takahashi N, Maruyama K, Motoyama S, Katayose Y, Ogawa JI: Inhibition of heat shock protein 90 sensitizes melanoma cells to thermosensitive ferromagnetic particle-mediated hyperthermia with low Curie temperature. Cancer Sci 2009;100:558–564.
31 Sato A, Tamura Y, Sato N, Yamashita T, Takada T, Sato M, Osai Y, Okura M, Ono I, Ito A, Honda H, Wakamatsu K, Ito S, Jimbow K: Melanoma-targeted chemo-thermo-immuno (CTI)-therapy using N-propionyl-4-S-cysteaminylphenol-magnetite nanoparticles elicits CTL response via heat shock protein-peptide complex release. Cancer Sci 2010;101:1939–1946.
32 Natarajan A, Xiong CY, Gruettner C, DeNardo GL, DeNardo SJ: Development of multivalent radioimmunonanoparticles for cancer imaging and therapy. Cancer Biother Radiopharm 2008;23: 82–91.
33 Ito A, Honda H, Kobayashi T: Cancer immunotherapy based on intracellular hyperthermia with magnetite nanoparticles: a novel concept of heat-controlled necrosis with heat shock protein expression. Cancer Immunol Immunother 2006;55:320–328.
34 Koshkina NV, Waldrep JC, Roberts LE, Golunski E, Melton S, Knight V: Paclitaxel liposome aerosol treatment induces inhibition of pulmonary metastases in murine renal carcinoma model. Clin Cancer Res 2001;7:3258–3262.
35 Thorek DL, Chen AK, Czupryna J, Tsourkas A: Superparamagnetic iron oxide nanoparticle probes for molecular imaging. Ann Biomed Eng 2006;34: 23–38.
36 MacFall JR, Soher BJ: From the RSNA refresher courses: MR imaging in hyperthermia. Radiographics 2007;27: 1809–1818.

Prof. Esther H. Chang, PhD
Department of Oncology, Lombardi Comprehensive Cancer Center, TRB/E420
Georgetown University Medical Center, 3970 Reservoir Rd., NW
Washington, DC 20057-1469 (USA)
Tel. +1 202 687 8418, E-Mail change@georgetown.edu

Alexiou C (ed): Nanomedicine – Basic and Clinical Applications in Diagnostics and Therapy.
Else Kröner-Fresenius Symp. Basel, Karger, 2011, vol 2, pp 154–163

Local Cancer Therapy with Magnetic Nanoparticles

Rainer Tietze · Stefan Lyer · Eveline Schreiber · Jenny Mann · Stephan Dürr · Christoph Alexiou

Department of Otorhinolaryngology, Head and Neck Surgery, Else Kröner-Fresenius-Stiftung Professorship for Nanomedicine, Section for Experimental Oncology and Nanomedicine (SEON), University Hospital Erlangen, Erlangen, Germany

Abstract

The application of nanotechnology for treatment, diagnosis, and monitoring of diseases is summarized under the term nanomedicine. A particularly promising application is attributed to nanoparticular drug delivery systems. The goal of these carrier systems is the selective enrichment of active substances into diseased tissue structures, an increase in bioavailability, a decrease in active substance degradation, and above all the reduction and/or avoidance of unwanted side effects. Apart from numerous nanosystems acting as carriers, the use of iron oxide nanoparticles has to be particularly emphasized. On the one hand those particles are the carriers of the active substance and on the other hand they are also visualizable with conventional imaging techniques (X-ray and magnetic resonance tomography) and can therefore be used for 'theranostic' purposes. Beyond that, they can be employed for hyperthermia, a further important supporting pillar of nanomedicine. Both procedures should lead to personalized and specific therapy, which has medical and socioeconomic relevance in view of the worldwide increasing number of cancer patients.

Cancer still persists as one of the leading causes of death around the world. Globally, in the year 2007 a total of 7.9 million deaths (13% of all deaths) were recorded and according to the WHO this number is still increasing [1]. In the European Union the annual number of cancer incidences was 1.6 million, and about 1 million people died from cancer. The corresponding data in Germany were 426,806 and 210,930, respectively, in the year 2006 [2]. The tendency is ascending in spite of all of the efforts that have been made and is mainly caused by an aging population. Half of all of the incidences that have been diagnosed are in the state of a local tumor extension and can be treated with established therapies like resection of the tumor and irradiation of the affected area plus systemic chemotherapy or even treatment with tumor antibodies. To the other half of the patients, no promising surgery can be offered. These locally far extended tumors are treated with a combined schedule of both systemic chemotherapy and irradiation. Conventional chemotherapy is always applied systemically and lacks sufficient enrichment of therapeutic agents in the tumor area and moreover impairs the nontargeted tissue in the whole organism. Severe side effects like suppression of bone marrow, dysfunction of the liver and kidneys, loss of hair, and nausea are some of the consequences of this treatment. Especially the effectiveness of cytostatic drugs correlates with their dose. A highly increased dose is needed for effective therapy but fails due to severe side effects. The chemotherapeutic dosage often represents a compromise between the toxicity for tumor cells and the tolerance of the patient's organism. One of the future challenges will be the development of directed therapy drafts, addressing the tumor specifically while sparing the remainder of the organism. Nanoparticles display unique physical and chemical properties due to their size of 1–100 nm. Also within this range are antibodies, receptors, nucleic acids, proteins, and other biological macromolecules. Biomimetic features as well as an extraordinary surface-to-volume ratio and the easy metamorphosis of properties lead to promising tools in the field of medicine, especially imaging, diagnosis, and therapy [3, 4].

Target-Oriented Tumor Therapy Using Nanomedicine

Nanomedicine offers a splendid prospect for directed drug application to avoid the negative side effects of conventional tumor therapy mentioned above. This is especially true for eliminating the severe side effects of systemic chemotherapy. As a consequence, the diseased body compartment is regarded as a biological unit which will be treated separately. This kind of draft is aimed at optimization of the local therapeutic effect. Moreover, transportation in nanosized ranges makes slightly soluble or instable agents available to address tumor cells [5]. A multitude of different antitumor drugs, radiotherapeutic nuclides, genetic material, and antibodies as macromolecular agents can use the nanosized delivery vehicle for a focused enrichment [6]. Thus, high potential substances can be applied that have not yet been applicable due to their high systemic toxicity or because of metabolic barriers. For oncology purposes, this means a maximized level of antineoplastic agents in the tumor area and a minimized level in the remainder of the organism.

Iron Oxide Nanoparticles for Tumor Therapy

Magnetic iron oxide nanoparticles have already found their way to medical fields. For diagnostic purposes, nanoparticles are used for magnetic cell separation whereby antibodies featuring a specific affinity are attached to particle surfaces. Hence, cells, DNA, or bacteria can be separated magnetically. Moreover, this technique is useful to verify the therapeutic effect after chemotherapeutic treatment [7]. Colloidal magnetic iron oxide particles (ferrofluids) are used already as contrast agents for MRI [8, 9]. Commercially available contrast media such as Resovist®, Endorem®, Sinerem®, and Combidex® are in use. Magnetite and maghemite particles are most commonly used in medicine and they are generally well tolerated [10]. They possess magnetic properties only during magnetic field application. After turning off the magnetic field they are not magnetic any more. One of the important parameters for in vivo application is the size of the particles. Particles at a size of about 100 nm seem to be especially useful. Bigger particles can easily be caught by mononuclear phagocyte system cells and can be enriched in the liver and spleen. This is especially true for systemic (intravenous) application [11]. Furthermore, bigger particles tend to lead to vascular obliterations in contrast to smaller ones [12]. One of the first clinical trials of magnetic particles took place in 1966 under the aegis of Alksne et al. [13] by generating a closure of intracranial aneurysms using ferromagnetic iron. By means of magnetic albumin microspheres as a vehicle, carrying magnetic particles and doxorubicin as the cytostatic drug, tumor remissions could be achieved in animal experiments with rats [14]. Carbon-coated iron oxide particles carrying doxorubicin have also been developed for local chemotherapy. These particles (MTC-Dox; Ferex, USA) were applied intravascularly followed by attraction of an external magnetic field in the scope of a phase I/II multicenter study on patients with hepatocellular carcinoma [15]. Previously, Lübbe et al. [16] conducted a first-time human trial with starch-coated magnetic nanoparticles in a phase I/II study whereby epirubicin was ionically bound to the particles. The formulation was applied intravenously and its tolerance could be demonstrated. For the concept of magnetic drug targeting (MDT) with intra-arterially administered magnetic nanoparticles, a sufficient particle size (80–150 nm) is important to attract them by means of an external magnetic field [17, 18]. The advantage of this kind of application is the major local enrichment of particles in the respective area due to bypassing of the metabolic barriers of the mononuclear phagocyte system. The magnetic force affecting the particles strongly depends on the radius (proportional to r^3). Besides the particle size, the charge and chemical features are relevant. For in vivo applications the nanoparticles are usually coated for colloidal stability and to prevent segregation into particles and carrier medium [11, 19]. A collapse of the colloidal stability would end in the emergence of a thrombus.

Figure 1 exhibits the principle of MDT and the histological evidence of particle enrichment in tumor tissue. For experimental studies with animals, ferrofluids were injected intra-arterially into the tumor-supplying vessel (arteria femoralis), and simultaneously a strong external magnetic field was applied over a VX2 squamous cell carcinoma placed at the hind limb. An enrichment of ferrofluids could be demonstrated using histology, MRI, and computed microtomography [20–22]. For nanoparticles acting therapeutically and for diagnostic purposes, in this approach they can also be designated as 'theranostics'. Further morphological investigations showed no pathological alteration in the liver, kidneys, spleen, lung, or brain. Metabolic decomposition of the ferrofluids proceeds in the liver and spleen in analogy to the physiological iron metabolism. In these organs ferrofluids can be detected histologically for 3 months after application. Radiotracer studies using ^{59}Fe have provided evidence for the embedding of iron arisen from ferrofluids into the hemoglobin biosynthesis. Furthermore, those studies have yielded quantitative results and have shown a 114-fold higher enrichment of iron oxide nanoparticles per gram of tissue in the case of attraction by an external magnetic field versus no external magnetic field [17, 23]. In the scope of a therapy study, intra-arterial application into the arteria femoralis of rabbits transplanted with a VX2 squamous cell carcinoma was performed. Therefore, ferrofluids loaded with the chemotherapeutic agent mitoxantrone (MTO) were used. As shown in this approach, 20% of the systemic dose (regular systemic dose 10 mg/m^2, equates to 100%) has been proven sufficient for complete tumor remission without any side effects, and the cellular uptake of iron oxide particles could also be shown [24, 25]. Furthermore, electron microscopy verified by energy-dispersive X-ray spectroscopy (EDX) analysis could visualize the iron oxide spots inside tumor cells, which is evidence of the therapeutic effect (fig. 2).

Hyperthermia

Hyperthermia is the intentional arrangement of an increased temperature in the organism and therefore probably the oldest therapeutic draft to treat tumors. Bacterial infections and in consequence bacterial induced pyrexia were the state of the art to treat cancer patients before World War II [26]. Magnetic hyperthermia with iron oxide nanoparticles is based on focused enrichment in the tumor area either by direct injection into the carcinogenic tissue (fig. 3) or by

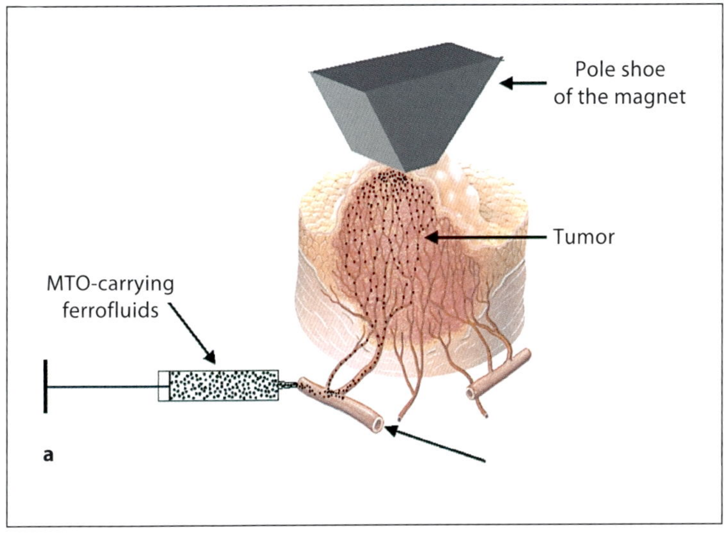

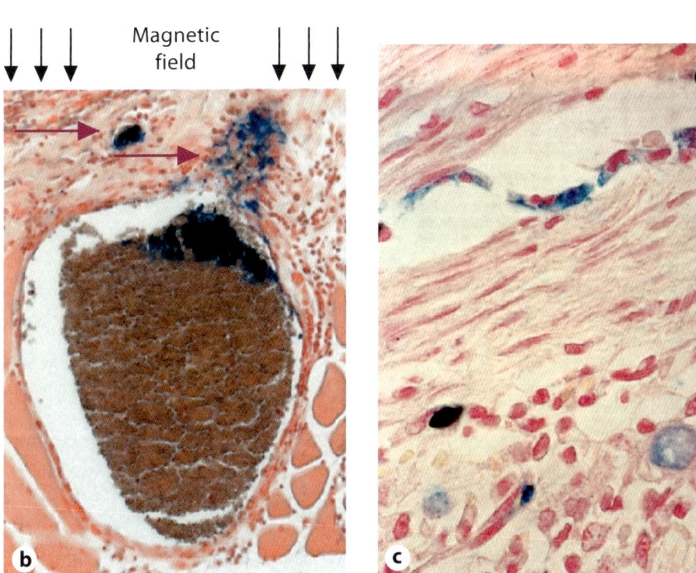

Fig. 1. **a** Principle of MDT [20]. **b** Particle infiltration into the tumor tissue after MDT. The tumor vessel and extravasation of particles into the tumor tissue (magnification 400-fold, Prussian blue staining) is visible; iron oxide appears blue (arrows). **c** Histologic investigation of the tumor tissue after MDT (magnification 1,000-fold, Prussian blue staining) blue stained iron oxide particles are highlighted by an arrow.

means of 'targeting' using tumor-supplying vessels. By exposure of the nanoparticle-spiked tumor to an alternating magnetic field, the induced material is heated up and destroys the malignant tissue. The first trendsetting experiments regarding hyperthermia took place in 1957 and were conducted by Gilchrist et al. [27]; the results of these experiments were approved by others [28]. Further studies affirm the effectiveness of this method for both conditions: hyperthermic treatment (applying temperatures of up to 47°C for a duration of 30 min) and thermoablation (temperatures over 55°C for a duration of 4–5 min) [29]. The physical background for this kind of therapy of magnetic nanoparticle-mediated heating is the dissipation of mag-

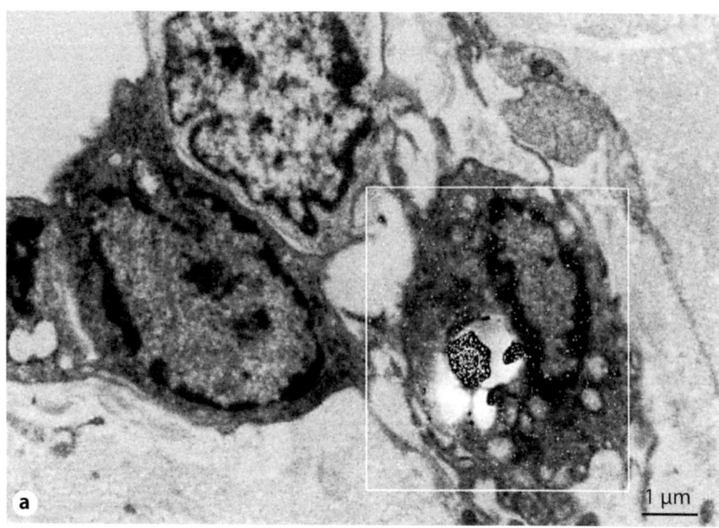

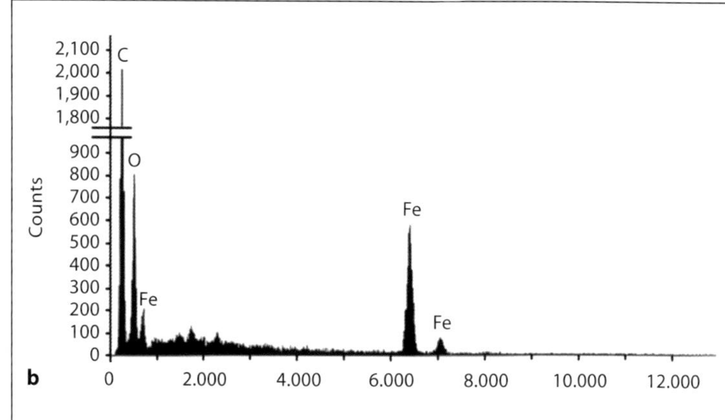

Fig. 2. a Electron microscopy of ultra-thin cross sections (thickness 70–100 nm) of tumor tissue after MDT. Focusing on the framed area, measurement of the Fe Kα signal (tumor cell) is shown. **b** Fe-EDX spectrum of ferrofluid aggregates within the tumor cell [25].

netic energy due to varying particle reversals of magnetization causing loss processes. The magnetic heat treatment of unifocal small carcinomas has been investigated extensively and has already reached the stage of clinical trials [30, 31]. Contrarily, there is no practice for the treatment of frequently occurring multicenter tumors that exclude organ preservation surgery. One of the most important questions will be how the combination of hyperthermic and chemotherapeutic modalities can be merged on the base of nanoparticles for successful and prospective therapy drafts [32].

Therapeutic Success of Magnetic Drug Targeting in Animal Studies

The first pilot studies evidenced the efficiency of MDT. In those drafts, experimental VX2 tumors were implanted subcutaneously into the left hind limb of experimental animals (female New Zealand white rabbits). After 4–6 weeks, the tumors had grown to a size that made intervention necessary. MTO was bound to superparamagnetic Fe_3O_4 nanoparticles by electrostatic interaction (hydrodynamic diameter ~100 nm) in an aqueous solution (ferrofluid). The applied doses were

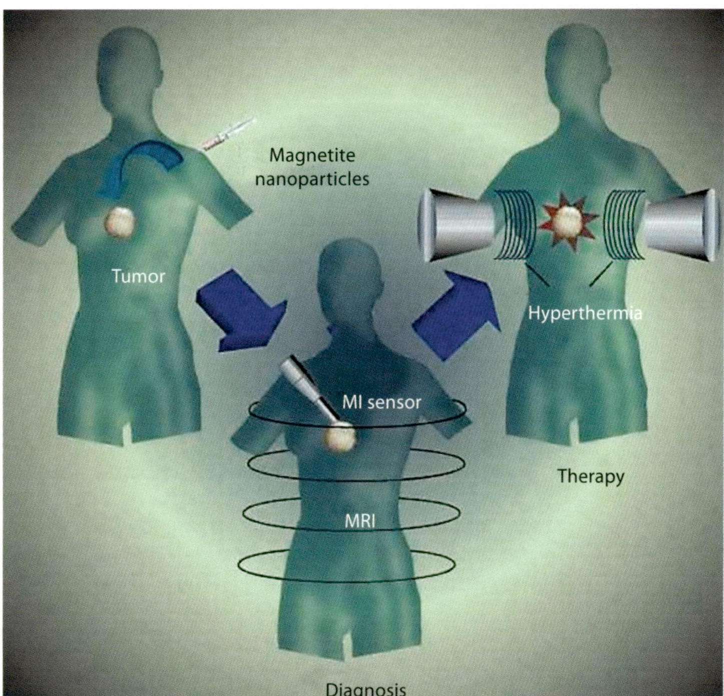

Fig. 3. Hyperthermia therapy scheme with magnetic nanoparticles [54]. **a** Application of magnetic nanoparticles. **b** Control of the accumulation of magnetic nanoparticles using MRI and a magnetoimpedance (MI) sensor. **c** An alternating magnetic field heats up the nanoparticles.

approximately 10% compared to the regular systemic dose (10 mg/m^2) of MTO. One of the major aspects of the MDT technique is the intra-arterial administration of the nanoparticles close to the tumor area. The drug-loaded nanoparticles were applied by injection through catheters and then attracted to the tumors by a focused external magnetic field during the application. Thus, no metabolic barrier could stop the ferrofluid from flushing the tumor vascular system. MDT was performed using just a single cycle of treatment. Widespread distribution of the particles is necessary for the therapeutic success that could be observed in these experiments [33]. Visualization of vascularization and tumor size before and after MDT is essential to control the therapeutic progress (fig. 4). A useful imaging tool for these studies was a biplane angiographic system with flat-detector computed tomography (FD-CT) [34]. Conventional angiographic series were obtained in order to visualize the vascularization in the tumor region. Additionally, FD-CT was performed to reconstruct cross-sectional CT images.

Current Developments in Nanosized Drug Carriers

Fullerenes and compounds derived from so-called carbon nanotubes as well as other condensed carbon clusters are the most popular species of a class of substances that were once implicated with 'nanotechnology'. Those roots date back to the 70s and early 80s of the past century [35]. Among many other scopes, using them as tiny carriers for therapeutic substances has become one of the most promising possibilities [36]. Meanwhile, a bulky number of structures in the nanosized scale have been developed to face this challenge. The

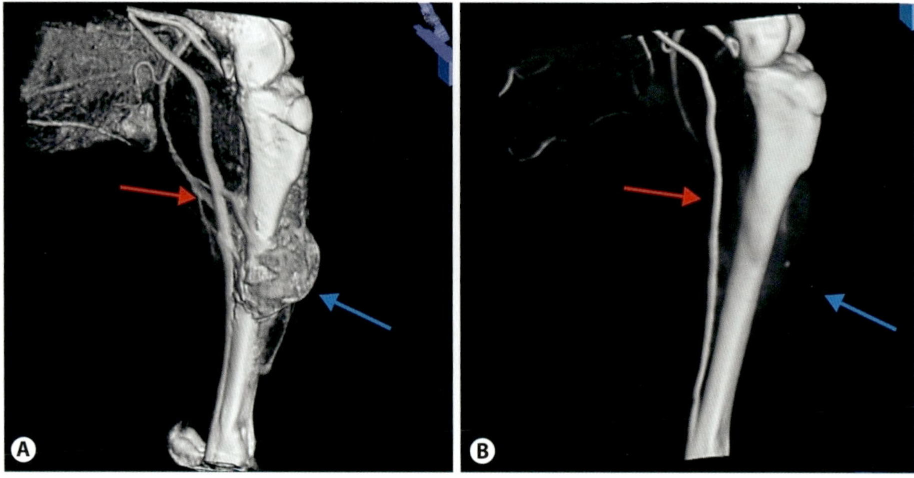

Fig. 4. Tumor remission after MDT. Angiographic imaging before (**A**) and after (**B**) MDT. Red arrow: arteria femoralis exhibiting tumor vessels. Blue arrow: tumor tissue. **B** Tumor tissue and vessels are not detectable any more after application of 1 course of MDT [33].

most important ones are well known as liposomes, micelles, encapsulations, emulsions, and metallic nanoparticles. Having acted at first as pure vehicles for carrying drugs effectively over metabolic barriers, now new nano-cargo systems able to release the agents in a controlled way in the targeted area are the focus of researchers. The release occurs by alternated physical and chemical parameters in the area of interest. Transportation and controlled release of therapeutic agents using nano- and micro-carrier systems in the area of interest is a promising approach in current medicine [6, 37]. The operation mode can thereby vary considerably. For example, it can be based on specific attributes of diseased tissues [38, 39]. Thus, inflamed tissue can be discerned from healthy tissue by a lower pH value, increased temperature, and a higher permeability of vessels [40]. These basic facts can be used for 'targeting' with temperature- and pH-responsive nanoparticles [41, 42]. Most of these thermoresponsive systems are based on polymer-coated, drug-loaded nanoparticles or capsules. Shen et al. [43] developed temperature-sensitive polymer-conjugated albumin nanospheres copolymerized with a poly(N-isopropylacrylamide) moiety. In this case the release occurs via dehiscing of the interspaces of the polymer network on the outside of the nanoparticles due to the increased temperature. As a consequence the nanospheres can be attacked and hydrolyzed easily by trypsin. Poly(N-isopropylacrylamide) is well known for facile alterations of conformation at a temperature level of about 32°C in aqueous media, and therefore it is often used as a coblock polymer for thermolabile surfaces [44]. Connecting alternative polymers featuring different conditions for switching their conformation, defined profiles for release on demand can be arranged. Due to multivalent properties of the surrounding polymer network, temperature- as well as pH-responsive release systems are possible [42, 45]. Nevertheless, thermosensitive polymer systems lack sharply differentiated thresholds, so the rate of drug release merely differs marginally. A different approach, in which ceramides triggering apoptosis are thereby used as effective antican- cer agents, has been used by Stover et al. [46]. Nanoparticles with bioactive sphingolipid-like second messenger sequences are incorporated depending on the temperature. Ceramides hardly penetrate the cell membrane and are degraded very easily. Linear dendritic nanoparticles contain a thermoresponsive poly(N-isopropylacrylamide) moiety that changes the whole polymer conformation above a threshold temperature. Due to the more hydrophobic property the particles now pass the membrane, become hydrolyzed, and release the effective ceramide. Other concepts, which are not yet at the clinical trial stage, perform drug release via the absorption of light. More precisely, the absorption of light causes heat development and leads to bond breakage between the drug and the coating substance. As proof of principle, substances were used to feature fluorescence only in the unbound state. These model compounds display fluorescence after bond breakage to the nanoparticles' surface [47]. Above a temperature of 60 °C a retro-Diels-Alder reaction divides the compound into a terminal maleimide linker and a furane moiety. A retro-Diels-Alder ring opening is also possible by alternating magnetic fields causing heat induction [48]. This kind of defined bond breaking enables an abrupt release of considerable amounts of therapeutic substances and offers new ways of achieving progressed drug release systems.

Tailor-Made Therapy Systems as Promising Concepts in Nanomedicine

By combining different approaches, application of different substances (classic drugs as well as nonclassic agents like nucleic acids, antibodies, etc.) and varying particle surfaces multifunctional particles can be enhanced. This is due to the fact that nanoparticles offer a platform for therapeutic agents but also linker for functional imaging purposes (^{18}F-PET) can be bound and so multifunctional 'theranostic' particles can be put on the stage [49–51]. For this reason, concepts that once expressed the hopes and utopias connected with nanotechnology at its beginning are not out of reach. At that time the dreams were about intelligent sensors and tools being able to fight fatal diseases inside human bodies as was written by Drexler [52] in his book *Engines of Creation – The Coming Era of Nanotechnology* in 1986. Medically, socially, and economically, cancer is considered a major topic. The biggest dilemma in cancer therapy is that despite all of the efforts being made and in spite of the increasing costs of therapeutic approaches, the number of victims has hardly decreased. At present and in the future, the sourcing of many new therapeutic approaches will get out of hand. Many new therapies will only be applied in a smaller number of patients. There is no doubt that new therapy drafts must possess significant medical progress, but also their costs must be evaluated. Keeping the increasing numbers of cancer incidents in mind, the need for action is highly visible. MDT and hyperthermia are promising and innovative therapy drafts in oncology. The main advantages are the focused tumor therapy and clearly reduced drug dose, whereby negative side effects connected with expensive costs for supporting measures for the patient can be reduced considerably. Taking into account these data and the increasing number of cancer patients, experts expect an appreciable economic potential for medical purposes in nanotechnology [53].

Acknowledgements

The authors would like to thank the Else Kröner-Fresenius-Stiftung for their support and the implementation of the Else Kröner-Fresenius-Stiftung Professorship for Nanomedicine at the University Hospital Erlangen. Furthermore, the authors would like to thank the DFG (PAK 151) and the German Ministry of Science and Education (BMBF, FKZ: 01EX1012B) for their advancement.

Disclosure Statement

The authors have nothing to disclose.

References

1 World Health Organization: Cancer. 2011. http://www.who.int/mediacentre/factsheets/fs297/en/index.html.

2 Robert Koch Institut: Krebs in Deutschland 2005/2006 – Häufigkeiten und Trends. Berlin, Robert Koch Institut, 2010, pp 121.

3 Wagner V, Dullaart A, Bock AK, Zweck A: The emerging nanomedicine landscape. Nat Biotechnol 2006;24:1211–1217.

4 Peer D, Karp JM, Hong S, Farokhzad OC, Margalit R, Langer R: Nanocarriers as an emerging platform for cancer therapy. Nat Nanotechnol 2007;2:751–760.

5 Perkins WR, Ahmad I, Li XG, Hirsh DJ, Masters GR, Fecko CJ, Lee JK, Ali S, Nguyen J, Schupsky J, Herbert C, Janoff AS, Mayhew E: Novel therapeutic nanoparticles (lipocores): trapping poorly water soluble compounds. Int J Pharm 2000;200:27–39.

6 Pankhurst QA, Thanh NK, Jones SK, Dobson J: Progress in applications of magnetic nanoparticles in biomedicine. J Phys D Appl Phys 2009;42: 224001.

7 Bilkenroth U, Taubert H, Riemann D, Rebmann U, Heynemann H, Meye A: Detection and enrichment of disseminated renal carcinoma cells from peripheral blood by immunomagnetic cell separation. Int J Cancer 2001;92: 577–582.

8 Taupitz M, Wagner S, Hamm B, Dienemann D, Lawaczeck R, Wolf KJ: MR lymphography using iron oxide particles: detection of lymph node metastases in the VX2 rabbit tumour model. Acta Radiol 1993;34:10–15.

9 Harisinghani MG, Barentsz J, Hahn PF, Deserno WM, Tabatabaei S, van de Kaa CH, de la Rosette J, Weissleder R: Noninvasive detection of clinically occult lymph-node metastases in prostate cancer. N Engl J Med 2003;348:2491–2499.

10 Gao Y: Biofunctionalization of magnetic nanoparticles; in Kumar C (ed): Biofunctionalization of Nanomaterials. Weinheim, Wiley-VCH, 2005, vol 72, p. 72–89.

11 Storm G, Belliot SO, Daemen T, Lasic DD: Surface modification of nanoparticles to oppose uptake by the mononuclear phagocyte system. Adv Drug Deliv Rev 1995;17:31–48.

12 Muller RH, Jacobs C, Kayser O: Nanosuspensions as particulate drug formulations in therapy: rationale for development and what we can expect for the future. Adv Drug Deliv Rev 2001;47: 3–19.

13 Alksne JF, Fingerhut AG, Rand R: Magnetically controlled metallic thrombosis of intracranial aneurysms. Surgery 1966;60:212–218.

14 Widder KJ, Marino PA, Morris RM, Howard DP, Poore GA, Senyei AE: Selective targeting of magnetic albumin microspheres to the Yoshida sarcoma: ultrastructural evaluation of microsphere distribution. Eur J Cancer Clin Oncol 1993;19:141–147.

15 Wilson MW, Kerlan RK Jr, Fidelman NA, Venook AP, LaBerge JM, Koda J, Gordon RL: Hepatocellular carcinoma: regional therapy with a magnetic targeted carrier bound to doxorubicin in a dual MR imaging/conventional angiography suite – initial experience with four patients. Radiology 2004;230:287–293.

16 Lübbe AS, Bergemann C, Riess H, Schriever F, Reichardt P, Possinger K, Matthias M, Dorken B, Herrmann F, Gurtler R, Hohenberger P, Haas N, Sohr R, Sander B, Lemke AJ, Ohlendorf D, Huhnt W, Huhn D: Clinical experiences with magnetic drag targeting: a phase I study with 4′-epidoxorubicin in 14 patients with advanced solid tumors. Cancer Res 1996;56:4686–4693.

17 Alexiou C, Jurgons R, Schmid RJ, Bergemann C, Henke J, Erhardt W, Huenges E, Parak F: Magnetic drug targeting – biodistribution of the magnetic carrier and the chemotherapeutic agent mitoxantrone after locoregional cancer treatment. J Drug Target 2003; 11:139–149.

18 Alexiou C, Diehl D, Henninger P, Iro H, Rockelein R, Schmidt W, Weber H: A high field gradient magnet for magnetic drug targeting. IEEE Trans Appl Supercond 2006;16:1527–1530.

19 Brigger I, Morizet J, Laudani L, Aubert G, Appel M, Velasco V, Terrier-Lacombe MJ, Desmaele D, d'Angelo J, Couvreur P, Vassal G: Negative preclinical results with stealth nanospheres-encapsulated doxorubicin in an orthotopic murine brain tumor model. J Control Release 2004;100:29–40.

20 Alexiou C, Jurgons R, Schmid R, Erhardt W, Parak F, Bergemann C, Iro H: Magnetic drug targeting – a new approach in locoregional tumor therapy with chemotherapeutic agents: experimental animal studies. HNO 2005;53: 618–622.

21 Alexiou C, Arnold W, Hulin P, Klein RJ, Renz H, Parak FG, Bergemann C, Lubbe AS: Magnetic mitoxantrone nanoparticle detection by histology, X-ray and MRI after magnetic tumor targeting. J Magn Magn Mater 2001; 225:187–193.

22 Rahn H, Gomez-Morilla I, Jurgons R, Alexiou C, Odenbach S: Microcomputed tomography analysis of ferrofluids used for cancer treatment. J Phys Condens Matter 2008;20:204152–204156.

23 Alexiou Ch, SA, Hulin P, Klein RJ, Bergemann Ch, Arnold W: Magnetic drug targeting: biodistribution and dependency on magnetic field strength. J Magn Magn Mater 2002;252:363–366.

24 Alexiou C, Arnold W, Klein RJ, Parak FG, Hulin P, Bergemann C, Erhardt W, Wagenpfeil S, Lubbe AS: Locoregional cancer treatment with magnetic drug targeting. Cancer Res 2000;60:6641–6648.

25 Alexiou C, Schmid RJ, Jurgons R, Kremer M, Wanner G, Bergemann C, Huenges E, Nawroth T, Arnold W, Parak FG: Targeting cancer cells: magnetic nanoparticles as drug carriers. Eur Biophys J 2006;35:446–450.

26 Nauts HC,Fowler GA, Bogatko FH: A review of the influence of bacterial infection and of bacterial products (Coley's toxins) on malignant tumors in man: a critical analysis of 30 inoperable cases treated by Coley's mixed toxins, in which diagnosis was confirmed by microscopic examination selected for special study. Acta Med Scand 1953;276: 1–103.

27 Gilchrist RK, Medal R, Shorey WD, Hanselman RC, Parrott JC, Taylor CB: Selective inductive heating of lymph nodes. Ann Surg 1957;146:596–606.

28 Gordon RT, Hines JR, Gordon D: Intracellular hyperthermia: a biophysical approach to cancer treatment via intracellular temperature and biophysical alterations. Med Hypotheses 1979;5: 83–102.

29 Jordan A, Scholz R, Wust P, Fahling H, Krause J, Wlodarczyk W, Sander B, Vogl T, Felix R: Effects of magnetic fluid hyperthermia (MFH) on C3H mammary carcinoma in vivo. Int J Hyperthermia 1997;13:587–605.
30 Johannsen M, Thiesen B, Jordan A, Taymoorian K, Gneveckow U, Waldofner N, Scholz R, Koch M, Lein M, Jung K, Loening SA: Magnetic fluid hyperthermia (MFH) reduces prostate cancer growth in the orthotopic Dunning R3327 rat model. Prostate 2005;64: 283–292.
31 Hilger I, Andra W, Hergt R, Hiergeist R, Schubert H, Kaiser WA: Electromagnetic heating of breast tumors in interventional radiology: in vitro and in vivo studies in human cadavers and mice. Radiology 2001;218:570–575.
32 Moskovitz B, Meyer G, Kravtzov A, Gross M, Kastin A, Biton K, Nativ O: Thermo-chemotherapy for intermediate or high-risk recurrent superficial bladder cancer patients. Ann Oncol 2005;16:585–589.
33 Lyer S, Tietze R, Jurgons R, Struffert T, Engelhorn T, Schreiber E, Dorfler A, Alexiou C: Visualisation of tumour regression after local chemotherapy with magnetic nanoparticles – a pilot study. Anticancer Res 2010;30:1553–1557.
34 Kalender WA, Kyriakou Y: Flat-detector computed tomography (FD-CT). Eur Radiol 2007;17:2767–2779.
35 Iijima S: Helical microtubules of graphitic carbon. Nature 1991;354:56–58.
36 Kostarelos K, Bianco A, Prato M: Promises, facts and challenges for carbon nanotubes in imaging and therapeutics. Nat Nanotechnol 2009;4:627–633.
37 Pankhurst QA, Connolly J, Jones SK, Dobson J: Applications of magnetic nanoparticles in biomedicine. J Phys D Appl Phys 2003;36:R167–R181.
38 Arap W, Haedicke W, Bernasconi M, Kain R, Rajotte D, Krajewski S, Ellerby HM, Bredesen DE, Pasqualini R, Ruoslahti E: Targeting the prostate for destruction through a vascular address. Proc Natl Acad Sci USA 2002;99: 1527–1531.
39 Torchilin VP: Drug targeting. Eur J Pharm Sci 2000;11:S81–S91.
40 Gerlowski LE, Jain RK: Microvascular permeability of normal and neoplastic tissues. Microvasc Res 1986;31:288–299.
41 Shen Z, Wei W, Zhao YJ, Ma GH, Dobashi T, Maki Y, Su ZG, Wan JP: Thermosensitive polymer-conjugated albumin nanospheres as thermal targeting anti-cancer drug carrier. Eur J Pharm Sci 2008;35:271–282.
42 Chuang CY, Don TM, Chiu WY: Synthesis of chitosan-based thermo- and pH-responsive porous nanoparticles by temperature-dependent self-assembly method and their application in drug release. J Polym Sci A Polym Chem 2009;47:5126–5136.
43 Shen ZY, Ma GH, Dobashi T, Maki Y, Su ZG: Preparation and characterization of thermo-responsive albumin nanospheres. Int J Pharm 2008;346: 133–142.
44 Lee CF, Wen CJ, Lin CL, Chiu WY: Morphology and temperature responsiveness-swelling relationship of poly(N-isopropylamide-chitosan) copolymers and their application to drug release. J Polym Sci A Polym Chem 2004;42: 3029–3037.
45 Chuang CY, Don TM, Chiu WY: Synthesis and properties of chitosan-based thermo- and pH-responsive nanoparticles and application in drug release. J Polym Sci A Polym Chem 2009;47: 2798–2810.
46 Stover TC, Kim YS, Lowe TL, Kester M: Thermoresponsive and biodegradable linear-dendritic nanoparticles for targeted and sustained release of a pro-apoptotic drug. Biomaterials 2008;29: 359–369.
47 Bakhtiari AB, Hsiao D, Jin GX, Gates BD, Branda NR: An efficient method based on the photothermal effect for the release of molecules from metal nanoparticle surfaces. Angew Chem Int Ed Engl 2009;48:4166–4169.
48 Purushotham S, Chang PE, Rumpel H, Kee IH, Ng RT, Chow PK, Tan CK, Ramanujan RV: Thermoresponsive core-shell magnetic nanoparticles for combined modalities of cancer therapy. Nanotechnology 2009;20: 305101.
49 Levy-Nissenbaum E, Radovic-Moreno AF, Wang AZ, Langer R, Farokhzad OC: Nanotechnology and aptamers: applications in drug delivery. Trends Biotechnol 2008;26:442–449.
50 Purushotham S, Chang PE, Rumpel H, Kee IH, Ng RT, Chow PK, Tan CK, Ramanujan RV: Thermoresponsive core-shell magnetic nanoparticles for combined modalities of cancer therapy. Nanotechnology 2009;20:305101–305112.
51 Welch MJ, Hawker CJ, Wooley KL: The advantages of nanoparticles for PET. J Nucl Med 2009;50:1743–1746.
52 Drexler KE: Engines of Creation – the Coming Era of Nanotechnology. New York, Anchor Books, 1986.
53 OECD: Opportunities and risks of nanotechnologies: report in co-operation with the OECD International Futures Programme. 2005. http://www.oecd.org/dataoecd/32/1/44108334.pdf
54 Barakat NS: Magnetically modulated nanosystems: a unique drug-delivery platform. Nanomedicine (Lond) 2009; 4:799–812.

Univ. Prof. Dr. med. Christoph Alexiou
Else Kröner-Fresenius-Stiftung Professorship for Nanomedicine
Section for Experimental Oncology and Nanomedicine
Department of Otorhinolaryngology, Head and Neck Surgery
University Hospital Erlangen
Glückstrasse 10, DE–91054 Erlangen (Germany)
Tel. +49 9131 85 34769, E-Mail c.alexiou@web.de

Alexiou C (ed): Nanomedicine – Basic and Clinical Applications in Diagnostics and Therapy.
Else Kröner-Fresenius Symp. Basel, Karger, 2011, vol 2, pp 164–176

Parameters Influencing the Efficacy of Magnetic Heating of Small Breast Tumors

Ingrid Hilger · Werner A. Kaiser

Department of Experimental Radiology, Institute of Diagnostic and Interventional Radiology I, Centrum of Radiology, University Hospital Jena, Jena, Germany

Abstract

Detection of tumors in the breast has been distinctly improved in the last years, which favors the development of minimally invasive therapeutic methods promising a high therapeutic efficacy. Moreover, the cosmetic outcome of these techniques is becoming a factor of increased significance. Herein, we propose the treatment of small tumors via magnetic heating, meaning the selective accumulation of magnetic materials, i.e. maghemite/magnetite, at the tumor area and exposure of the breast to an alternating magnetic field for several minutes. This produces a selective heating spot which allows for a localized elimination of tumor cells. Even though the principle seems simple at first glance, the efficacy of the methodology strongly depends on how the different biomedical, physical, and technical parameters are combined and adjusted between each other. Therefore, in the present study we will consider the different aspects that have to be taken into account to fulfill these requirements.

Image-based detection of tumors in the breast has distinctly improved in the last years. Increasingly, smaller tumors with a good prognosis (diameters less than 10 mm) can be detected, particularly with sophisticated methods such as magnetic resonance mammography [1]. These developments favor early detection and – consequently – the efficacy and outcome of the therapy.

Whereas traditional concepts in the early 20th century have been based on the assumption that breast cancer is a local disease to be treated by radical mastectomy, today there is increasing interest in strategies with lower invasiveness, similar survival rates, and encouraging cosmetic outcomes [2]. Radical mastectomy is currently restricted to cases with multicentricity, multifocality, tumor spreading over several quadrants, or extensive infiltration.

Most of the minimally invasive techniques are based on different tools which have been previously addressed to tumors in regions of the body other than the breast. In recent years, increasing evidence of successful treatment of breast tumors using interstitial laser photocoagulation, radio-

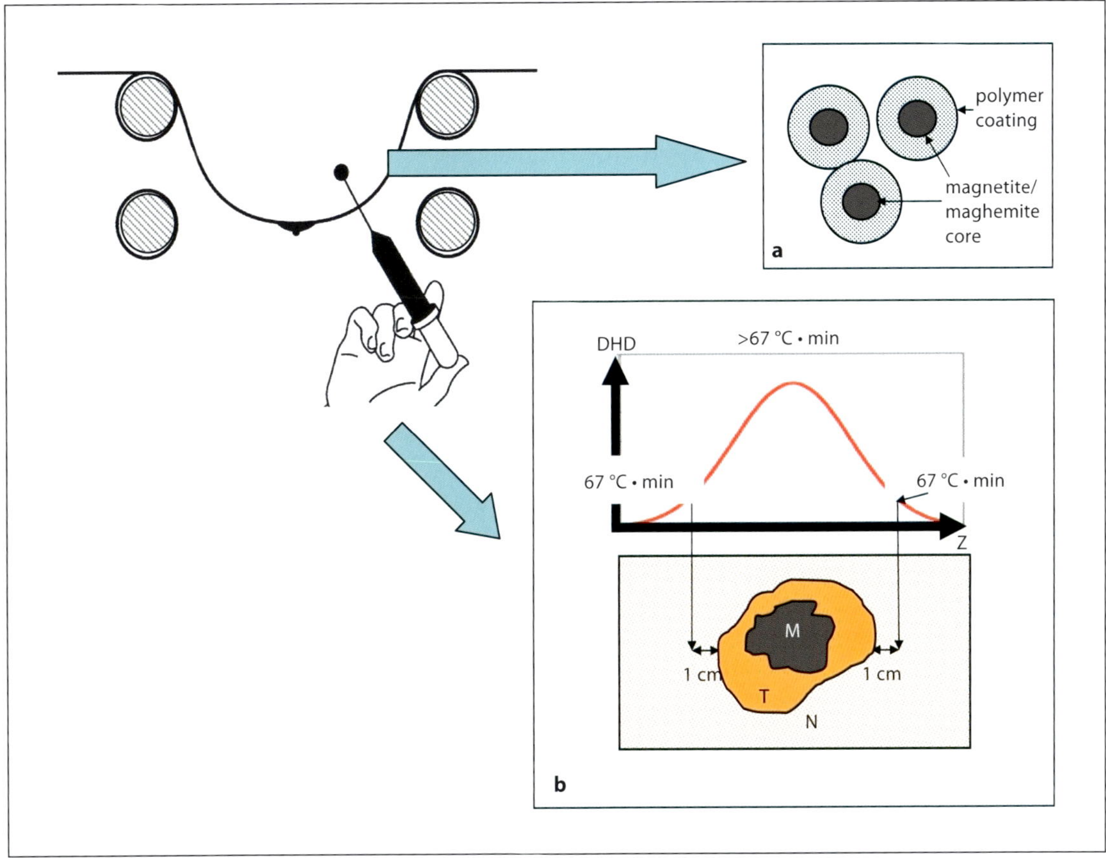

Fig. 1. Scheme showing the principle of the magnetic heating procedure to treat small tumors in the breast. A magnetic material is selectively accumulated at the tumor area. After exposure to an AC magnetic field, heating is produced to inactivate tumor cells. **a** The magnetic material is typically composed of spherical magnetite/maghemite nanoparticles and polymer coatings. **b** Parameters could be adjusted to induce thermoablative temperatures with critical deposited temperature dosages (DHD) ideally at 1 cm from the tumor border in analogy to surgical procedures (for more details see text). Bottom: tumor (T) with centrally deposited magnetic (M) material (black). Top: bell-shaped curve of the expected DHD at the median axis of the tumor (z-direction).

frequency, focused ultrasound, or cryotherapy has been reported [3]. Nevertheless, a particularly high accuracy for energy deposition at the target could be achieved by the use of a magnetic material (iron oxide, e.g. magnetite/maghemite) and exposure of the breast to an alternating magnetic field so that the deposited magnetic material absorbs energy from the field, converting it into heat (fig. 1). Whereas initial approaches were first studied in dogs and several tumor models [4, 5] with encouraging results, the method has been tested in the clinical setting for selected tumor entities only, such as prostate and glioblastoma [6, 7]. For breast cancer, corresponding applications are lacking to date, mainly due to the lack of systematic examinations, the diversity of relevant parameters, and the high requirements for good therapeutic and cosmetic outcomes. On the other hand, the breast is easily accessible to magnetic fields because of its topographic location which is distant from vital organs. Therefore, the present article will describe the relevant parameters to be considered when implementing the proposed method for clinical practice.

Magnetic Materials

One of the basic prerequisites for effective magnetic heating treatments is that the nanoparticles have adequate magnetization values. A tiny regulation of the temperatures to be deposited into the target tissue is accomplished by the selection of appropriate magnetic materials. Iron oxides (magnetite/maghemite) are particularly preferred since they have been used as contrast agents in radiology for many years, e.g. in the detection of cancers in the liver and spleen. It has been shown that 250 mg of magnetite/maghemite per kg of body weight does not induce any hepatotoxicity in mice, and 1–3 mg of magnetite/maghemite per kg of body weight was tolerable for human beings [8–10].

The generation of heat using magnetic particles is based on energy dissipation in the magnetization processes of the particles. The corresponding mechanisms are highly dependent upon the particle morphology [11, 12]. Whereas the magnetization reversal of particles with core diameters larger than 100 nm is governed by the motion of magnetic domain walls, smaller magnetite/maghemite cores (below 100 nm) behave like single-domain particles where no domain wall exists at all. Importantly, the magnetic behavior of particles with core diameters below 50 nm is strongly influenced by thermal agitation already at room temperature. The magnetization reversal of these so-called superparamagnetic particles proceeds either by rotation of the magnetic moment inside the core (Néel relaxation) or by mechanical rotation of the whole particle (Brown relaxation) [13] during exposure to an alternating magnetic field. The particular features of different loss processes have been considered in detail in relation to the principal limitations that might occur when addressing hyperthermic applications with fine magnetic particles [11]. The different loss processes in magnetic particle systems that might be of interest for magnetic heating applications have been compiled by Hergt et al. [14], whereas some of the experimental data on specific heating power reported in the literature was given by Andrä [15]. Heating potentials for maghemite particles as high as 1 kW/g (amplitude 8.8 kA/m, frequency 410 kHz) have been determined [16]. In the meantime, several investigations have shown that comparatively good thermal energies could be obtained with nanoparticles with sizes between 10 and 50 nm [16–19].

The fact that magnetite/maghemite nanoparticles containing contrast media Endorem® (Guerbet, Villepinte, France) reveal any heating effects (unpublished data) implies that distinct efforts have been made to regulate synthesis parameters in an adequate manner to yield new magnetic nanoparticles with the desired structural features [20]. A particular challenge is the production and stabilization of nanoparticles with a relatively larger core size distribution when using magnetic field frequencies in the kilohertz range and magnetic field amplitudes of up to 15 kA/m.

Not only the heating potential but also their interaction with target structures will depend on the nanoparticle morphology. The nanoparticle uptake by cells could be of interest for local deposition of the material and may depend on the nature of the coating, the surrounding medium, and their aggregation behavior. If the nanoparticles get into the blood compartment, aggregation might occur in the presence of plasma proteins. This particular process seems to be highly dependant on the nature of the coating, as shown via magnetorelaxometry analysis [21] (see also below).

To enhance treatment efficacy, the use of multifunctional nanoparticles was suggested. For example, the adsorptive binding of cisplatin to magnetic nanoparticles could allow for a dedicated chemotherapeutic release at the tumor site [22]. In combination with magnetic heating, it is expected to exert additive cytotoxic effects [23] particularly when it reaches its target site of action forming DNA adducts after its chloride ions have been displaced by hydroxyl groups. This process is thought to inhibit DNA replication and chain elongation [24]. The absorptive binding is beneficial from the viewpoint that cisplatin shows better chemotherapeutic properties when it is not adsorbed to other molecules [25].

Nanoparticle Loading of Tumors

Different procedures for nanoparticle loading of tumors have been proposed. Among them, intratumoral application allows for easy control of the nanoparticle deposition at the target region. Due to deposition of comparatively large nanoparticle amounts, the induction of thermoablative temperatures is feasible. Nonetheless, there is a need for defined stereotactic and imaging procedures for adequate inoculation of the material into the target tissue, and the homogeneous nanoparticle distribution is still a challenging issue. Inhomogeneous distributions make the occurrence of regions of underdosage very likely to occur, particularly when using hyperthermic treatments (see below).

Systemic nanoparticle application into the blood system is less invasive as compared to intratumoral application but requires adequate nanoparticle formulations to allow for a selective accumulation at the tumor area. In most cases, more than 90% of the intravenously injected nanoparticle dose will not be available for accumulation at the tumor region due to the occurrence of opsonization processes and consequently high sequestration rates by the mononuclear phagocyte system (MPO). In this context, the mechanical properties of the polymer coating have an important role in determining protein adsorption. PEGylated nanoparticles turned out to exhibit beneficial properties by preventing protein binding on surfaces [26] and therefore increasing the blood half-life, which also increases the availability for nanoparticle extravasation into the tumor interstitium. Another interesting approach is the use of high-affinity nanoparticles by coupling antibodies, antibody fragments, peptides, and aptamers with high binding affinity to molecular structures at the tumor region. Several targeting approaches have been published to date [27–29] showing the basic feasibility of the approach. Nevertheless, it has become clear that no thermoablative temperatures but rather hyperthermic temperatures [30] will be achieved by this methodology, requiring additional oncologic modalities to ensure adequate inactivation of tumor cells.

More sophisticated techniques exploit the capture of nanoparticles after systemic application via the use of external magnets. In this case, appropriate magnetic field gradients have to be taken into account. The present experience revealed that more or less superficial tumors could be loaded with this technology. Initial in vitro experiments showed that distinctly higher temperature increases are obtained when labeling tumor cells via magnetic targeting prior to magnetic heating applications [31]. Nanoparticles are mainly localized in endosomal and lysosomal compartments as is already known for cells incubated with nanoparticles only.

Magnetic Field Application

The AC magnetic field amplitude H and frequency distinctly influence the heating power of magnetic nanoparticles. The frequency dependence of heating power may be different for different size distributions of nanoparticles, varying with frequency from the square law to near independence.

In general, superparamagnetic particles show a square dependence on the field amplitude [32].

Concerning therapeutic applications, the frequency dependence of the specific loss of power of the particle type to be applied should be known to enable a valuable frequency choice, but simply increasing the heating power by increasing the field amplitude and field frequency is not beneficial. It has to be considered that inductive heating of tissue due to eddy currents might occur. This heating effect reasonably should not cause any damage to the healthy tissue. Calculations for a cylindrical tissue model [33] showed that heating power is proportional to the square of the product ($H \times f \times R$) where R is the radius of the exposed region. Experimental investigations [34] performed with a test person confirmed these theoretical results and became the base of a widely used limit for exposure of the human torso: $(H \times f)_{Brez} = 4.85 \times 10^8$ A/(m × s). This number is related to the radius R = 15 cm. For field applications to smaller parts of the body the allowed value ($H \times f$) may be higher.

The described limitations of the AC field parameters demonstrate that the specific heating power of the nanoparticles should be optimized with respect to the field parameters. In this context we preferably apply field amplitudes of up to 11 kA/m and a frequency of 410 kHz. Herein, circular coils with diameters of 9 and 15 cm in combination with high frequency generators of 5 and 15 kW of power, respectively, were used for animal experiments. In contrast, for treatment of glioblastomas a ferrite core applicator working at 100 kHz and a maximum field amplitude of 15 kA/m has been suggested [35].

For the range of field parameters discussed above and the field coil apertures needed for human patients, it is technically difficult to provide a homogeneous field within the target region. Since the heat generation and consequently the temperature elevation in the tissue depends on the square of the field amplitude, the remaining field inhomogeneity of the applicator has to be taken into account. The position of the tumor within the field applicator and the field amplitude at this position should be exactly known for a reliable therapy. To assure that the temperature elevation is within the therapeutic range, a minimally invasive temperature measurement in the tumor tissue during field application is necessary and the field amplitude may be used to control the temperature.

Temperature Regime

The ultimately deposited temperature regimes will be crucial for the outcome of the proposed minimally invasive therapy method. They are the equivalent of the amount, time of residence, and activity of the deposited chemotherapeutic drugs at the tumor region in routine oncologic procedures. Temperature regimes do reflect the specific temperature dosages and the time prevalence at a specific tumor volume. Even though the necessary time-dependent temperature dosages have been widely investigated for hyperthermic treatments (treatments up to temperatures of 42°C for approximately 60 min), few data is available regarding thermoablative tumor treatments. From water bath experiments, we postulated that the critical temperature leading to cell death should be between 51 and 55°C for an exposure time of 4 min. The corresponding deposited heating dose (DHD) as calculated by estimating the area between the time-dependent temperature curves and the temperature level without heating is between 47 and 61°C × min [36]. Of course, both the course of the temperature rise and decay during and after treatment will have a distinct impact on the corresponding heat dosage. This data is in quite good agreement with findings related to the thermostability of several vital proteins [37–39] and the occurrence of tissue coagulation [40]. This process can be clearly demonstrated by the treatment of muscle tissues with an inherent color change due to degradation of myoglobin and the occurrence of hemolysis at temperatures higher than 50°C [41, 42].

The induction of DNA damage is another crucial factor for inactivation of tumor cells. Experiments on human tumor cell cultures revealed a strong correlation between DNA damage, cell survival, and temperature, as determined via the cloning assay and trypan blue exclusion. Herein, the threshold thermoablative temperature for tumor cell elimination was found to be between 55 and 60°C for 5 min. Particularly, cells seem to be able to repair induced DNA damage up temperatures of 55°C since the DNA repair machinery is viable in these conditions. Instead, after inactivation of the DNA repair machinery at 60°C and higher temperatures, the damage is not repaired any more, leading to cell death via apoptosis. A similar situation was also encountered in vivo after magnetic heating of tumor-bearing mice [43].

Taking these data together, in the expected clinical setting, temperatures of approximately 55°C should allow for reliable inactivation of tumor cells. In analogy to operative procedures and considering the problems of reliable detection of tumor borders – particularly with X-ray mammography – the range of the critical temperatures should be extended to a security rim of 1 cm around the tumor border (fig. 1). Since thermoablative effects are irreversible, the proposed methodology could ideally be applied as a single therapy modality in defined patient groups with tumors of good prognosis (e.g. smaller than 10 mm and with a negative lymph node status). Future developments will show to which extent this procedure is feasible and reasonable in each individual case. It implies, indeed, that the whole tumor should be completely covered with a homogeneous DHD. This procedure remains a challenging issue, particularly in connection with the intratumoral application of the magnetic material.

Assuming low heat conduction due to low blood flow – as it is the case in the breast – the maximum temperature rise ΔT to be achieved in the target region depends on the square of the target radius as well as on the heat conductivity of the environment [11]. Estimations have shown that for presently available nanoparticles the therapy is limited to a target radius of more than 1 mm [14]. This effect was investigated in detail by Rabin [44] with the conclusion that heating of single tissue cells filled with nanoparticles ('intracellular hyperthermia') is completely ineffective [45].

To monitor temperature developments and consequently the DHD, only invasive procedures are available to date. For this purpose, the implantation of catheters for thermometry remains necessary with the intrinsic disadvantages. Further developments of noninvasive thermometry or sophisticated computer simulations are on the way [46] and future work will show whether new developments are applicable in relation to the proposed method.

Nanoparticle Retention at the Tumor Site

To ensure the induction of highly selective heat spots at the tumor region, adequate nanoparticle retention times are crucial. Our investigations have shown that at least for a time period of 30 min after application of the magnetic material no distinct variation in the location and configuration of the intratumoral particle agglomerates seems to take place. The nanoparticle release from the tumor is rather low. Spectrometric determination of iron amounts in selected organs of treated mice indicated that there seems to be a tendency for a preferential wash out of nanoparticles with a cationic surfactant as compared with those having an anionic one (table 1). The different behavior of the aforementioned nanoparticles could depend on the different extent of particle absorption through blood proteins (opsonization). A higher particle accumulation of the cationic nanoparticles was observed in the animal's liver, spleen, and lung. The iron accumulation in the liver and spleen is known to be related to se-

Table 1. Iron mass per gram of dried tissue mass in different tissues of tumor-bearing immunodeficient mice measured 50 min after the intratumoral application of charged magnetic nanoparticles

Samples	Spleen, %	Lung, %	Liver, %	Blood, %
I	6.7 ± 2.3 n = 5	0.0 ± 0.0 n = 3	2.8 ± 0.9 n = 5	2.4 ± 0.1 n = 3
II	14.0 ± 2.4 n = 6	16.8 ± 8.4 n = 4	6.6 ± 0.7 n = 6	22.3 ± 8.5 n = 4

Values are given as a percentage of the total injected iron mass. Corrected values are in relation to tissue-specific iron contents that were determined using non-treated controls.

questration by phagocytic cells of the MPO. These observations indicate that the local immobilization of nanoparticles at the tumor area depends on the features of the nanoparticle surface and that, indeed, systemic interrelations are likely to take place even after intratumoral application. In the expected clinical setting, the probability of the occurrence of thermal damage in organs of the MPO should be low due to its topographic distance from the breast (e.g. liver and spleen) where exposure to AC magnetic fields per se takes place.

Formation of Localized Heat Spots in vivo

The observations from experiments with tumor-bearing mice show that intratumoral nanoparticle application and exposure of the animal to an alternating magnetic field leads to the induction of highly localized temperature spots (fig. 2). We could demonstrate that – on average – temperatures up to around 71 °C [12] can be induced at the tumor center via magnetic heating. At the peripheries, the increase in temperature seems to be more variable, ranging between 13 and 73 °C, with corresponding DHD between 40 and 262 °C × min [47]. Regions of DHD underdosage can be observed, e.g. in 8 of 36 tumors with regions of DHD lower than the critical heating dosage between 47 and 61 °C × min (see above and Hilger et al. [36]). Similar findings were reported when treating experimental tumors with magnetic fluid hyperthermia (temperatures of up to 47 °C for 30 min [48]). The occurrence of regions of temperature underdosage is the result of low local concentrations of magnetic nanoparticles within the tumor region. At this point, there is an increased demand for appropriate fitting of the magnetic particle distribution to the shape of the tumor.

The presently available data also indicate that interindividual variations in the induced tumoral temperatures are likely to occur [47]. This observation implies the need for individual temperature monitoring. Such interindividual variations in temperature development are related to the distinct tumor tissue morphology which determines the nature of the nanoparticle distribution after intratumoral application. Also the use of inhomogeneous magnetic fields could induce these effects where variations in the positioning of the tumors within the inductor coil lead to slightly altered temperature increases.

Effects of Thermal Washout

Thermal washout has been known to counteract the local induction of heat, independently of the methodology used. Nevertheless, the situation in the breast is expected to be comparatively favorable in terms of blood flow-based thermal washout influences. In general, blood flow in tumors is chaotic in nature [49–51] as a consequence

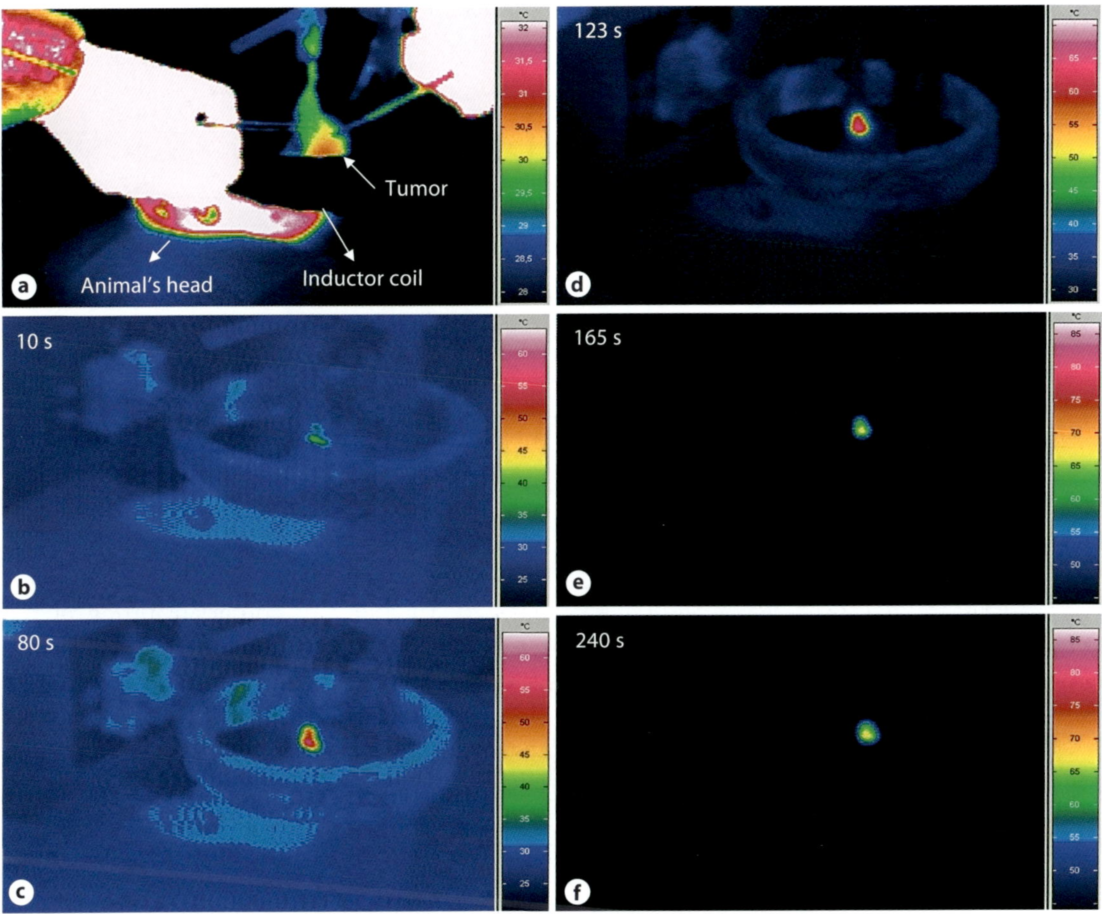

Fig. 2. Magnetic heating induces localized heat spots at the tumor region. Infrared images of tumor-bearing rats after application of magnetic material (15 mg magnetite/maghemite nanoparticles per tumor) and exposure to an AC magnetic field for 4 min. The tumor-bearing leg of a rat is placed in the center of the coil. **a** Reference image (magnetic field off). **b–f** Magnetic field on (amplitude 7 kA/m; frequency 325 kHz). Color-coded pictures with respective temperature scalings as indicated at the right side of each image.

of the nonorganized and heterogeneous vessel structure, e.g. the absence of basal lamina in the vessel wall, the presence of sinusoids, etc. [52, 53]. The rate of blood flow in tumors of the breast is nearly comparable to that of other tumors, e.g. tumors in the liver or the central nervous system (up to approximately 30 ml/100 g/min [54]) but lower in relation to the normal host tissue [54]. Particularly, when compared to the liver, the breast is a low vascularized body region [55]. Moreover, the presence of fatty tissue in the breast should also be beneficial for the induction of localized heat spots. In contrast, it is already known that for heating of tumors in the highly vascularized liver, even the occlusion of single vessels is necessary when situated in the vicinity of the portal vein.

In general, the effects of heat transport are also clearly visible when using subcutaneous tumor models. Here, temperatures in the proximal periphery were shown to be lower compared to

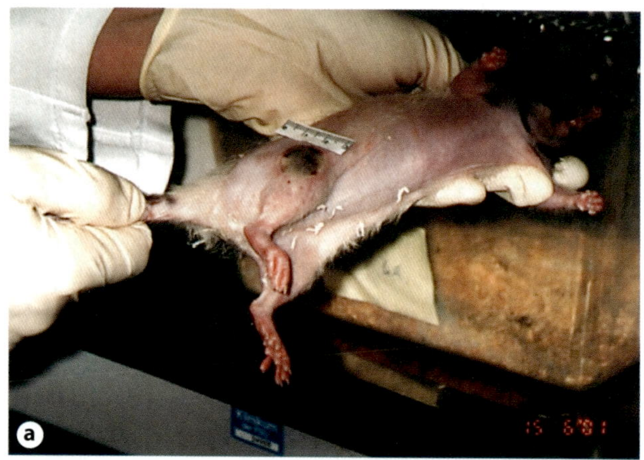

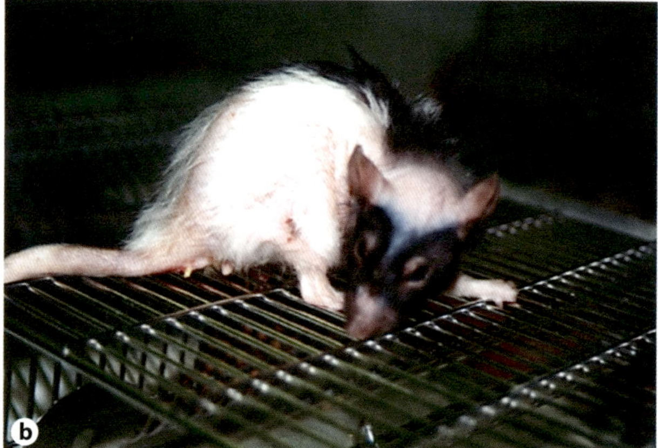

Fig. 3. Magnetic heating allows for selective treatment of the tumor region. Representative pictures of treated rats after application of the magnetic material (15 mg magnetite/maghemite nanoparticles per tumor) and exposure of the animals to an AC magnetic field (amplitude 7 kA/m; frequency 325 kHz) for 4 min. **a** At 2 days after treatment, only the area where the magnetic material was deposited is visible. **b** At 56 days after treatment, animals typically present a scar at the former tumor area.

the other peripheries [47]. This is caused by the fact that only the proximal tumor periphery was surrounded by normal tissue; the other ones were surrounded by air. Air is known to have higher insulation features than biological tissue mainly composed of water. Additionally, blood flow in the proximal tumor surroundings might have reinforced the aforementioned effects to some extent.

Particularly for hyperthermic treatments, adaptive regulations from the vascular system could additionally influence the extent of the thermal washout. This is attributed to the fact that normal (innervated) blood vessels within or in the tumor surroundings are capable of reacting against a heating stimulus as opposed to newly formed noninnervated vessels in the tumor. In this context, it was shown that blood flow increases after hyperthermic treatment of tumors for 20–50 min to temperatures between 41 and 45°C [56] due to vasodilatation. These effects are influenced not only by the tumor type but also by the temperature and treatment duration [57, 58]. The response to the hyperthermic stimuli consists of a 'reactive' phase followed by a 'destructive' phase as a consequence of ischemia and necrosis [59]. Ideally, using short exposure times (several minutes) for tumor treatments at thermoablative temperatures, the reactive phase could be prevented by promoting the destructive one.

Heating Effects on Tissues

Particularly, the utilization of thermoablative temperature-treated tumor cells leads to chromatin aggregation along the inner nuclear membrane and the occurrence of pycnotic cell nuclei. Also distinct DNA damage takes place as observed after processing of tumor cells from treated animals for single-cell electrophoresis [43]. These are clear signs for the selective induction of coagulative necrosis. In relation to the effects associated with hyperthermic treatments, changes in the association of nuclear proteins involved in DNA replication leading to the induction of DNA damage have been reported. Other cellular effects are related to an altered plasma membrane protein distribution which influences membrane permeability, calcium uptake, and disruption of mitochondrial membrane potentials. In general, effects on protein folding affecting the activation of programmed cell death have been observed [for review see 60].

In the context of the expected systemic effects, heat treatment of tumors in the breast is likely to show a beneficial tolerability similar to that which has already been shown for the treatment of prostate tumors [7]. Moreover, in our preclinical experiments with tumor-bearing rats a good tolerability was observed (fig. 3). Even though iron oxide nanoparticles seem to be physiologically well tolerated [61], there is some evidence that they are capable of inducing oxidative stress [62]. Moreover, an antitumor immune response might be induced after hyperthermia, together with the acquisition of host immunity [63]. Accordingly, it was demonstrated that when a tumor was transplanted in the left or right side of an animal and only one of them was treated, the tumor disappeared [64]. Nevertheless, the data acquired so far are rather ambiguous, showing that there is a demand for further systematic work to shed more light on this issue.

In relation to the metabolism and clearance of nanoparticles particularly after intratumoral application, no clear data is available so far. Basically, the pharmacokinetics should be nearly comparable to the results of applications as contrast media in diagnostics. For example, after intravenous application of magnetite/maghemite containing Fe^{59} isotopes in rats, a half-life for whole-body clearance of iron of 45 days was found [65]. It is expected that, after treatments, nanoparticles could be taken up by phagocytes. Degradation of individual nanoparticles into iron ions seem to occur in intracellular lysosomes as a result of the presence of different hydrolytic enzymes. Utilization of iron ions by natural iron metabolism pathways [66] is very likely to occur.

Possible limitations of the method consist of the treatment of breast tumors in proximity to the chest wall as they are suboptimally accessible to alternating magnetic fields. Moreover, the treatment of multifocal tumors could be problematic when applying the magnetic material directly into the tumor. Therefore, investigations considering tumor-specific targeting of the magnetic particles could be very helpful.

Outlook

Minimally invasive treatment of breast cancer seems to be feasible in view of the present knowledge. Even though treatment of this tumor entity is related to high expectations concerning both the cosmetic and the therapeutic outcomes, magnetic heating could be a valuable technology for the treatment of solitary tumors with diameters of up to 1–2 cm. There is good potential for a highly controlled temperature development right at the tumor area. Even an individual adaptation of the parameters for each patient's situation with respect to the tumor location and diameter is feasible.

Acknowledgement

Most of our investigations reported here have been conducted in different projects funded by the German Research Foundation (DFG).

Disclosure Statement

The authors have nothing to disclose.

References

1 Sardanelli F, Podo F, Santoro F, Manoukian S, Bergonzi S, Trecate G, Vergnaghi D, Federico M, Cortesi L, Corcione S, Morassut S, Di Maggio C, Cilotti A, Martincich L, Calabrese M, Zuiani C, Preda L, Bonanni B, Carbonaro LA, Contegiacomo A, Panizza P, Di Cesare E, Savarese A, Crecco M, Turchetti D, Tonutti M, Belli P, Maschio AD: Multicenter surveillance of women at high genetic breast cancer risk using mammography, ultrasonography, and contrast-enhanced magnetic resonance imaging (the High Breast Cancer Risk Italian 1 Study): final results. Invest Radiol 2011;46:94–105.

2 Van Dongen JA, Barteling H, Fentiman IS, et al: Factors influencing local relapse and survival and results of salvage treatment after breast-conserving therapy in operable breast cancer: EORTC Trial 1081, breast conservation compared with mastectomy in TNM stage I and II breast cancer. Eur J Cancer 1992;28A:801–805.

3 Zhao Z, Wu F: Minimally-invasive thermal ablation of early-stage breast cancer: a systemic review. Eur J Surg Oncol 2010;36:1149–1155.

4 Gilchrist RK, Medal R, Shorey WD, Hanselman RC, Parrott JC, Taylor CB: Selective inductive heating of lymph nodes. Ann Surg 1957;146:596–606.

5 Gilchrist RK, Shorey WD, Russel M, Hanselman RC, dePeyster FA, Yang J, Medal R: Effects of electromagnetic heating on internal viscera: a preliminary to the treatment of human tumors. Ann Surg 1965;161:890–896.

6 Johannsen M, Gneveckow U, Taymoorian K, Thiesen B, Waldofner N, Scholz R, Jung K, Jordan A, Wust P, Loening SA: Morbidity and quality of life during thermotherapy using magnetic nanoparticles in locally recurrent prostate cancer: results of a prospective phase I trial. Int J Hyperthermia 2007; 23:315–323.

7 Johannsen M, Thiesen B, Wust P, Jordan A: Magnetic nanoparticle hyperthermia for prostate cancer. Int J Hyperthermia 2010;26:790–795.

8 Bacon BR, Stark DD, Park CH, Saini S, Gromann EV, Hahn PF, Compton CC, Ferrucci JT: Ferrite particles: a new magnetic resonance imaging contrast agent. Lack of acute or chronic hepatoxicity after intravenous administration. J Lab Clin Med 1987;110:164–171.

9 Marchal G, van Heke P, Demaerel P, Decrop E, Kennis C, Baert AL, van der Schuerren E: Detection of liver metastases with superparamagnetic iron oxide in 15 patients: results of MR imaging at 1.5 T. Am J Radiol 1988;152: 771–775.

10 Rummeny E, Weissleder R, Stark DD, Elizondo G, Ferrucci JT: Kernspintomographie fokaler Leber- und Milzläsionen. Radiologe 1988;28:380–386.

11 Hergt R, Andrä W, d'Ambly CG, Hilger I, Kaiser WA, Richter U, Schmidt HG: Physical limits of hyperthermia using magnetite fine particles. IEEE Trans Magn 1998;34:3745–3754.

12 Hilger I, Andrä W, Hergt R, Hiergeist R, Schubert H, Kaiser WA: Electromagnetic heating of breast tumors in interventional radiology: in vitro and in vivo studies in human cadavers and mice. Radiology 2001;218:570–575.

13 Blums E, Cebers A, Maiorov MM: Magnetic Fluids. New York, de Gruyter, 1997.

14 Hergt R, Hiergeist R, Hilger I, Kaiser WA: Magnetic Nanoparticles for Thermoablation. Recent Res Dev Mater Sci 2002;3:723–742.

15 Andrä W: Magnetic hyperthermia; in Andrä W, Nowak H (eds): Magnetism in Medicine. Berlin, Wiley-VCH, 1998, pp 455–470.

16 Hergt R, Hiergeist R, Hilger I, Kaiser WA, Lapatnikov Y, Margel S, Richter U: Maghemite nanoparticles with very high AC-losses for application in RF-magnetic hyperthermia. J Magn Magn Mater 2004;270:345–357.

17 Fricker J: Drugs with a magnetic attraction to tumours. Drug Discov Today 2001;6:387–389.

18 Hergt R, Hiergeist R, Zeisberger M, Glockl G, Weitschies W, Ramirez P, Hilger I, Kaiser WA: Enhancement of AC-losses of magnetic nanoparticles for heating applications. J Magn Magn Mater 2004;280:358–368.

19 Hergt R, Hiergeist R, Zeisberger M, Schüler D, Heyen U, Hilger I, Kaiser W: Magnetic properties of bacterial magnetosomes as potential diagnostic and therapeutic tools. J Magn Magn Mater 2005;293:80–86.

20 Frimpong RA, Hilt JZ: Magnetic nanoparticles in biomedicine: synthesis, functionalization and applications. Nanomedicine 2010;5:1401–1414.

21 Eberbeck D, Kettering M, Bergemann C, Zirpel P, Hilger I, Trahms L: Quantification of the aggregation of magnetic nanoparticles with different polymeric coatings in cell culture medium. J Phys D Appl Phys 2010;43:405002.

22 Kettering M, Zorn H, Bremer-Streck S, Oehring H, Zeisberger M, Bergemann C, Hergt R, Halbhuber KJ, Kaiser WA, Hilger I: Characterization of iron oxide nanoparticles adsorbed with cisplatin for biomedical applications. Phys Med Biol 2009;54:5109–5121.

23 Raaphorst GP, Li LF, Yang DP, LeBlanc JM: Cisplatin sensitization by concurrent mild hyperthermia in parental and mutant cell lines deficient in homologous recombination and non-homologous endjoining repair. Oncol Rep 2005;14:281–285.

24 Wang D, Lippard SJ: Cellular processing of platinum anticancer drugs. Nat Rev Drug Discov 2005;4:307–320.

25 Xu P, Van Kirk EA, Li S, Murdoch WJ, Ren J, Hussain MD, Radosz M, Shen Y: Highly stable core-surface-crosslinked nanoparticles as cisplatin carriers for cancer chemotherapy. Colloids Surf B Biointerfaces 2006;48:50–57.
26 Kim HR, Andrieux K, Delomenie C, Chacun H, Appel M, Desmaële D, Taran F, Georgin D, Couvreur P, Taverna M: Analysis of plasma protein adsorption onto PEGylated nanoparticles by complementary methods: 2-DE, CE and Protein Lab-on-chip system. Electrophoresis 2007;28:2252–2261.
27 Hilger I, Leistner Y, Berndt A, Fritsche C, Haas KM, Kosmehl H, Kaiser WA: Near-infrared fluorescence imaging of HER-2 protein over-expression in tumour cells. Eur Radiol 2004;14:1124–1129.
28 Hilger I, Trost R, Reichenbach JR, Linss W, Lisy MR, Berndt A, Kaiser WA: MR imaging of Her-2/neu protein using magnetic nanoparticles. Nanotechnology 2007;18:135103.
29 Lisy MR, Goermar A, Thomas C, Pauli J, Resch-Genger U, Kaiser WA, Hilger I: In vivo near-infrared fluorescence imaging of carcinoembryonic antigen-expressing tumor cells in mice. Radiology 2008;247:779–787.
30 Hilger I, Dietmar E, Linss W, Streck S, Kaiser WA: Developments for the minimally invasive treatment of tumours by targeted magnetic heating. J PhysCondens Matter 2006;18:S2951–S2958.
31 Kettering M, Winter J, Zeisberger M, Alexiou C, Bremer-Streck S, Bergemann C, Kaiser WA, Hilger I: Magnetically based enhancement of nanoparticle uptake in tumor cells: combination of magnetically induced cell labeling and magnetic heating (in German). Rofo 2006;178:1255–1260.
32 Hiergeist R, Andrä W, Buske N, Hergt R, Hilger I, Richter U, Kaiser WA: Application of magnetite ferrofluids for hyperthermia. J Magn Magn Mater 1999;201:420–422.
33 Atkinson W, Brezovich I, Chakraborty D: Usable frequencies in hyperthermia with thermal seeds. IEEE Trans Biomed Eng 1984;31:70–75.
34 Brezovich I: Low frequency hyperthermia: capacitive and ferromagnetic thermoseed methods; in Palival P, Hetzel FW (eds): Medical Physics Monograph. New York, American Institute of Physics, 1988, pp 82–111.
35 Jordan A, Scholz R, Maier-Hauff K, Johannsen M, Wust P, Nadobny J, Schirra H, Schmidt H, Deger S, Loening S, Lanksch W, Felix R: Presentation of a new magnetic field therapy system for the treatment of human solid tumors with magnetic fluid hyperthermia. J Magn Magn Mater 2001;225:118–126.
36 Hilger I, Frühauf S, Andrä W, Hiergeist R, Hergt R, Kaiser WA: Magnetic heating as a therapeutic tool. Thermology Int 2001;11:130–136.
37 Ritchie KP, Keller BM, Syed KM, Lepock JR: Hyperthermia (heat shock)-induced protein denaturation in liver, muscle and lens tissue as determined by differential scanning calorimetry. Int J Hyperthermia 1994;10:605–618.
38 Joly MA: A Physico-Chemical Approach to Denaturation of Proteins. New York, Academic Press, 1965.
39 Xu Y, Qian R: Analysis of thermal injury process based on enzyme deactivation mechanisms. J Biomech Eng 1995; 117:462–465.
40 Heisterkamp J, van Hillegersberg R, IJzermans JN: Critical temperature and heating time for coagulation damage: implications for interstitial laser coagulation (ILC) of tumors. Lasers Surg Med 1999;25:257–262.
41 Hilger I, Andrä W, Bähring R, Daum A, Hergt R, Kaiser WA: Evaluation of temperature increase with different amounts of magnetite in liver tissue samples. Invest Radiol 1997;32:705–712.
42 Hunt MC, Sørheim O, Slinde E: Color and heat denaturation of myoglobin forms in ground beef. J Food Sci 1999; 64:847–851.
43 Hilger I, Rapp A, Greulich KO, Kaiser WA: Assessment of DNA damage in target tumor cells after thermoablation in mice. Radiology 2005;237:500–506.
44 Rabin Y: Is intracellular hyperthermia superior to extracellular hyperthermia in the thermal sense? Int J Hyperthermia 2002;18:194–202.
45 Jordan A, Scholz R, Wust P, Schirra H, Schiestel T, Schmidt H, Felix R: Endocytosis of dextran and silan-coated magnetite nanoparticles and the effect of intracellular hyperthermia on human mammary carcinoma cells in vitro. J Magn Magn Mater 1999;194:185–196.
46 Li Z, Vogel M, Maccarini P, Stakhursky V, Soher B, Craciunescu O, Das S, Arabe O, Joines W, Stauffer P: Improved hyperthermia treatment control using SAR/temperature simulation and PRFS magnetic resonance thermal imaging. Int J Hyperthermia 2011;27:86–99.
47 Hilger I, Hiergeist R, Hergt R, Winnefeld K, Schubert H, Kaiser WA: Thermoablation of tumors using magnetic nanoparticles: an in vivo feasibility study. Invest Radiol 2002;37:580–586.
48 Jordan A, Scholz R, Wust P, Fähling H, Krause J, Wlodarczyk W, Sander B, Vogl T, Felix R: Effects of magnetic fluid hyperthermia (MFH) on C3H mammary carcinoma in vivo. Int J Hyperthermia 1997;13:587–605.
49 Tannock IF, Steel GG: Quantitative techniques for study of the anatomy and function of small blood vessels in tumors. J Natl Cancer Inst 1969;42: 771–782.
50 Solesvik OV, Rofstad EK, Brustad T: Vascular structure of five human malignant melanomas grown in athymic nude mice. Br J Cancer 1982;46:557–567.
51 Hirano A, Matsui T: Vascular structure in brain tumors. Hum Pathol 1975;6: 611–621.
52 Warren BA: The vascular morphology of tumors; in Peterson HI (ed): Tumor Blood Circulation. Boca Raton, CRC Press, 1979, pp 1–47.
53 Hammersen F, Endrich B, Messmer K: The fine structure of tumor blood vessels. 1. Participation of non-endothelial cells in tumor angiogenesis. Int J Microcirc Clin Exp 1985;4:31–43.
54 Reinhold HS: Physiological effects of hyperthermia. Recent Results Cancer Res 1988;107:32–43.
55 Lu XQ, Burdette EC, Bornstein BA, Hansen JL, Svensson GK: Design of an ultrasonic therapy system for breast cancer treatment. Int J Hyperthermia 1996;12:375–399.
56 Waterman FM, Nerlinger RE, Moylan DJ, Leeper DB: Response of human tumor blood flow to local hyperthermia. Int J Radiat Oncol Biol Phys 1987;13: 75–82.
57 Dewhirst M, Gross JF, Sim D, Arnold P, Boyer D: The effect of rate of heating or cooling prior to heating on tumour and normal tissue microcirculatory blood flow. Biorheology 1984a; 21:539–558.

58 Dewhirst MW, Sim DA, Gross J, Kundrat MA: Effect of heating rate on tumour and normal tissue microcirculatory function; in Overgaard J (ed): Hyperthermic Oncology. London, Taylor and Francis, 1984, pp 177–180.

59 Reinhold HS, Endrich B: Invited review: tumour microcirculation as a target for hyperthermia. Int J Hyperthermia 1986;2:11–137.

60 Roti JL: Cellular responses to hyperthermia (40–46 degrees C): cell killing and molecular events. Int J Hyperthermia 2008;24:3–15.

61 Babincová M, Leszczynska D, Sourivong P, Babinec P: Selective treatment of neoplastic cells using ferritin-mediated electromagnetic hyperthermia. Med Hypotheses 2000 54:177–179.

62 Buyukhatipoglu K, Clyne AM: Superparamagnetic iron oxide nanoparticles change endothelial cell morphology and mechanics via reactive oxygen species formation. J Biomed Mater Res A 2011;96:186–195.

63 Multhoff G, Gaipl U: Molecular and immunological effects of hyperthermia on tumor progression and metastasis. Onkologe 2010;16:1043–1050.

64 Yanase M, Shinkai M, Honda H, Wakabayashi T, Yoshida J, Kobayashi T: Antitumor immunity induction by intracellular hyperthermia using magnetite cationic liposomes. Jpn J Cancer Res 1998;89:775–782.

65 Weissleder R, Stark DD, Engelstad BL, Bacon BR, Compton CC, White DL, Jacobs P, Lewis J: Superparamagnetic iron oxide: pharmacokinetics and toxicity. AJR Am J Roentgenol 1989;152: 167–173.

66 Klausner RD, Ronault TA, Harford JB: Regulating the fate of mRNA: the control of cellular iron metabolism. Cell 1993;72:12–28.

Prof. Dr. Ingrid Hilger
Institut für Diagnostische und Interventionelle Radiologie I
Zentrum für Radiologie, Universitätsklinikum Jena
Erlanger Allee 101, DE–07747 Jena (Germany)
Tel. +49 3641 9325 921, E-Mail ingrid.hilger@med.uni-jena.de

Alexiou C (ed): Nanomedicine – Basic and Clinical Applications in Diagnostics and Therapy.
Else Kröner-Fresenius Symp. Basel, Karger, 2011, vol 2, pp 177–184

Galectins, Glycans, and Mucins as Targets for Novel and Specific Antibody Therapies in Gynecologic Cancer Therapies

Udo Jeschke · Irmgard R.M. Wiest · Christian Schindlbeck · Darius Dian · Klaus Friese

Klinik und Poliklinik für Frauenheilkunde und Geburtshilfe – Campus Innenstadt, Ludwig-Maximilians-Universität München, München, Deutschland

Abstract

Galectin-1 (Gal-1), a member of the mammalian lectins, recognizes galactose-containing sequences of oligosaccharides associated with several cell surface glycoconjugates. Gal-1 recognizes appropriate glycoepitopes on human breast cancer cells. Gal-1 is expressed in many malignant and normal tissues. Furthermore, it is known that Gal-1 can initiate T cell apoptosis. The Thomsen-Friedenreich (TF) antigen is a specific oncofetal carbohydrate epitope (Galβ1–3GalNAcα-O epitope) expressed on the surface of various carcinomas. It mediates endothelium adhesion and tumor invasion and could thus be a marker of aggressiveness. As it also causes an immune response, it is an interesting target for the development of therapeutic antibodies. Mucin derives from the word mucus and designates the structural component of it. The O-glycans are involved in antigen binding and/or signal transduction. Mucins are divided into 2 groups: membrane-bound ones such as mucin-1 (MUC1) and secreted mucins. MUC1 plays a role in the barrier function of the mucous membranes and helps regulate cell adhesion. In gynecologic cancer tissue, MUC1 is glycosylated with the TF epitope. Gal-1 is able to bind to the TF epitope attached to MUC1. Gal-1, TF, and MUC1 are interesting targets for novel and specific antibody therapies in gynecologic cancer therapies.

A number of humanized and chimeric monoclonal antibodies (MAbs) against human epidermal receptor 2 (HER2), epidermal growth factor receptor (EGFR), and vascular endothelial growth factor (VEGF) have been developed for the treatment of tumors, including breast, colorectal, lung, head and neck, and ovarian cancer, over the last years [1].

Trastuzumab [2, 3], bevacizumab [4, 5], cetuximab [6, 7], and panitumumab [8, 9] are MAbs that are used in clinical practice with acceptable rates of advancement in the overall survival of patients [5, 7, 10].

Unfortunately, the number of further tumor-specific and sufficient antigen targets for the development of new potential therapeutic antibodies is limited.

Therefore, the search for targets on the primary tumors or the corresponding metastatic cells to offer tailored therapy is of major importance in future tumor therapies.

Galectins

The galectin family members are defined by a conserved amino acid sequence motif in the carbohydrate recognition domain (CRD) and an affinity for β-galactosides. Galectin-1 (Gal-1), the earliest described member of the family, is a homodimeric protein with a CRD of 134 amino acids [11]. Expression of Gal-1 has been specially identified in lymphoid organs such as the thymus, lymph nodes, activated macrophages, and T cells and in immune-privileged sites such as the placenta and cornea, suggesting an important role in generating and maintaining immune tolerance [12]. Although LacNAc is the basic ligand recognized by Gal-1, the prototype galectin binds with increased avidity to multiple Galβ1–4GlcNAc sequences presented on branched N-linked or repeating LacNAc residues on N- and O-linked glycans. Gal-1 with a single CRD forms a noncovalently associated homodimer to become functionally bivalent under physiological conditions. The bivalent nature of Gal-1 facilitates glycan-mediated cell surface receptor cross-linking believed to be essential in inducing signaling events [13, 14]. Extracellularly, via recognition of glycan ligands, Gal-1 exerts distinct biological effects in various tissues and on cells, including cell adhesion [15, 16], metastasis [17], cell growth regulation [18, 19], immunosuppression [20], and apoptosis [13].

Binding experiments of breast cancer tumor cells to laminin and plasma or placental fibronectin showed that binding was generally reduced by treatment of cells with Gal-1 [21]. Cell binding of Gal-1 proved to be very effective in blocking cell association to fibronectins after preincubation with cell suspensions. Reduced expression of Gal-1 seemed to be associated with a positive lymph node status in the breast cancer group. These results can be interpreted to reflect cell-type-dependent requirements of galectin ligand presentation during the metastatic cascade [21].

Gal-1 is also expressed in the placenta. The placenta plays a key role in the maintenance of local tolerance and allows the mother to accept the embryo untilcompletion of the pregnancy. The complex process of tolerance accompanying the survival of the fetus is controlled at the embryo-maternal interface by factors deriving from decidualized endometrium and from the trophoblast itself. Trophoblasts develop various strategies to evade the damaging attack by the maternal immune response including expression of nonclassical MHC class I antigens and of complement regulatory proteins [22, 23]. Choriocarcinoma cell lines were evaluated as an experimental model of trophoblast-derived immunoregulation [24]. With these cells we found a moderate staining for mucin-1 (MUC1) but a strong expression of the Thomsen-Friedenreich (TF) tumor antigen [25, 26]. This pattern very likely represents the syncytiotrophoblast-like phenotype because the syncytiotrophoblast in vivo also strongly expresses TF and MUC1 [27]. The TF antigen and galectins have already been implicated in tumor cell adhesion and tissue invasion. In a previous study the involvement of TF was demonstrated in both homotypic aggregation of MDA-MB-435 human breast carcinoma cells and their adhesion to the endothelium [28].

The Thomsen-Friedenreich Epitope

The TF antigen (or, more precisely, epitope) could be a suitable candidate for the development of new therapeutic antibodies. The TF epitope has been known as a tumor-associated antigen [29] for a long time and was found to be a specific marker of malignant breast carcinomas [30]. The presence of TF during an early fetal phase, its near absence in noncarcinomatous postfetal tissues, and its association with carcinomas suggest that TF is a stage-specific oncofetal carbohydrate antigen. The TF epitope occurs only in very limited amounts at a few locations in normal adult human tissues [31] but is expressed by fetal epithelia and mesothelia [32]. In addition, it is also found on transferrin isolated from human amniotic fluid [33] and is expressed by the syncytiotrophoblast and extravillous trophoblast [27]. Moreover, it is expressed by the surface and glandular epithelial tissue of the receptive endometrium [34]. The TF antigen oligosaccharide is related to blood group antigens and consists of galactose-β1–3N-acetylgalactosamine (Galβ1–3GalNAcα-O-Ser/Thr). In several tumor entities, like gastric [35], colon [36], or lung cancer [37], or in cancer of the cervix uteri [38], a correlation between TF expression and a negative prognosis could be demonstrated. Its prognostic impact in other tumor locations, especially in breast cancer, was also investigated. A correlation between a high tumor stage and TF expression was observed [38] nevertheless, whereas in another study high TF expression predicted improved survival [39]. In previous studies we could demonstrate that TF is also expressed nearly completely on disseminated tumor cells in the bone marrow of breast cancer patients [40]. In another study, patients with TF-positive breast tumors showed a favorable prognosis. This contrasts with studies on gastrointestinal tumors. We hypothesized that the immunogenicity of TF could lead to an immune response and therefore to a reduction of TF-positive tumor cells in breast cancer patients [41].

Mucin-1

In epithelial cells, TF is a carbohydrate epitope of MUC1, which belongs to a family of highly glycosylated proteins present on the apical surface of many glandular epithelial cells. On tumor cells MUC1 is posttranslationally modified, resulting in incomplete O-glycosylation and exposure of the TF epitope. Gal-1 is able to bind to the MUC1-attached TF epitope (fig. 1). MUC1, also named CD227, EMA, PAS-O, PEM, MAM-6, or episialin, plays a role in the barrier function of the mucous membranes and helps regulate cell adhesion. The chromosomal localization was defined on 1q21 [42]. It is a high molecular weight transmembrane glycoprotein with unique properties. It is extremely long and its extracellular part consists of identical repeats of a 20-amino acid peptide, the number of which varies in the population. The repeats are heavily O-glycosylated, with characteristic glycan chains. This can be monitored directly with MAbs to the glycans as well as indirectly with MAbs to peptide epitopes because their binding is often hindered by the glycans [43].

Cancer cells, especially adenocarcinomas, express aberrant forms and amounts of mucins. The expression of distinct exposed oligosaccharide structures confers on tumor cells an enormous range of potential ligands for interaction with other receptors at the cell surface [44]. An increased and often depolarized expression throughout the entire cell cytoplasm has been recorded [45]. It is generally believed that MUC1 overexpression by tumor cells facilitates invasive growth and metastasis [46, 47]. It is also thought that overexpression of MUC1 disrupts cell-cell and cell-extracellular matrix adhesions [48–50]. In addition, MUC1 was shown to contribute to cancer cell escape from immune surveillance as MUC1-expressing cells are less susceptible to T cell- and NK cell-mediated lysis [51]. On the other hand, absence of MUC1 expression also occurred and was associated with the absence of estrogen receptor and progesterone receptor in several breast cancer studies. These findings support the observation by Luna-More et al. [52] that tumors of the human breast negative for MUC1 are high grade, are estrogen receptor and progesterone receptor negative, and are associated with positive axillary lymph nodes. Other studies also

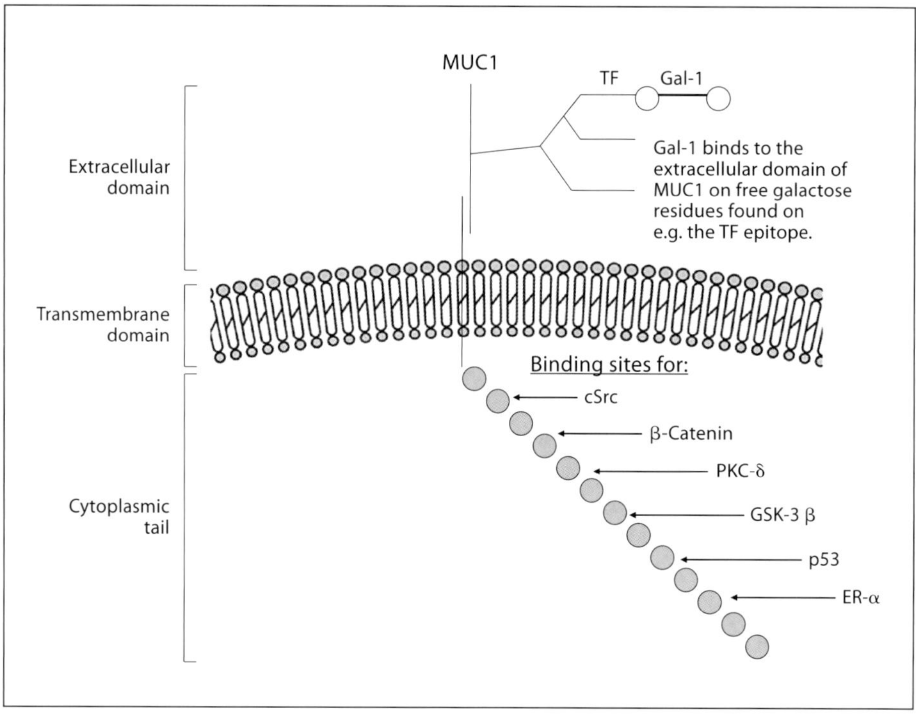

Fig. 1. Gal-1 as well as MAbs against the TF epitope and/or MUC1 could bind to the extracellular domain of MUC1 and induce signal transduction processes in tumor cells leading to inhibition of proliferation as well as the onset of apoptosis.

related low or negative MUC1 expression to a higher tumor grade and poor prognosis [53]. MUC1 may also participate in cellular signaling through the ability of its cytoplasmic tail to interact with β-catenin [54, 55]. MUC1 has 6 conserved tyrosines which are potential sites for phosphorylation and signaling [56]. All 4 ErbBs have been reported to bind to the cytoplasmic tail of MUC1 [57]. A YEKV site is phosphorylated by either Src or EGF-activated ErbB1. A second site (YTNP) binds the cytoplasmic Grb-Sos complex and can initiate mitogenic signaling through the canonical Ras-Erk pathway [54]. Trastuzumab (Herceptin) is a humanized antibody that is used to treat breast cancers that overexpress the cell surface receptor HER2. Unfortunately, many breast cancer patients do not respond to trastuzumab therapy, even though their tumors overexpress HER2. In a recent study the involvement of MUC1 in trastuzumab resistance was investigated [58]. Fessler et al. [58] showed that in a cancer cell line with trastuzumab resistance a dramatic increase in the amount of a cleaved form of the MUC1 protein was identified.

In a recent study we investigated a novel anti-MUC1 antibody called PankoMab as a potential new diagnostic tool and compared its ability to define prognostically relevant groups of patients with 2 established antibodies (DF3 and VU-4-H5) nominally directed against the same antigen [59]. In detail, we examined the relationship between MUC1 expression patterns according to the 3 antibodies in invasive ductal carcinoma of the breast and established tumor characteristics like lymph node involvement, hormone receptor status, and grading. To avoid ambiguous results due to the heterogeneity of breast cancer, we focused on invasive ductal carcinomas, the most common type of breast cancer [59].

All 3 anti-MUC1 antibodies were of the same isotype (mouse IgG1). The antibody DF3 was first described in 1984 by Abe and Kufe et al. [60]. It was generated with a membrane fraction of breast cancer tissue as the immunogen. DF3 is part of the CA15-3 sandwich assay for serum MUC1 [61]. The antibody VU-4-H5 was generated with a synthetic MUC1 peptide consisting of 3 tandem repeats as the immunogen. Both antibodies, DF3 and VU-4-H5, were evaluated during the ISOBM TD-4 International Workshop on Monoclonal Antibodies against MUC1 [62]. As a result of this workshop, they were confirmed in their MUC1 specificity and the epitope sequence determined as APDTRPAP for both antibodies. Nonetheless, the fine specificities were not identical. In immunohistochemistry DF3 did react with normal gastrointestinal and breast tissues without pretreatment, whereas VU-4-H5 became positive with normal tissues only after partial deglycosylation by periodate oxidation. In addition, the cellular localization of the antigen analyzed with a MUC1-positive tumor cell line was different [43].

PankoMab [63] is a novel MUC1 antibody explicitly tailored to recognize a tumor-associated MUC1 epitope (TA-MUC1) described earlier [64, 65]. This epitope consists of a special carbohydrate-induced conformation of the PDTRP motif. PankoMab has an improved tumor selectivity, making it very attractive as a potential therapeutic antibody. PankoMab may also be suited as a diagnostic antibody. In contrast to the antibodies DF3 and VU-4-H5, PankoMab reactivity revealed a strong correlation with the expression of the estrogen receptor. These findings are in accordance with the observation by Luna-More et al. [52] that tumors of the human breast negative for MUC1 are estrogen receptor and progesterone receptor negative.

Conclusion

Galectins are a family of carbohydrate-binding proteins that interact with β-galactosides (e.g. the TF epitope). Gal-1 is involved in binding cell surface glycoconjugates, cell adhesion, and migration processes and is an important regulator of tumor angiogenesis. If Gal-1 binds to the TF epitope of breast cancer cells and shows apoptotic potential in breast tumor cell lines together with inhibition of proliferation on TF-positive tumor cells it could be a useful therapeutic tool. PankoMab recognizes tumor-associated MUC1 epitopes in gynecologic cancer cells. In vitro experiments with PankoMab can be used as a model for inhibition of cell proliferation experiments independently of antibody-dependent cell-mediated cytotoxicity or complement-dependent cytotoxicity. Antibody-mediated cell killing is called antibody-dependent cell-mediated cytotoxicity. MAbs bind complement which leads to direct cell toxicity known as complement-dependent cytotoxicity. Antibodies against the TF epitope (NM-TF1) could also have potential as therapeutic tools in breast cancer patients.

Disclosure Statement

The authors have nothing to disclose.

References

1 Argyriou AA, Kalofonos HP: Recent advances relating to the clinical application of naked monoclonal antibodies in solid tumors. Mol Med 2009;15:183–191.

2 Chiu SJ, Ueno NT, Lee RJ: Tumor-targeted gene delivery via anti-HER2 antibody (trastuzumab, Herceptin) conjugated polyethylenimine. J Control Release 2004;97:357–369.

3 zum Buschenfelde CM, Hermann C, Schmidt B, Peschel C, Bernhard H: Antihuman epidermal growth factor receptor 2 (HER2) monoclonal antibody trastuzumab enhances cytolytic activity of class I-restricted HER2-specific T lymphocytes against HER2-overexpressing tumor cells. Cancer Res 2002; 62:2244–2247.

4 Andriolo RB, Puga ME, Belfort R Jr, Atallah AN: Bevacizumab for ocular neovascular diseases: a systematic review. Sao Paulo Med J 2009;127:84–91.

5 Ardavanis A, Doufexis D, Kountourakis P, Malliou S, Karagiannis A, Kardara E, Sykoutri D, Charalampia M, Rigatos G: Salvage therapy of pretreated advanced breast cancer with bevacizumab and paclitaxel every two weeks: a retrospective case review study. BMC Cancer 2009;9:338.

6 Schwartzberg LS, Stepanski EJ, Fortner BV, Houts AC: Retrospective chart review of severe infusion reactions with rituximab, cetuximab, and bevacizumab in community oncology practices: assessment of clinical consequences. Support Care Cancer 2008; 16:393–398.

7 Tappenden P, Jones R, Paisley S, Carroll C: Systematic review and economic evaluation of bevacizumab and cetuximab for the treatment of metastatic colorectal cancer. Health Technol Assess 2007;11:1–128, iii–iv.

8 Giusti RM, Cohen MH, Keegan P, Pazdur R: FDA review of a panitumumab (Vectibix) clinical trial for first-line treatment of metastatic colorectal cancer. Oncologist 2009;14:284–290.

9 Power DG, Shah MA, Asmis TR, Garcia JJ, Kemeny NE: Safety and efficacy of panitumumab following cetuximab: retrospective review of the Memorial Sloan-Kettering experience. Invest New Drugs 2009;28:353–360.

10 Barros Costa RL: Targeted therapy: comprehensive review. Am J Hosp Palliat Care 2009;26:137–146.

11 Barondes SH, Castronovo V, Cooper DM, Cummings RD, Drickamer K, Feizi T, Gitt MA, Hirabayashi J, Hughes C, Kasai K, et al: Galectins: a family of animal beta-galactoside-binding lectins. Cell 1994;76:597–598.

12 La M, Cao TV, Cerchiaro G, Chilton K, Hirabayashi J, Kasai K, Oliani SM, Chernajovsky Y, Perretti M: A novel biological activity for galectin-1: inhibition of leukocyte-endothelial cell interactions in experimental inflammation. Am J Pathol 2003;163:1505–1515.

13 Perillo NL, Pace KE, Seilhamer JJ, Baum LG: Apoptosis of T cells mediated by galectin-1. Nature 1995;378: 736–739.

14 Walzel H, Neels P, Bremer H, Kohler H, Raab N, Barten M, Brock J: Immunohistochemical and glycohistochemical localization of the beta-galactoside-binding S-type lectin in human placenta. Acta Histochem 1995;97:33–42.

15 Hughes RC: The galectin family of mammalian carbohydrate-binding molecules. Biochem Soc Trans 1997;25: 1194–1198.

16 Hafer-Macko C, Pang M, Seilhamer JJ, Baum LG: Galectin-1 is expressed by thymic epithelial cells in myasthenia gravis. Glycoconj J 1996;13:591–597.

17 Raz A, Lotan R: Endogenous galactoside-binding lectins: a new class of functional tumor cell surface molecules related to metastasis. Cancer Metastasis Rev 1987:6:433–452.

18 Wells V, Mallucci L: Identification of an autocrine negative growth factor: mouse beta-galactoside-binding protein is a cytostatic factor and cell growth regulator. Cell 1991;64:91–97.

19 Adams L, Scott GK, Weinberg CS: Biphasic modulation of cell growth by recombinant human galectin-1. Biochim Biophys Acta 1996;1312:137–144.

20 Offner H, Celnik B, Bringman TS, Casentini-Borocz D, Nedwin GE, Vandenbark AA: Recombinant human beta-galactoside binding lectin suppresses clinical and histological signs of experimental autoimmune encephalomyelitis. J Neuroimmunol 1990;28: 177–184.

21 Andre S, Kojima S, Yamazaki N, Fink C, Kaltner H, Kayser K, Gabius HJ: Galectins-1 and -3 and their ligands in tumor biology: non-uniform properties in cell-surface presentation and modulation of adhesion to matrix glycoproteins for various tumor cell lines, in biodistribution of free and liposome-bound galectins and in their expression by breast and colorectal carcinomas with/without metastatic propensity. J Cancer Res Clin Oncol 1999;125:461–474.

22 Proll J, Blaschitz A, Hutter H, Dohr G: First trimester human endovascular trophoblast cells express both HLA-C and HLA-G. Am J Reprod Immunol, 1999;42:30–36.

23 Bulmer JN, Johnson PM: Antigen expression by trophoblast populations in the human placenta and their possible immunobiological relevance. Placenta 1985;6:127–140.

24 Grummer R, Hohn HP, Mareel MM, Denker HW: Adhesion and invasion of three human choriocarcinoma cell lines into human endometrium in a three-dimensional organ culture system. Placenta 1994;15:411–429.

25 Jeschke U, Reimer T, Bergemann C, Wiest I, Schulze S, Friese K, Walzel H: Binding of galectin-1 (gal-1) on trophoblast cells and inhibition of hormone production of trophoblast tumor cells in vitro by gal-1. Histochem Cell Biol 2004;121:501–508.

26 Jeschke U, Karsten U, Wiest I, Schulze S, Kuhn C, Friese K, Walzel H: Binding of galectin-1 (gal-1) to the Thomsen-Friedenreich (TF) antigen on trophoblast cells and inhibition of proliferation of trophoblast tumor cells in vitro by gal-1 or an anti-TF antibody. Histochem Cell Biol 2006;126:437–444.

27 Jeschke U, Richter DU, Hammer A, Briese V, Friese K, Karsten U: Expression of the Thomsen-Friedenreich antigen and of its putative carrier protein mucin 1 in the human placenta and in trophoblast cells in vitro. Histochem Cell Biol 2002;117:219–226.

28 Glinsky VV, Huflejt ME, Glinsky GV, Deutscher SL, Quinn TP: Effects of Thomsen-Friedenreich antigen-specific peptide P-30 on beta-galactoside-mediated homotypic aggregation and adhesion to the endothelium of MDA-MB-435 human breast carcinoma cells. Cancer Res 2000;60:2584–2588.

29 Springer GF, Desai PR, Murthy MS: Histochemical methods for the demonstration of Thomsen-Friedenreich antigen in cell suspensions and tissue sections [Klin. Wochenschr. 56, 761–765 (1978)]. Klin Wochenschr 1978;56: 1225–1226.

30 Springer GF, Desai PR, Scanlon EF: Blood group MN precursors as human breast carcinoma-associated antigens and 'naturally' occurring human cytotoxins against them. Cancer 1976;37: 169–176.

31 Cao Y, Stosiek P, Springer GF, Karsten U: Thomsen-Friedenreich-related carbohydrate antigens in normal adult human tissues: a systematic and comparative study. Histochem Cell Biol 1996;106:197–207.

32 Barr N, Taylor CR, Young T, Springer GF: Are pancarcinoma T and Tn differentiation antigens? Cancer 1989;64: 834–841.

33 van Rooijen JJ, Jeschke U, Kamerling JP, Vliegenthart JF: Expression of N-linked sialyl Le(x) determinants and O-glycans in the carbohydrate moiety of human amniotic fluid transferrin during pregnancy. Glycobiology 1998; 8:1053–1064.

34 Jeschke U, Walzel H, Mylonas I, Papadopoulos P, Shabani N, Kuhn C, Schulze S, Friese K, Karsten U, Anz D, Kupka MS: The human endometrium expresses the glycoprotein mucin-1 and shows positive correlation for Thomsen-Friedenreich epitope expression and galectin-1 binding. J Histochem Cytochem 2009;57:871–881.

35 Baldus SE, Zirbes TK, Glossmann J, Fromm S, Hanisch FG, Monig SP, Schroder W, Schneider PM, Flucke U, Karsten U, Thiele J, Holscher AH, Dienes HP: Immunoreactivity of monoclonal antibody BW835 represents a marker of progression and prognosis in early gastric cancer. Oncology 2001;61: 147–155.

36 Baldus SE, Zirbes TK, Hanisch FG, Kunze D, Shafizadeh ST, Nolden S, Monig SP, Schneider PM, Karsten U, Thiele J, Holscher AH, Dienes HP: Thomsen-Friedenreich antigen presents as a prognostic factor in colorectal carcinoma: a clinicopathologic study of 264 patients. Cancer 2000;88:1536–1543.

37 Takanami I: Expression of Thomsen-Friedenreich antigen as a marker of poor prognosis in pulmonary adenocarcinoma. Oncol Rep 1999;6:341–344.

38 Wolf MF, Ludwig A, Fritz P, Schumacher K: Increased expression of Thomsen-Friedenreich antigens during tumor progression in breast cancer patients. Tumour Biol 1988;9:190–194.

39 Imai J, Ghazizadeh M, Naito Z, Asano G: Immunohistochemical expression of T, Tn and sialyl-Tn antigens and clinical outcome in human breast carcinoma. Anticancer Res 2001;21:1327–1334.

40 Schindlbeck C, Jeschke U, Schulze S, Karsten U, Janni W, Rack B, Sommer H, Friese K: Characterisation of disseminated tumor cells in the bone marrow of breast cancer patients by the Thomsen-Friedenreich tumor antigen. Histochem Cell Biol 2005;123:631–637.

41 Schindlbeck C, Jeschke U, Schulze S, Karsten U, Janni W, Rack B, Krajewski S, Sommer H, Friese K: Prognostic impact of Thomsen-Friedenreich tumor antigen and disseminated tumor cells in the bone marrow of breast cancer patients. Breast Cancer Res Treat 2007; 101:17–25.

42 Gendler SJ, Lancaster CA, Taylor-Papadimitriou J, Duhig T, Peat N, Burchell J, Pemberton L, Lalani EN, Wilson D: Molecular cloning and expression of human tumor-associated polymorphic epithelial mucin. J Biol Chem 1990;265:15286–15293.

43 Cao Y, Karsten U, Hilgers J: Immunohistochemical characterization of a panel of 56 antibodies with normal human small intestine, colon, and breast tissues. Tumour Biol 1998; 19(suppl 1):88–99.

44 Hollingsworth MA, Swanson BJ: Mucins in cancer: protection and control of the cell surface. Nat Rev Cancer 2004;4:45–60.

45 Guddo F, Giatromanolaki A, Koukourakis MI, Reina C, Vignola AM, Chlouverakis G, Hilkens J, Gatter KC, Harris AL, Bonsignore G: MUC1 (episialin) expression in non-small cell lung cancer is independent of EGFR and c-erbB-2 expression and correlates with poor survival in node positive patients. J Clin Pathol 1998;51:667–671.

46 Gaemers IC, Vos HL, Volders HH, van der Valk SW, Hilkens J: A stat-responsive element in the promoter of the episialin/MUC1 gene is involved in its overexpression in carcinoma cells. J Biol Chem 2001;276:6191–6199.

47 Mommers EC, Leonhart AM, von Mensdorff-Pouilly S, Schol DJ, Hilgers J, Meijer CJ, Baak JP, van Diest PJ: Aberrant expression of MUC1 mucin in ductal hyperplasia and ductal carcinoma in situ of the breast. Int J Cancer 1999;84:466–469.

48 Huang L, Chen D, Liu D, Yin L, Kharbanda S, Kufe D: MUC1 oncoprotein blocks glycogen synthase kinase 3beta-mediated phosphorylation and degradation of beta-catenin. Cancer Res 2005;65:10413–10422.

49 Li Y, Bharti A, Chen D, Gong J, Kufe D: Interaction of glycogen synthase kinase 3beta with the DF3/MUC1 carcinoma-associated antigen and beta-catenin. Mol Cell Biol 1998;18:7216–7224.

50 Li Y, Kuwahara H, Ren J, Wen G, Kufe D: The c-Src tyrosine kinase regulates signaling of the human DF3/MUC1 carcinoma-associated antigen with GSK3 beta and beta-catenin. J Biol Chem 2001;276:6061–6064.

51 van de Wiel-van Kemenade E, Ligtenberg MJ, de Boer AJ, Buijs F, Vos HL, Melief CJ, Hilkens J, Figdor CG: Episialin (MUC1) inhibits cytotoxic lymphocyte-target cell interaction. J Immunol 1993;151:767–776.

52 Luna-More S, Rius R, Weil B, Jimenez A, Bautista MD, Perez-Mellado A: EMA: a differentiation antigen related to node metastatic capacity of breast carcinomas. Pathol Res Pract 2001;197: 419–425.

53 Rahn JJ, Dabbagh L, Pasdar M, Hugh JC: The importance of MUC1 cellular localization in patients with breast carcinoma: an immunohistologic study of 71 patients and review of the literature. Cancer 2001;91:1973–1982.

54 Carraway KL, Ramsauer VP, Haq B, Carothers Carraway CA: Cell signaling through membrane mucins. Bioessays 2003;25:66–71.

55 Carraway KL, Price-Schiavi SA, Zhu X, Komatsu M: Membrane mucins and breast cancer. Cancer Control 1999;6: 613–614.

56 Carraway KL 3rd, Funes M, Workman HC, Sweeney C: Contribution of membrane mucins to tumor progression through modulation of cellular growth signaling pathways. Curr Top Dev Biol 2007;78:1–22.

57 Carraway KL, Price-Schiavi SA, Komatsu M, Idris N, Perez A, Li P, Jepson S, Zhu X, Carvajal ME, Carraway CA: Multiple facets of sialomucin complex/MUC4, a membrane mucin and erbb2 ligand, in tumors and tissues (Y2K update). Front Biosci 2000;5:D95–D107.

58 Fessler SP, Wotkowicz MT, Mahanta SK, Bamdad C: MUC1* is a determinant of trastuzumab (Herceptin) resistance in breast cancer cells. Breast Cancer Res Treat 2009;118:113–124.

59 Dian D, Janni W, Kuhn C, Mayr D, Karsten U, Mylonas I, Friese K, Jeschke U: Evaluation of a novel anti-mucin 1 (MUC1) antibody (PankoMab) as a potential diagnostic tool in human ductal breast cancer; comparison with two established antibodies. Onkologie 2009;32:238–244.

60 Abe M, Kufe DW: Sodium butyrate induction of milk-related antigens in human MCF-7 breast carcinoma cells. Cancer Res 1984;44:4574–4577.

61 Hayes DF, Zurawski VR Jr, Kufe DW: Comparison of circulating CA15-3 and carcinoembryonic antigen levels in patients with breast cancer. J Clin Oncol 1986;4:1542–1550.

62 Price MR, Rye PD, Petrakou E, Murray A, Brady K, Imai S, Haga S, Kiyozuka Y, Schol D, Meulenbroek MF, Snijdewint FG, von Mensdorff-Pouilly S, Verstraeten RA, Kenemans P, Blockzjil A, Nilsson K, Nilsson O, Reddish M, Suresh MR, Koganty RR, Fortier S, Baronic L, Berg A, Longenecker MB, Hilgers J, et al: Summary report on the ISOBM TD-4 Workshop: analysis of 56 monoclonal antibodies against the MUC1 mucin. San Diego, Calif., November 17–23, 1996. Tumour Biol 1998;19 (suppl 1):1–20.

63 Danielczyk A, Stahn R, Faulstich D, Loffler A, Marten A, Karsten U, Goletz S: PankoMab: a potent new generation anti-tumour MUC1 antibody. Cancer Immunol Immunother 2006;55:1337–1347.

64 Karsten U, Serttas N, Paulsen H, Danielczyk A, Goletz S: Binding patterns of DTR-specific antibodies reveal a glycosylation-conditioned tumor-specific epitope of the epithelial mucin (MUC1). Glycobiology 2004;14:681–692.

65 Karsten U, von Mensdorff-Pouilly S, Goletz S: What makes MUC1 a tumor antigen? Tumour Biol 2005;26:217–220.

Prof. Dr. Udo Jeschke
Department of Obstetrics and Gynecology – Maistrasse
Ludwig Maximilians University, Maistrasse 11
DE–80377 Munich (Germany)
Tel. +49 89 5160 4266, E-Mail udo.jeschke@med.uni-muenchen.de

Alexiou C (ed): Nanomedicine – Basic and Clinical Applications in Diagnostics and Therapy.
Else Kröner-Fresenius Symp. Basel, Karger, 2011, vol 2, pp 185–196

The Commercial Development of Antibodies as Drugs

Michael A.W. Eaton[a] · John R. Adair[b]

[a]ETP Nanomedicine Secretariat, VDI/VDE Innovation + Technik GmbH, Department Innovation Europe, Berlin, Germany;
[b]Ithaka Life Sciences Ltd, NETPark Incubator, Sedgefield, UK

Abstract

The introduction of a new therapeutic modality, as happened with the arrival of monoclonal antibodies on the market in the 1980s, can change the pharmaceutical landscape. The introduction of such a new product class inevitably created some technical challenges, e.g. overcoming in vivo immunogenicity, but solutions were eventually found and recombinant antibodies are now a major class of biopharmaceuticals. The high specificities and affinities of antibodies have led to their use both as therapeutics in their own right and, to a lesser extent, as a targeting molecule for other effector molecules such as radioisotopes and cytotoxic drugs. The linkage of chemical effector entities to proteins creates major development problems in a regulated environment, many of which remain to be solved. Nanomedicine has and will produce major benefits to patients, but it is important that researchers understand the essential requirements for translation to drugs.

The pharmaceutical industry in the 20th century has helped patients, supported the science base, and created wealth. It has done this by understanding and improving the drug discovery process but it has also adopted, slowly in some cases, new therapeutic modalities such as recombinant antibodies [1]. Recombinant antibodies are the product of advances in the 1970s and 1980s in chemistry, computing, and molecular biology. Antibodies had previously been used to a very limited extent because of the problems associated with specificity, efficacy, scale-up, and purity.

All of this changed with the advent of the biotechnology revolution: the discovery of restriction enzymes and isolation of other DNA-modifying enzymes enabled nucleic acid sequences to be isolated and precisely manipulated for the first time. Together with methods for introducing these nucleic acids into heterologous cells such as bacteria, and later mammalian cells, this allowed the cloning of these sequences and an explosion of information on gene sequences and the control of gene expression. The technologies to sequence nucleic acids rapidly, developed in the late 1970s by Maxam and Gilbert [2] and by Sanger et al. [3], have led to today's highly automated processes that enable the entire genomes of man and many other organisms to be sequenced. In addition to the manipulation of isolated natural DNA sequences, developments in the assembly of synthetic nucleotide sequences in vitro from individual oligonucleotides allowed the development not

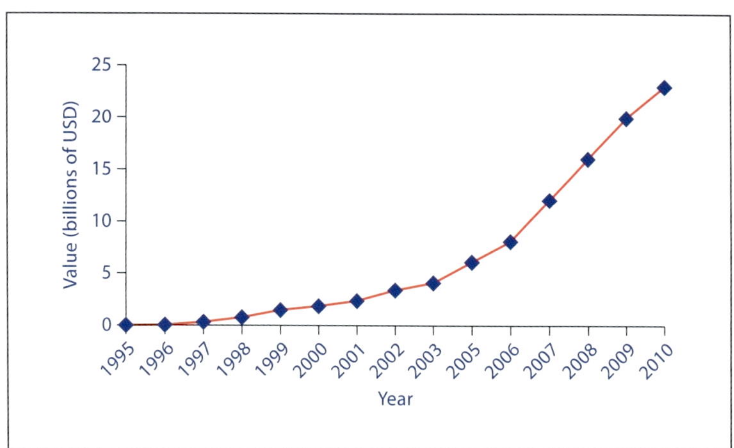

Fig. 1. Development of the therapeutic antibody market size – revenue (from company annual reports).

only of synthetic whole genes but also of novel and modified gene sequences [for a summary of the development of early techniques see 4]. The development of the now ubiquitous polymerase chain reaction [5] in the mid-1980s further accelerated the pace of genetic manipulation. In parallel, developments in computing provided the ability to manipulate the growing volume of sequence and structural information, becoming an invaluable and essential tool for the biotechnologist. These technical advances provided a basis for the better understanding of biological processes and systems allowing the commercial development of novel medicinal products: chemical, biological, and more recently cellular.

The initial output of the infant biotechnology industry was the ability to produce in large quantities pure medicinal proteins, such as growth hormone and insulin, and to provide enzymes for industrial processes. The ability to manipulate natural sequences by gene synthesis and mutagenesis allowed novel proteins to be developed; early examples included investigations into the development of novel interferons [6–8]. However, the discovery by Köhler and Milstein [9] in 1975 of the hybridoma procedure for generating monoclonal antibodies provided the starting point for the huge development of recombinant monoclonal antibodies which today dominates biotechnology sales volumes. The groundbreaking work of Köhler and Milstein (fig. 1) [9] enabled significant quantities of monoclonal antibodies to be obtained, initially by growing the hybridoma cells in mice and recovering the antibody from the resulting ascites fluid. The first commercial product of the technology, the antibody OKT3, came on the market in 1986 as an immunosuppression treatment for kidney transplantation rejection. By the end of the 1970s immunoglobulin genes were being cloned and the genetic materials for producing recombinant antibodies became available. Some scientists at the time thought that the cloning of antibody genes into bacteria was unnecessary and eccentric! However, with increasing knowledge of how antibody genes were organized and able to be manipulated, cloning into prokaryotic and eukaryotic cells became the new standard. One of the other features of the industry began to become apparent: intellectual property started to become valuable [10], none more so than the so-called Boss/Cabilly patents which described methods for putting the antibody heavy and light chains into heterologous cells such as *Escherichia coli*.

One of the driving forces for producing recombinant forms of antibodies was the observation that the murine antibodies, initially pro-

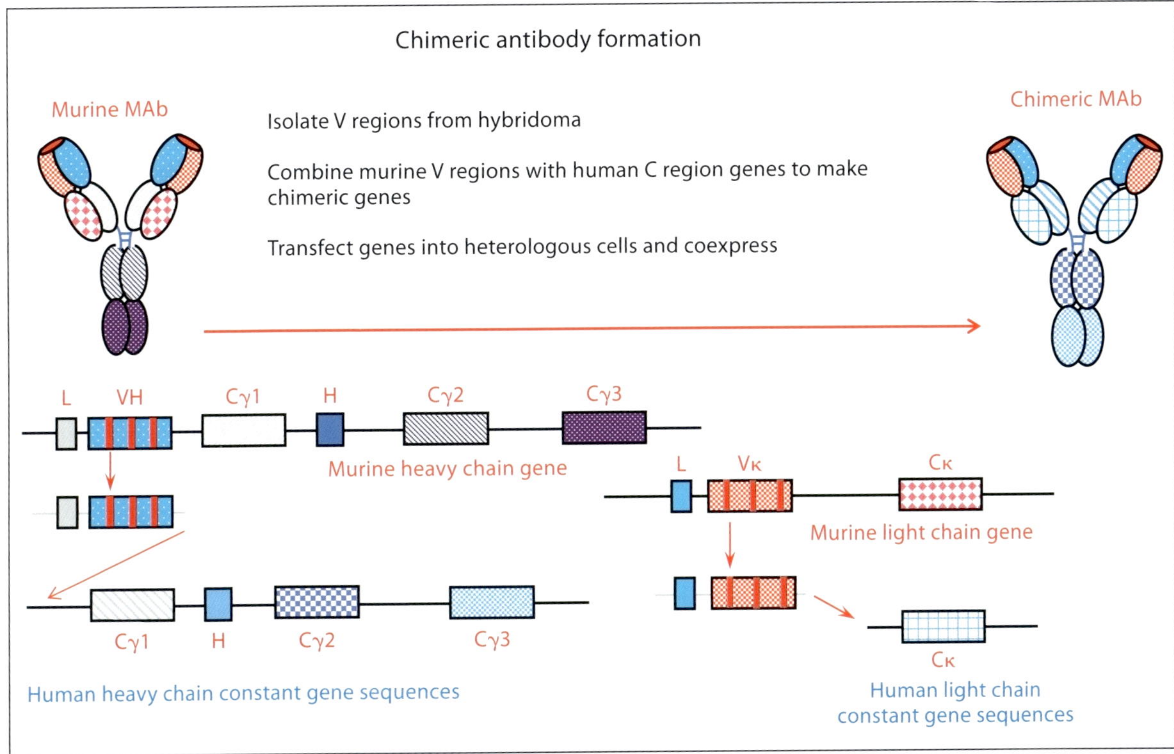

Fig. 2. Diagram of chimeric antibody structure.

duced by the hybridoma technology, were generally immunogenic when injected into man, and for technical reasons it was proving difficult to produce human monoclonal antibodies. Therefore, as a result of the understanding of antibody structures and gene sequences, in the early 1980s the regions of murine antibodies that recognized and bound to their cognate antigens were isolated and combined with human antibody sequences to produce hybrid antibodies [reviewed in 11]. Initially, the murine binding domains were combined with the remaining regions from human antibodies to produce chimeric antibodies (fig. 2). A weakness of this approach is that the murine sequences in the chimeric antibody still led to immunogenicity in many, but not all, cases. Nevertheless, by the late 1990s a number of chimeric antibodies were on the market (table 1).

In the absence of a useful and generally applicable method of making human hybridomas, an eventual solution was to 'fully' humanize the antibody. The initial description of a method for making 'humanized' or 'reshaped human' antibodies came in 1986 from the Medical Research Council's laboratories in Cambridge, UK [12, 13]. This involved obtaining the sequences for the binding site of a nonhuman antibody, also known as the complementarity determining region, and combining these with the remaining sequence of a human antibody and expressing the hybrid gene in a suitable host cell. Subsequently, a number of other groups worldwide disclosed methods to improve on the initial methodology to reliably regenerate the binding affinity and specificity of the original nonhuman antibody in the human antibody framework [11]. The first products aris-

Table 1. Approved therapeutic monoclonal antibodies

Antibody	International nonproprietary name	Target	Company	Igg format	First approval year	Indications
Orthoclone OKT3	Muromonab-CD3	CD3	J&J	Murine	1986	Transplant rejection
ReoPro	Abciximab	GpIIb/IIIa	Eli Lilly	Chimeric Fab fragment	1994	Adjunct to percutaneous coronary intervention for the prevention of cardiac ischemic complications
Panorex	Edrecolomab	EpCAM, CD226	GSK	Murine	1995	Colorectal cancer (now withdrawn)
Rituxan	Rituximab	CD20	Genentech/Roche	Chimeric	1997	NHL, RA
Zenapax	Daclizumab	CD25	PDL Biopharm/ Roche	Humanized	1997	Transplant rejection (withdrawn for commercial reasons)
Remicade	Infliximab	TNF	Centocor (J&J)/ Eli Lilly	Chimeric	1998	Crohn's disease, ankylosing spondylitis, psoriatic arthritis, ulcerative colitis, RA, psoriasis
Simulect	Basiliximab	CD25	Novartis	Chimeric	1998	Transplant rejection
Synagis	Palivizumab	RSV F	MedImmune	Humanized	1998	Prevention of serious lower respiratory tract disease caused by respiratory syncytial virus
Herceptin	Trastuzumab	CD340 p185HER2, c-erb-B2, human epidermal growth factor receptor 2	Genentech/Roche	Humanized	1998	Breast cancer
Mylotarg	Gemtuzumab ozogamicin	CD33	Pfizer/UCB	Humanized	2000	Acute myeloid leukemia (approval withdrawn in the USA due to lack of efficacy in follow-on trials)
Campath	Alemtuzumab	CD52	Bayer Schering	Humanized	2001	B cell chronic lymphocytic leukemia
Zevalin	Ibritumomab tiuxetan	CD20	BiogenIdec/ Bayer Schering	Murine ^{90}Y-labeled	2002	NHL
Humira	Adalimumab	TNF	Abbott Labs/ AstraZeneca	Human (phage)	2002	RA, psoriatic arthritis, psoriasis, Crohn's disease, ankylosing spondylitis, juvenile RA
Bexxar	Tositumomab	CD20	GlaxoSmithKline	Murine ^{131}I-labeled	2003	NHL
TNT	n/a	n/a	Peregrine	Murine ^{131}I-labeled	2003	Lung cancer
Xolair	Omalizumab	IgE	Genentech/Roche	Humanized	2003	Asthma
Raptiva	Efalizumab	CD11a	Genentech/Roche	Humanized	2003	Plaque psoriasis (now withdrawn)
Erbitux	Cetuximab	EGFR	Bristol-Myers Squibb/ImClone	Chimeric	2004	Colorectal cancer; head and neck cancer
Avastin	Bevacizumab	VEGF	Genentech/Roche	Humanized	2004	Colorectal cancer, non-small cell lung cancer, breast cancer, renal cancer
Tysabri	Natalizumab	α4 subunit of $\alpha 4\beta 1$ (VLA-4) and $\alpha 4\beta 7$ integrins	BiogenIdec/Elan	Humanized	2004	Multiple sclerosis, Crohn's disease
Actemra	Tocilizumab	IL-6R (CD126)	Chugai/Roche	Humanized	2005	Castleman's disease, RA, systemic onset juvenile idiopathic arthritis, polyarticular-course juvenile idiopathic arthritis
TheraCim	Nimotuzumab	EGFR	YM Biosciences	Humanized	2005	Head and neck cancer

Table 1 (continued)

Antibody	International nonproprietary name	Target	Company	Igg format	First approval year	Indications
Lucentis	Ranibizumab	VEGF	Genentech/Roche	Humanized Fab fragment	2006	Wet age-related macular degeneration, macular edema
Vectibix	Panitumumab	EGFR	Amgen	Human (transgenic mouse)	2006	Metastatic colorectal cancer
Soliris	Eculizumab	C5a	Alexion Pharma	Humanized	2007	Paroxysmal nocturnal hemoglobinuria
Cimzia	Certolizumab pegol	TNF	UCB Pharma	Humanized PEGylated Fab fragment	2007	Crohn's disease, RA
Removab	Catumaxomab	CD3 and EpCAM (CD226)	Fresenius	Rat/murine hybrid	2009	Malignant ascites
STELARA™	Ustekinumab	IL12/IL23	Centocor	Human (transgenic mouse)	2009	Plaque psoriasis
Symponi™	Golimumab	TNF	Centocor/ Schering Plough	Human (transgenic mouse)	2009	RA, psoriatic arthritis, ankylosing spondylitis
Ilaris®	Canakinumab	IL-1β	Novartis	Human (transgenic mouse)	2009	Cryopyrin-associated periodic syndromes
Arzerra	Ofatumumab	CD20	GSK/Genmab	Human (transgenic mouse)	2009	Refractory chronic lymphocytic leukemia (i.e. where other treatments have failed)
Prolia/Xgeva	Denosumab	RANKL	Amgen	Human (transgenic mouse)	2010	Osteoporosis, fracture prevention in prostate cancer

RA = Rheumatoid arthritis; NHL = non-Hodgkin's lymphoma.

ing from this technology were approved in the late 1990s and since then over a dozen humanized antibodies have been approved for human use (table 1). A cartoon of the resulting antibody is shown in figure 3.

During this period other methods were developed for making recombinant antibodies that used binding domains from human antibody genes, and these technologies led to a further set of approved products (table 1). These included methods of generation of libraries of specificities and selection of desired binding molecules from phage display libraries [14, 15], or making transgenic mice in which the murine immunoglobulin loci were disabled and human immunoglobulin genes inserted into their genomes [16]. The original hybridoma technology can then be used to make human monoclonal antibodies from these transgenic animals. In the last decade, several antibody products have arisen from these more recent technologies, providing a range of robust methods for making these antibody products [17].

The market for antibodies has increased dramatically since their arrival in the mid-1980s; the market for anti-TNF antibodies for the treatment of inflammatory diseases accounts for sales of more than USD 10 billion. The largest single market is rheumatoid arthritis where anti-TNFs have had a dramatic impact on patients' lives, often enabling them to lead a pain-free life for the first time in many years. Witty from GSK was quoted in Scrip in 2008 [18] as saying biologicals would account for 50% of drugs in development in 10 years; it looks like this will be achieved earlier.

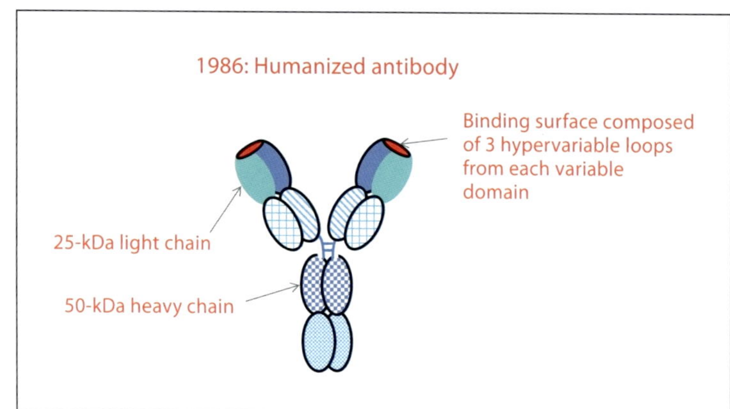

Fig. 3. Diagram of humanized antibody structure.

Antibodies bind to their target antigen and in many cases function by blocking the normal interaction of the target with its cognate ligand; in the case of soluble targets this acts to reduce the effective concentration of the target in the system. In some cases, binding may lead to intracellular signaling, which may induce cell death. However, in addition to this ability to bind to a target antigen with high affinity, antibodies have several other functions. Through sequences in the C terminal domains of the immunoglobulin (the 'Fc' region; see fig. 4) antibodies interact with receptors on certain immune cells and also with the first elements of the complement cascade and so can recruit these effector elements to kill cells to which the antibody is directed. These effector functions broaden the utility of antibodies and in the case of anticancer antibodies are useful means of cell killing [19, 20]. The constant region also has a less well-known function as an enzyme, producing reactive oxygen species to kill bacteria, etc. [21]. These innate functions are useful, as evidenced by the many products in the market, but the antibody is also capable of being used as a targeting platform for the delivery of other effector agents, which can include radioisotopes, drugs, toxins, cytokines, and enzymes. Indeed, one of the earliest examples of reengineering antibodies involved the formation of an antibody-enzyme fusion [22, 23]. While a wide range of 'payloads' have been examined in animal models and in the clinical setting, examples of approved products that are the result of conjugation of an antibody to a chemical effector are much less common in the market: Zevalin™, Bexxar™, and TNT™ target radioisotopes to tumors, while Mylotarg™ (prior to its withdrawal from the market in 2010) targeted the drug calicheamicin to tumors.

The use of antibodies as a targeting platform involves the consideration of and solution to a variety of issues, both biological and commercial.

These issues include:

- A high potency of payload. Normally this must have a drug potency in the picomolar range.
- Linkage of the payload to the antibody (method and number per antibody). Acid-labile linkages generally prove extremely difficult to manufacture to Good Manufacturing Practice (GMP) specifications.
- Delivery to the target cell of therapeutic concentrations of payload.
- Requirement of internalization of the payload into cells in some situations.
- Stability of the product. The product should have a shelf life greater than 1 year and it should ideally, for marketing reasons, be in solution.

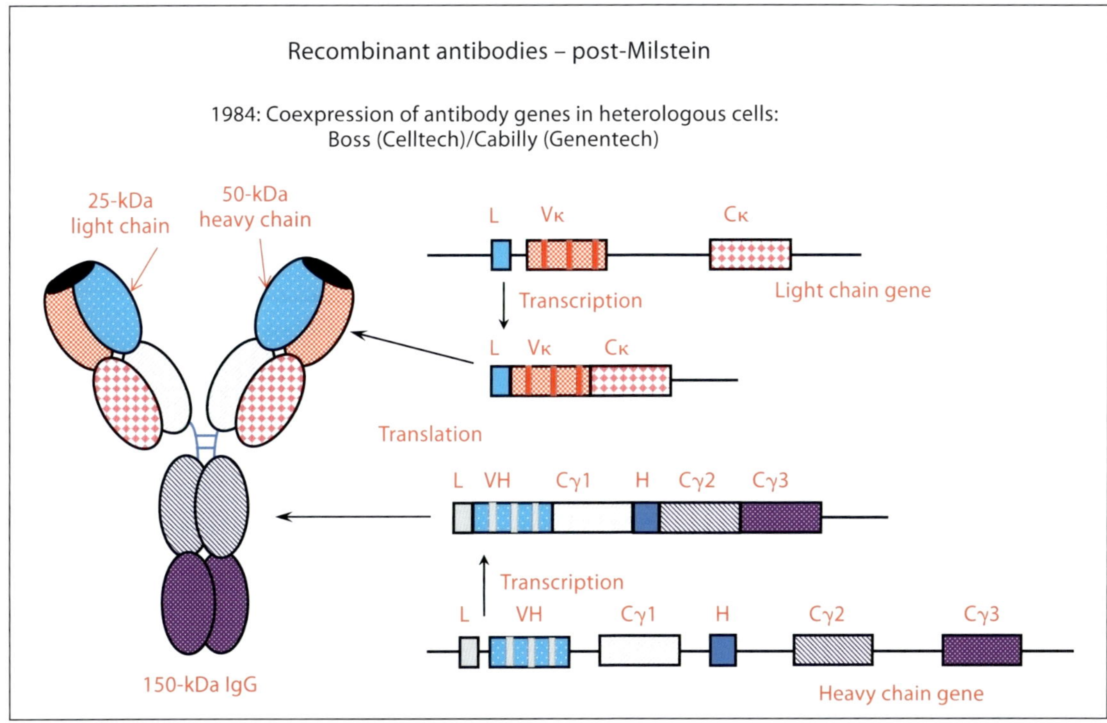

Fig. 4. Production of antibodies.

- Manufacturing issues. Random attachment of drugs generates a number of problems such as the consistency and regio-control of the conjugation. Site-specific addition removes this problem but yield remains important. Disulphide linkages are notoriously difficult in manufacturing and the reduction of thiols in proteins is difficult to regio-specifically control to the specification required by a regulator.

An issue for radioisotopically labeled antibodies is delivery of an effective dose to a tumor; this involves consideration of the number of isotopes that can be attached to an antibody, the kind of isotope (energy and path length), the mode of attachment (for stability and to reduce nonspecific toxicity), the kinetics of penetration into the tumor volume (which is affected by antibody size, affinity, tumor volume, and vascularization), and the antibody half-life (which affects serum residence time and therefore time for accumulation in the tumor). In addition, numbers of antigens per cell and heterogeneity of expression in a tumor all add to the complex equation. Radioimmunotherapy is consequentially not a favorite technology of pharmaceutical companies because of its complexity on many fronts.

Antibody-drug conjugates have been researched academically for decades; Mylotarg's entry to the market in 2000 stimulated further research, the products of which are only now in late clinical studies (see table 2). For targeting of drugs several cytotoxins are known, including those that bind to and damage DNA (such as calicheamicin) and those that bind to cellular proteins (such as the tubulin-binding maytansines and auristatins) [24, 25]. In general, the industry prefers the synthetic cytotoxins because of supply and cost issues, but the cytotoxins must be ex-

Table 2. Antibody drug conjugates on the market and in clinical development

Product	International nonproprietary name	Product type	Antigen	Application or indication	Approval year or latest clinical phase	Company/ licensee
Mylotarg™	Gemtuzumab ozogamicin	Humanized IgG-calicheamicin conjugate	CD33	Approved under accelerated approval provisions for the treatment of patients 60 years and older in first relapse with CD33-positive AML who are not considered candidates for cytotoxic chemotherapy	US 2000 approval withdrawn June 2010	Pfizer/UCB Pharma
				Oncology	Phase III	
Trastuzumab-DM1		Humanized IgG-DM1 (maytansinoid) drug conjugate	CD340 p185HER2, c-erb-B2, human epidermal growth factor receptor 2	Advanced HER2-positive breast cancer patients who have previously received multiple HER2-targeted medicines and chemotherapies	BLA submitted July 2010	Genentech/ Roche/ Immunogen
Adcetris (SGN35)	Brentuximab Vedotin	Chimeric antibody-MMAE conjugate	CD30	Hodgkin's lymphoma	BLA submitted Feb 2011	Seattle Genetics
CMC544	Inotuzumab ozogamicin	Humanized IgG-calicheamicin drug conjugate	CD22	NHL	Phase III	Pfizer/UCB Pharma
CDX-011	Glembatumumab vedotin	Human IgG MMAE antibody drug	GPNMB	Breast cancer	Phase II	Curagen (now part of Celldex)
BT-062		Humanized IgG-DM4 drug conjugate	CD138	Relapsed or refractory multiple myeloma	Phase I/II	Biotest
Milatuzumab		Humanized IgG-doxorubicin drug conjugate	CD74	Multiple myeloma	Phase I/II	Immunomedics, Inc.
MLN2704		Deimmunized IgG-DM1 drug conjugate	PSMA	Prostate cancer	Phase I/II	Millennium
AGS-5ME		Human IgG-drug conjugate	n/a	Cancers	Phase I	Astellas Pharma
BIIB015 (anti-Cripto Mab)		Humanized IgG-DM4 drug conjugate	Cripto	Cripto-positive solid tumors	Phase I	Biogenldec
AVE9633		Humanized IgG-DM4 drug conjugate	CD33	Relapsed/refractory CD33-positive AML	Phase I	Sanofi-Aventis
BIWI1	Bivatuzumab mertansine	Humanized IgG-DM1 drug conjugate	CD44v6	Head and neck squamous cell carcinoma	Phase I	Boehringer Ingelheim/ Immunogen
huC242-DM1	Cantuzumab mertansine	Humanized IgG-DM1 drug conjugate	CanAg	Colorectal, pancreatic, gastric, and other gastrointestinal cancers as well as certain non-small cell lung cancer	I	Immunogen
IMGN242		Humanized IgG-DM4 drug conjugate	CanAg (CD227)	Gastric cancer	Phase I	Immunogen
IMGN388		Humanized IgG-DM1 drug conjugate	anti-integrin	Solid tumors	Phase I	Immunogen
IMGN901	Lorvotuzumab mertansine	Humanized IgG-DM1 drug conjugate	CD56	Relapsed or refractory solid tumors	Phase I	Immunogen
MDX-1203		Human IgG-CC-1065 (rachelmycin) drug conjugate	CD70	Advanced/recurrent ccRCC or relapsed/refractory B-NHL	Phase I	Medarex (now part of Bristol-Myers Squibb)

Table 2 (continued)

Product	International nonproprietary name	Product type	Antigen	Application or indication	Approval year or latest clinical phase	Company/ licensee
MEDI-547		Human IgG-MMAF drug conjugate	EphA2	Oncology	Phase I	MedImmune (now part of AstraZeneca)
PSMA ADC		Fully human IgG-MMAE drug conjugate	PSMA	Prostate cancer	Phase I	Progenics
SAR3419		Humanized IgG-DM4 drug conjugate	CD19	Relapsed or refractory B-NHL	Phase I	Sanofi-Aventis
SAR566658		Humanized IgG-DM4 drug conjugate	DS6	DS6-positive and refractory solid tumors	Phase I	Sanofi-Aventis
SGN-75		Humanized IgG-MMAF drug conjugate	CD70	CD70-positive NHL or renal cell carcinoma	Phase I	Seattle Genetics

MMAE = Monomethyl-auristatin-E, a derivative of auristatin; DM4 = maytansinoid, a derivative of DM1; MMAF = monomethyl auristatin phenylalanine; PSMA = prostate-specific membrane antigen; AML = acute myeloid leukemia; NHL = non-Hodgkin's lymphoma; ccRCC = clear cell renal cell carcinoma; B-NHL = B cell non-Hodgkin's lymphoma; BLA = Biologics Licence Application.

tremely potent and importantly have a good solubility to stop conjugate aggregation. Here, the efficient internalization of the payload into the cells at efficacious concentrations is a major issue. The manufacture of these antibody-drug conjugates presents challenges for their commercial development, not the least of which is the linkage between the antibody and the drug, which requires stability during delivery to the target cell and during internalization but which may require subtle chemistry to allow correct release from the targeting element inside the cell.

The strategic challenge of how to interface biologicals with chemical entities has surprisingly not been solved despite the fact that this technology is important both in the therapeutic field and in the diagnostic field. The key factors are speed, as the molar concentrations of reactants are generally low and in water, and the stability of both the reagent and the adduct. The most commonly used chemistry is to react a functionalized maleimide reagent with a protein ideally containing an engineered cysteine residue. While competing oxidation of this residue is a problem, the Michael addition of the thiol to the strained maleimide ring is an extremely fast reaction with no evidence of a reverse Michael reaction under normal conditions (fig. 5). The problem is that the starting maleimidyl reagent and the resulting succinimidyl derivative are both subjected to a slow pH-dependent hydrolysis, which can limit the shelf life of a product depending on its formulation [26]. This ring opening, albeit slow, can change the specification of a regulated product; the cross-linking will still be functional but its chemical composition will change. While this problem has stimulated patenting, it has not been recognized by academia. Nevertheless, a range of antibody-drug conjugates are in clinical development (table 2).

Site-specific attachment to proteins remains a challenge and it may well be that this will await the development of 'click chemistry' where the expression of the modified 'click amino acid' does not lower expression levels. A few laboratories are currently reengineering production organisms to ensure that the process is economic even when an unnatural amino acid is included in the fermentation.

In addition to drugs and radionuclides, another entity that has been used extensively as a

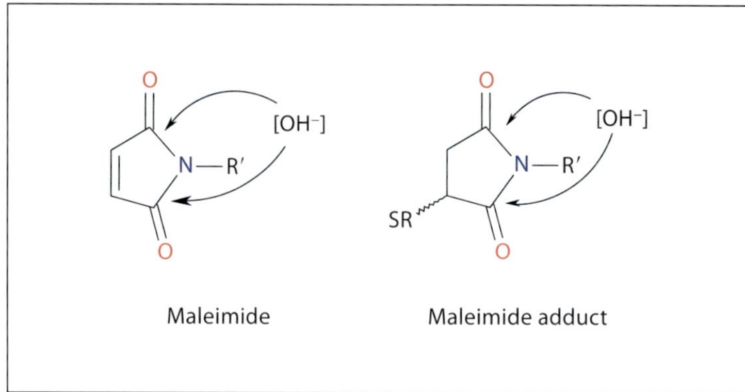

Fig. 5. Slow pH-dependent hydrolysis of maleimide reagents.

chemical conjugate is the polymer polyethylene glycol, otherwise known as PEG [27]. It is used to extend the biological half-life of therapeutic recombinant proteins with a size of less than 65 kDa (renal retention size). This size class includes certain antibody fragments, and the anti-TNF product CIMZIA™ is the first PEG-conjugated antibody Fab fragment (see fig. 4). Polymers can also target inflamed tissues by the so-called enhanced permeability and retention (EPR) effect [28–30]; this effect is modest, giving only a 2–3 enhancement, a level which can be difficult to see in the clinical setting. Polymer conjugates in development present special challenges for the following reasons:

- Analytically challenging in a GMP product.
- Linker chemistry – DMPK is difficult to analyze in vivo.
- Polymer metabolism and pharmacokinetics are analytically challenging. This may preclude the development of many polymer-based systems unless it is addressed.
- Polymers may have a pharmacology of their own.
- GMP polymers have an extremely high cost.
- Some polymers are not biodegradable, posing environmental challenges.
- Polymers are viscose at high concentrations (bear in mind a maximum of 1 ml for subcutaneous injection).
- Polymer metabolism is generally not researched (unfunded by grants).
- Analytical methods: it is extremely difficult to quantitate polymer residues in feces and urine.
- Immunogenicity and pseudo allergy can be enhanced by forming arrays/multimers [31]. Natural antibodies to polymers exist in naive patients.

PEG is the polymer most commonly conjugated with biologicals but it is a most uncommon polymer for the following reasons:

- It is weakly structured in water depending on the temperature but not as structured as a protein.
- It interacts with SDS but at higher concentrations than proteins and can be run on PAGE.
- It is soluble in both aqueous and organic solvent.
- It is extensively hydrated; the extent depends on the analytical method (2–10 waters per monomer).
- It is stable and relatively inert.
- It slows renal clearance of antibody fragments.
- It has a very low interaction with biological components.

- Two polymer chains give slower clearance than 1 due to entropy factors.
- Natural antibodies to polymers can present problems in the clinical setting, e.g. PEG-asparaginase.
- Its osmolality is very unusually a function of temperature.
- PEG is not metabolized in mammals, it does not cross biological barriers easily, and it is excreted unchanged.

Other compounds, including starches [32], have been considered as alternatives to PEG for improving in vivo half-life, but none of the alternatives had progressed to the clinical setting at the time of writing. Indeed, the difficulty and cost of the conjugation step in a GMP environment has weakened the position of antibody fragments with respect to whole antibodies.

Prospects

Antibodies are now a major product class for pharmaceutical and biotechnology companies. However, the large number of formats available, as well as their tunable half-lives as a result of PEGylation and similar strategies, means they are versatile tools for the delivery of a range of other molecules and structures within the expanding sphere of nanomedicine. Antibodies can be linked as targeting agents to other larger structures, such as liposomes, for the delivery of larger quantities of drugs to cells [33], for the delivery of nucleic acids [34], or for carrying inert molecules such as magnetic particles for diagnostic purposes [35] or therapy. Future developments will require further integration of a variety of disciplines to take optimal advantage of the exquisite targeting that antibodies can provide. Nanomedicine will provide the tools to improve the delivery of both protein and nucleic acid drugs; the challenge is a process to transfer translational knowledge from the industrial sector to researchers; without this, progress will be slow and inefficient.

Antibodies currently have a major marketing disadvantage compared with NCEs: in most cases they have to be injected. This produces a grand challenge for the sector: how to enlarge the market for antibodies to compete with NCEs by nanoformulation to permit nonparenteral delivery. The cost of infrequently dosed antibodies will fall further to the point where the effective cost of goods is similar to daily dosed NCEs. Given the increasing ease with which very specific antibodies and antibody fragments can be manufactured, it is not too farfetched to imagine an even wider role for nature's drugs in the future.

Disclosure Statement

The authors have nothing to disclose.

References

1 Reichert JM: Global antibody development trends. mAbs 2009;1:86–87.
2 Maxam AM, Gilbert W: A new method for sequencing DNA. Proc Natl Acad Sci USA 1977;74:560–564.
3 Sanger F, Nicklen S, Coulson AR: DNA sequencing with chain-terminating inhibitors. Proc Natl Acad Sci USA 1977;74:5463–5467.
4 Primrose SB, Twyman RM, Old RW: Principles of Gene Manipulation, ed 6. Oxford, Blackwell, 2001.
5 Saiki RK, Scharf S, Faloona F, Mullis KB, Horn GT, Erlich HA, Arnheim N: Enzymatic amplification of beta-globin genomic sequences and restriction site analysis for diagnosis of sickle cell anemia. Science 1985;230:1350.
6 Edge MD, Green AR, Heathcliffe GR, Meacock PA, Schuch W, Scanlon DB, Atkinson TC, Newton CR, Markham AF: Total synthesis of a human leukocyte interferon gene. Nature 1981;292:756–762.

7 Mark DF, Lu SD, Creasey AA, Yamamoto R, Lin LS: Site-specific mutagenesis of the human fibroblast interferon gene. Proc Natl Acad Sci USA 1984;81: 5662–5666.
8 Porter AG, Bell D, Adair J, Catlin GH, Clarke J, Davies JA, Dawson K, Derbyshire R, Doel SM, Dunthorne L, Finlay M, Hall J, Houghton M, Hynes C, Lindley I, Nugent M, O'Neil GJ, Smith JC, Stewart A, Tacon W, Viney J, Warburton N, Boseley PG, McCullagh KG: Novel modified beta-interferons: gene cloning, expression, and biological activity in bacterial extracts. DNA 1986;5:137–148.
9 Köhler G, Milstein C: Continuous cultures of fused cells secreting antibody of predefined specificity. Nature 1975; 256:495–497.
10 Storz U: IP issues in the therapeutic antibody industry; in Kontermann R, Dubel S (eds): Antibody Engineering. Berlin, Springer, 2010, vol 2, pp 517–581
11 Mountain A, Adair JR: Engineering antibodies for therapy. Biotechol Genet Eng Rev 1992;10:1–142
12 Jones PT, Dear PH, Foote J, Neuberger MS, Winter G: Replacing the complementarity-determining regions in a human antibody with those from a mouse. Nature 1986;321:522–525.
13 Riechmann L, Clark M, Waldmann H, Winter G: Reshaping human antibodies for therapy. Nature 1988;332:323–327.
14 Vaughan TJ, Williams AJ, Pritchard K, Osbourn JK, Pope AR, Earnshaw JC, McCafferty J, Hodits RA, Wilton J, Johnson KS: Human antibodies with sub-nanomolar affinities isolated from a large non-immunized phage display library. Nat Biotechnol 1996;14:309–314.
15 Hoogenboom H: Selecting and screening recombinant antibody libraries. Nat Biotechnol 2005;23:1105–1116.
16 Lonberg N: Human antibodies from transgenic animals. Nat Biotechnol 2005;23 1117–1125.
17 Nelson AL, Dhimolea E, Reichert JM: Development trends for human monoclonal antibody therapeutics. Nat Rev Drug Discov 2010;9:767–774.
18 Eaton M: The importance of the industrial academic interface for innovation in the pharmaceutical sector. Eur Ind Pharm 2010;6:8–11.
19 Desjarlais JR, Lazar GA, Zhukovsky EA, Chu SY: Optimizing engagement of the immune system by anti-tumor antibodies: an engineer's perspective. Drug Discov Today 2007;12:898–910.
20 Jefferis R: Glycosylation as a strategy to improve antibody-based therapeutics. Nat Rev Drug Discov 2009;8:226–233.
21 Wentworth P Jr, Jones LH, Wentworth AD, Zhu X, Larsen NA, Wilson IA, Xu X, Goddard WA 3rd, Janda KD, Eschenmoser A, Lerner RA: Antibody catalysis of the oxidation of water. Science 2001;293:1806–1809.
22 Neuberger MS, Williams GT, Mitchell EB, Jouhal SS, Flanagan JG, Rabbitts TH: A hapten-specific chimaeric IgE antibody with human physiological effector function. Nature 1986;314: 268–270.
23 Williams GT, Neuberger MS: Production of antibody-tagged enzymes by myeloma cells: application to DNA polymerase I Klenow fragment. Gene 1986;43:319–324.
24 Senter P: Potent antibody drug conjugates for cancer therapy. Curr Opin Chem Biol 2009;13:235–244.
25 Hughes B: Antibody drug conjugates for cancer: poised to deliver? Nat Rev Drug Discov 2010;9:665–667.
26 Khan MN: Kinetics and mechanism of the alkaline-hydrolysis of maleimide. J Pharm Sci 1984;12:1767–1771.
27 Payne RW, Murphy BM, Manning MC: Product development issues for PEGylated proteins. Pharm Dev Technol 2010, E-pub ahead of print.
28 Constantinou A, Chen C, Deonarain MP: Modulating the pharmacokinetics of therapeutic antibodies. Biotechnol Lett 2010;32:609–622.
29 Matsumura Y, Maeda H: A new concept for macromolecular therapeutics in cancer chemotherapy: mechanism of tumoritropic accumulation of proteins and the antitumor agent smancs. Cancer Res 1986;46:6387–6392.
30 Greish K: Enhanced permeability and retention (EPR) effect for anticancer nanomedicine drug targeting. Methods Mol Biol 2010;624:25–37.
31 Szebeni J: Complement activation-related pseudoallergy: a new class of drug-induced immune toxicity. Toxicology 2005;216:106–121.
32 Fresenius Kabi: HESylation – introduction. 2011. http://www.HESylation.com.
33 Yu B, Tai HC, Xue W, Lee LJ, Lee RJ: Receptor-targeted nanocarriers for therapeutic delivery to cancer. Mol Membr Biol 2010;27:286–298.
34 Liu B: Exploring cell type-specific internalizing antibodies for targeted delivery of siRNA. Brief Funct Genomic Proteomic 2007;6:112–119.
35 Vigor KL, Kyrtatos PG, Minogue S, Al-Jamal KT, Kogelberg H, Tolner B, Kostarelos K, Begent RH, Pankhurst QA, Lythgoe MF, Chester KA: Nanoparticles functionalized with recombinant single chain Fv antibody fragments (scFv) for the magnetic resonance imaging of cancer cells. Biomaterials 2010;31:1307–1315.

Prof. Mike Eaton
ETP Nanomedicine
Nethercote, Chinnor Road
Aston Rowant OX49 5SH (UK)
Tel. +44 1844 351 238, E-Mail nanomedicine@btinternet.com

Author Index

Subject Index